Bird Tracks & Sign

Bird Tracks & Sign

A Guide to North American Species

Mark Elbroch
and Eleanor Marks, Ph.D.
skull chapter by C. Diane Boretos

STACKPOLE
BOOKS

Published by
STACKPOLE BOOKS
5067 Ritter Road
Mechanicsburg, PA 17055
www.stackpolebooks.com

Printed in China

First edition

Cover design by Caroline Stover
Cover photo of northern flicker by Tom Vezo; cover photos of merlin feathers, Baltimore oriole nest, mourning dove tracks, red-naped sapsucker wells in willows, and ruffed grouse roost mark in snow by Mark Elbroch and Eleanor Marks

Illustrations and photographs by Mark Elbroch unless otherwise credited. Feather photographs in chapter six by Eleanor Marks unless otherwise credited.

Library of Congress Cataloging-in-Publication Data

Elbroch, Mark.
Bird tracks and sign : a guide to North American species / Mark Elbroch, Eleanor Marks, and C. Diane Boretos.—1st ed.
p. cm.
Includes bibliographical references (p.) and index.
ISBN 0-8117-2696-7
1. Birds—North America—Identification. 2. Animal tracks—North America—Identification. I. Marks, Eleanor (Eleanor Marie) II. Boretos, C. Diane. III. Title.

QL681.E42 2001
598′.097—dc21

2001020602

ISBN 978-0-8117-2696-2

To all the birds and bears,
and to my father, Larry Elbroch,
and my grandfather, Charles Gorst

M.E.

To my beloved co-inspirator,
Linda Crane (1946–2000),
genius tracker, especially of souls and seaweeds;
consummate confidante of birds and animals,
especially sandhill cranes, bears, snakes,
reindeer, and skunks;
bodhisattva to all.

E.M.

CONTENTS

Editor's Note: Abbreviations for the states in which the photographs were taken follow the captions.

Introduction

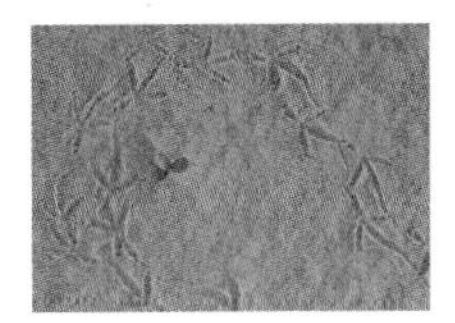

A thing is right when it tends to preserve the integrity, stability, and beauty of the biotic community. It is wrong when it tends otherwise.
ALDO LEOPOLD, ***A Sand County Almanac***

E.M.

There are so many fine field guides to the birds—why another? This book, while complementing those field guides, is a uniquely different footpath into the lifeways of birds, into their world of survival, into their habitats, and perhaps most importantly, into knowing them intimately by slowly coming to understand their homes and lives. Tracking will enable you to learn about the day-to-day lives of birds in that challenging world of survival, even when you can't see or hear them.

The other field guides and tapes will help you learn field marks and birdsong, and they are excellent tools for recognizing birds when you see a flash of color or hear a snippet of song or an abrupt warning call. But there are other ways to learn of birds' presence even when they are being secretive or have moved on, perhaps for their seasonal migration. This book provides you with ways to recognize and identify the presence of birds from other clues they leave that are far more enduring than a wisp of song or the brief glimpse of a tail flared for landing. Birds leave other clues to their presence in the form of tracks and sign. Sign refers to all the possible signs of their passing: sign of feeding, gathering materials for nesting, the nests or cavity holes themselves, pellets, droppings, feathers lost during molt, or kill sites.

Feigning a broken wing, a killdeer distracts intruders from her nest. (ME)

A great blue heron stalks a stream behind a thin veil of phragmites. (VA)

For all of us who enjoy tracking, beginner or advanced, becoming sensitive and connected enough to notice these kinds of sign vastly increases our awareness of living relationships in the landscape around us. In this field guide, each chapter focuses on one similar group of signs used to detect the presence of birds; these signs are classed together based on their relationship to particular biological functions of the bird (its physical processes), like feeding sign, or the scat sign of elimination. Just as learning the tracks of birds takes you far beyond relying only on field identification marks or birdsong (which can help only when the bird is present), learning about nests, scat, pellets, feeding sign, feathers, and skulls gives you a richer language for interpreting each possible clue as it reveals the presence of

Cultural Tracking

M.E.

If we all followed our lineages back far enough, we would find the trackers from which we all came. For there was an era in which the art and science of tracking was a necessary skill for survival. A glance at a track, and a story of incredible richness and complexity unfolded. Listening to birdsongs was not for mere identification; a note on the wind told of sneaking humans, large predators, or the location of a nest.

Real tracking is bigger than one lifetime. Tracking, as our ancestors knew it, was a body of knowledge handed down from generation to generation. Each person added to this knowledge base during his or her lifetime, and the source ever expanded. I call this cultural tracking.

Today, around the globe, there is renewed enthusiasm for tracking. Wildlife research is beginning to use tracking. One study in Montana collects data on bear sign, thereby avoiding having to capture bears, pull their teeth, drug them, collar them, or habituate them to human scent. Wildlife inventories are beginning to rely on tracks and sign to provide an accurate snapshot of wildlife in an area instead of on theoretical assumptions of inhabitants based upon vegetative analysis. Jon Young is teaching ancient bird language skills. And Louis Liebenberg has created a visual palm pilot for GPS units so that bushmen trackers in Africa are able to contribute to wildlife research. Individuals are also learning tracking on their own, taking to the woods to reconnect with wilderness, to engage in real relationships with wild creatures, and to learn a sense of place by reading the signs left in the environments in which they live. Tracking is really about a greater awareness, and it is truly a way of living more fully. Tracking centers us in a world that continues to move at a head-spinning speed. It slows us down.

We who are passionate about this subject are also beginning to rebuild what was lost in our country. May tracking and trackers be preserved and celebrated for all time. We humbly present the material in this book as another stone from which to rebuild our cultural knowledge of tracking. We have recorded a part of the art, a piece of the science, with the faith that those who read this work will add to it and share it with others, be inspired to continue to learn, and aid in the conservation of tracking as a whole. May you all add to this knowledge and push it further.

American crow tracks. (MA)

A crow skull lies in a bed of its own feathers, picked clean by the red-tailed hawk that captured it. (NY)

a particular bird in the habitat you are getting to know, even if that bird has already moved on or is resolutely silent and hidden in the trees.

As you look at, listen to, smell, touch, and sense the clues to the bird stories all around you, you will become not only a proficient tracker of birds, but a tracker of the birds' life cycles and habitat, of their prey and their hideouts, of weather and seeds, nuts and trees, death and life—in short, a tracker of ecology. As you become intimate with the lives of the birds, you will grow increasingly aware of the need for conservation to protect the habitats of these sometimes threatened inhabitants.

Birding Beginners

If you've never birded before, get a copy of the *National Geographic Field Guide to the Birds of North America* or Kenn Kaufman's *Focus Guide to the Birds of North America.* Kaufman arranges birds by size, which is the easiest way for someone to get started. Once you've been afield for a while, you will want to know a way to organize all the field signs you are learning into workable groups; the National Geographic guide will help you with that: You also don't need to know which birds are in the East or West because this book covers the entire continent and, like all field guides, organizes birds taxonomically by families that typically have common characteristics based on their evolutionary emergence and physical similarities.

As a tracker, you'll also find the National Geographic guide invaluable for its succinct descriptions of family characteristics at the beginning of each section. Reading those will help you understand how track and sign characteristics have been grouped in this book. At the end of each individual species description, there is information about the birds' habitat and seasonal range, which you can combine with your own field observations and the photographs and data in the book to help you make positive identifications. Other field guides provide this information, and Roger Tory Peterson's and *The Sibley Guide to Birds,* as well as local and regional field guides, are also excellent aids.

It will help in your identification if you take field notes, sketch anything that intrigues you, and measure significant things like tracks or the width of owl pellets. We highly recommend Clare Walker Leslie's *The Art of Field Sketching* for getting started with your field notes. Even if you are convinced you cannot draw, her book will prove you wrong.

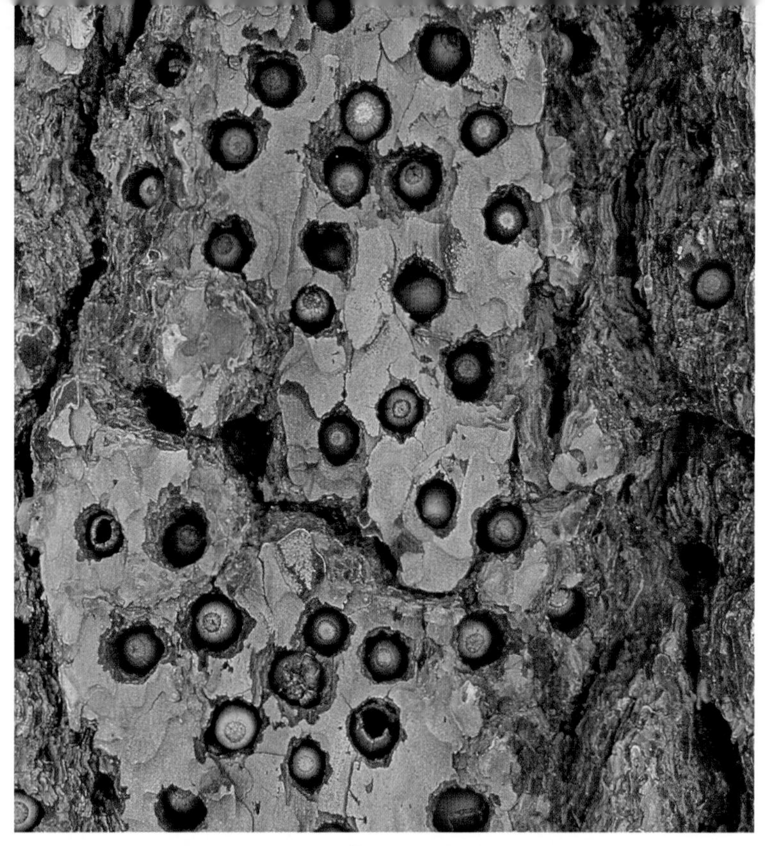

An acorn woodpecker's larder in a Jeffrey pine. (CA)

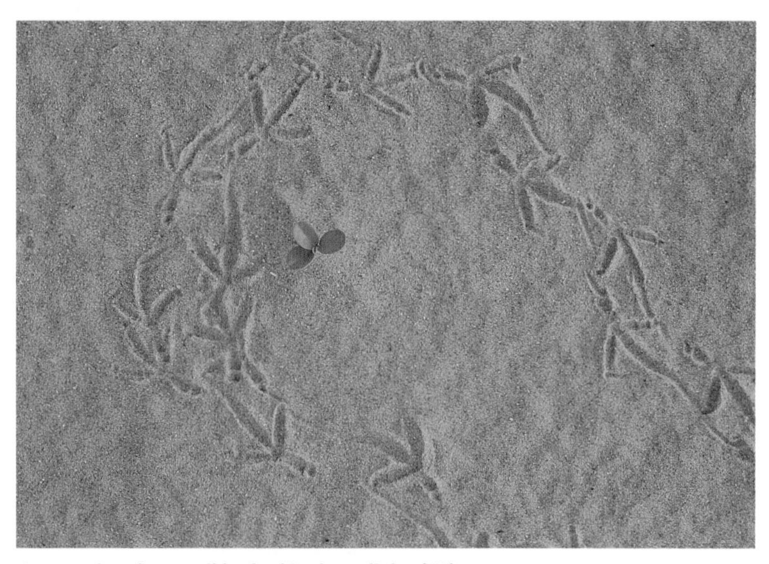

A mourning dove trail bathed in dawn light. (TX)

Start in Your Own Backyard

One of the easiest ways to get started in both tracking and birding is in your own backyard. This is a microhabitat you can get to know intimately, and you can begin to make improvements in your backyard habitat to promote visits by wildlife. *Building a Backyard Bird Habitat,* by Scott Shalaway, is a good comprehensive guide. The National Wildlife Federation's Backyard Wildlife Habitat Program (see Resources) offers advice about gardening for wildlife; nonchemical lawn care; using plants to attract pollinators, such as bees and hummingbirds; putting up nest boxes for birds and roosting boxes for bats; creating brush piles for roosts; providing winter water sources for birds; and many other wildlife-friendly ideas that will increase the number of tracks and species you will be able to see near your home.

You can put out track plates or track boxes overnight or for several days to see what species come by and observe how weather affects the aging of tracks. Position a board covered with a layer of fine sand or dirt where you've observed trails of tracks in your yard. Write down and sketch your observations, noting the weather. Try moving your track plate to different locations in your yard and see what happens.

If you want to get a better understanding of large-scale habitat in your area, and of ecosystems in general, there are four fantastic and comprehensive introductory books on this topic devoted to different regions of the United States, all Peterson Field Guides by John Kricher and Gordon Morrison: *Ecology of Eastern Forests, California and Pacific Northwest Forests, Rocky Mountain and Southwest Forests,* and *A Neotropical Companion.* The last is crucial reading if you want to know what is happening to "your" birds when many of them have migrated to their wintering grounds in Central America, the Caribbean, and South America.

Going on bird walks or taking a birding class are also excellent ways to get started in the world of birding. Once you get hooked on tracking, consider courses such as those offered by Jon Young's Wilderness Awareness School, especially the Kamana program, an intensive study-at-home course in tracking and ecology. Instructors at his home-base school in Washington are always available by phone to help with your questions, and they will connect you with others in your area who are taking the course too.

Endangered old-growth forest. (WA)

When you are birding or tracking, there are some important ethical considerations to protect the birds, other animals, their habitat, and the rights of other people. The American Birding Association has come up with some guidelines.

American Birding Association's Principles of Birding Ethics

Everyone who enjoys birds, birding, and tracking must always respect wildlife, its environment, and the rights of others. In any conflict of interest between birds and birders/trackers, the welfare of the birds and their environment comes first.

Code of Birding Ethics

Promote the welfare of birds and their environment.

1(a) Support the protection of important bird habitat.

1(b) To avoid stressing birds or exposing them to danger, exercise restraint and caution during observation, photography, sound recording, or filming.

Limit the use of recordings and other methods of attracting birds, and never use such methods in heavily birded areas or for attracting any species that is Threatened, Endangered, or of Special Concern, or is rare in your area.

Keep well back from nests and nesting colonies, roosts, display areas, and important feeding sites. In such sensitive areas, if there is a need for extended observation, photography, filming, or recording, try to use a blind or hide, and take advantage of natural cover.

Use artificial light sparingly for filming or photography, especially for close-ups.

1(c) Before advertising the presence of a rare bird, evaluate the potential for disturbance to the bird, its surroundings, and other people in the area, and proceed only if access can be controlled, disturbance can be minimized, and permission has been obtained from private land-owners. The sites of rare nesting birds should only be divulged to the proper conservation authorities.

1(d) Stay on roads, trails, and paths where they exist; otherwise keep habitat disturbance to a minimum.

(continued on page 10)

Respect the law and the rights of others.
2(a) Do not enter private property without the owner's explicit permission.
2(b) Follow all laws, rules, and regulations governing use of roads and public areas, both at home and abroad.
2(c) Practice common courtesy in contacts with other people. Your exemplary behavior will generate goodwill with birders/trackers and non-birders alike.

Ensure that feeders, nest structures, and other artificial bird environments are safe.
3(a) Keep dispensers, water, and food clean and free of decay and disease. It is important to feed birds continually during harsh weather.
3(b) Maintain and clean nest structures regularly.
3(c) If you are attracting birds to an area, ensure the birds are not exposed to predation from cats and other domestic animals, or dangers posed by artificial hazards.

Group birding, whether organized or impromptu, requires special care.
Each individual in the group, in addition to the obligations spelled out in Items #1 and #2, has responsibilities as a Group Member.
4(a) Respect the interests, rights, and skills of fellow birders/trackers, as well as those people participating in other legitimate outdoor activities. Freely share your knowledge and experience, except where code 1(c) applies. Be especially helpful to beginning birders/trackers.
4(b) If you witness unethical birding/tracking, assess the situation and intervene if you think it prudent. When interceding, inform the person(s) of the inappropriate action and attempt, within reason, to have it stopped. If the behavior continues, document it and notify appropriate individuals or organizations.

Group Leader Responsibilities:
4(c) Be an exemplary ethical role model for the group. Teach through word and example.
4(d) Keep groups to a size that limits impact on the environment and does not interfere with others using the same area.

(continued on page 11)

4(e) Ensure that everyone in the group knows and practices this code.
4(f) Learn and inform the group of any special circumstances applicable to the areas being visited (e.g. no tape recorders allowed).
4(g) Acknowledge that professional tour companies bear special responsibility to place the welfare of birds and the benefits of public knowledge ahead of the company's commercial interests. Ideally, leaders should keep track of tour sightings, document unusual occurrences, and submit records to appropriate organizations.

Additional copies of the Code of Birding Ethics can be obtained from: ABA, PO Box 6599, Colorado Springs, CO 80934-6599; (800) 850-2473 or (719) 578-1614; fax: (800) 247-3329 or (719) 578-1480; e-mail: member@aba.org; or on their website: www.americanbirding.org.

Tracks and Trails

1

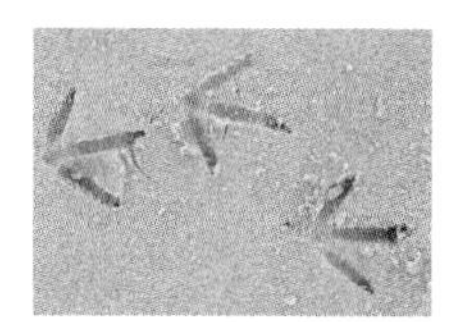

M.E.

It was a bit shocking to arrive in Death Valley National Park and be told that in addition to the fancy resort, there was a golf course. We never went to look at the thing—some places aren't meant to have green turf. But unintended good did come of this golf course. A pipe broke and water flowed freely down the valley past the campground. It crossed the main park road, where an orange "Flood" sign was posted.

People slammed on their brakes, dropping to 5 miles an hour to ford the torrent, which was actually just a slick of water crossing the pavement—so shallow you could stand in the center wearing sneakers and your feet would stay dry. But people slammed on their brakes anyway. We were definitely in the desert.

After the water passed over the road's edge, it dropped and cut a shallow course in the sand—a miniature Grand Canyon a foot deep and 10 feet wide at most. This narrow stream with muddy banks flowed right past our campground. We had arrived there well after dark, and I used the high beams of our rental car to assess the situation more fully. A coyote walked along one edge of the stream, leaving perfect tracks. Electricity raced through me and fueled my dreams until morning.

Before the sun had risen, I was off to explore the new creek. In the thick, gray predawn, I strained to see the activity recorded along the banks. So much lay before me. The miraculous water source meant the even greater miracle of mud in the midst of an otherwise dry desert, and mud is the medium upon which stories are written in tracks and trails.

Coyotes had left trails everywhere, as they socialized, drank, and hunted along the banks. Several kit foxes had come to drink, nervously walking up to the creek's edge—no doubt cautious due to all the coyote activity. A gray fox also came to drink, then crossed the creek and entered the desert beyond. (According to park literature, gray foxes do not inhabit the valley.) Ravens patrolled the edges, drinking from time to time. A great blue heron walked upstream, and Brewer's blackbirds, grackles, and sparrows all came to drink as well. Two great horned owls, one larger than the other, dropped in. Both birds walked straight into the water, where they stood and took gulps from the stream, raising their beaks to the stars to swallow. White-tailed antelope squirrels ran back and forth, and a wood rat made an appearance on one bank. Several band-tailed lizards patrolled an edge, and roadrunners leaped from one side to the other, not needing to stop to drink, as their physiology is adapted to dry climates. And all this in a distance less than half the length of a football field! I wonder, if not for the improbable gift of these tracks, how many lives I would have missed here, under the baking sun, on the floor of Death Valley.

The term *tracking* brings to mind footprints, or tracks. Birds certainly figure prominently among the wild creatures that leave tracks

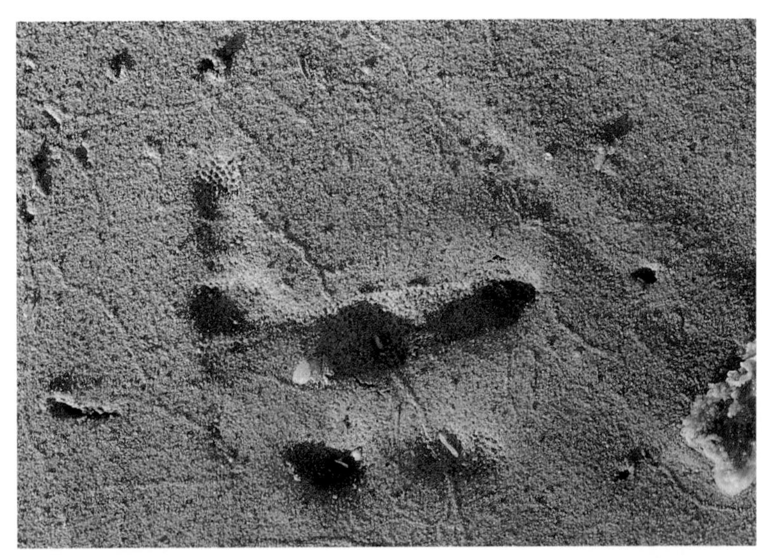

A great horned owl track at the edge of a water source. (The white substance is a salt deposit.) (CA)

behind, yet, for some reason, this approach to identification has been little studied, at least in the United States, perhaps due to the high visibility and audibility of many bird species. Nevertheless, this is a valuable means of determining the presence of particular birds. With practice and the right diagnostic questions, you can quickly narrow down the identification possibilities to bird families or even species.

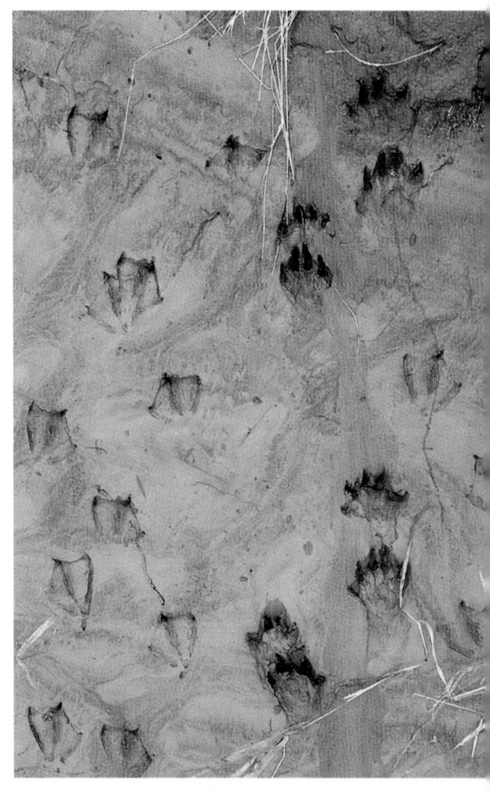

An otter shares a riverbank with several wood ducks (the smaller tracks). (MA)

A track should never be examined in isolation; it is always related to and speaks about an ecosystem. When understood in ecological relation to its environment, a track is bursting with information and clues to species identification. The location and environment surrounding a track provide information on behavior, habits, and more. So take note. The more you study the ecology of your local area, including which bird species are present each season, their behaviors, and their preferred habitats, the easier track identification becomes. You can't become a student of tracking without also becoming a well-rounded ecologist—a naturalist, if you will. Are you aware of which birds overwinter in your area, staying throughout the year, which summer around your home, and which only pass through during migrations? Do you know which birds feed on the ground, which in the air, which eat insects, and which eat seeds? The more ecological information you learn, the easier track identification becomes.

Certain bird species rarely touch the ground at all. Grebes, loons, swifts, flycatchers, and kingfishers are among those that are least comfortable on the ground. Nightjars also rarely leave tracks, as they touch down only to roost and rest, often on roads and rocks rather than soft substrates. Because it's rare to encounter the tracks of such species under natural circumstances, they are not covered in the following pages. Other bird families, whose tracks are common in the field in North America, are examined in great detail.

Track Analysis

A practical introductory method to the identification of bird tracks is to follow a three-step process I call the "3 perspectives": First, look at the ecology of the environment. Next, look at the track pattern. Then, look at the track. When you find a track in the field, stop, step back, and first take in the big picture, studying the environment surrounding the track. Determine the ecological framework—lowland or upland forest, wetland, grassland, or other type of ecosystem. Then approach the tracks themselves. Try to determine the pattern of a small group of tracks, which offers much information: A bird that frequents the ground tends to walk, whereas strong perching birds tend to hop. Now narrow your focus further, studying just one track. Don't try to label the track with a species name. Rather, carefully examine the track, and acknowledge what it's sharing with you. A bird track tells you a great deal about the ecology of the bird, giving a lot of clues as to how it exists, which aids in the identification of the species. Studying the tracks answers many questions: How much time does this bird spend on the ground? How comfortable is it on the ground? It's all there: A bird that spends most of its time perched has very narrow tracks, whereas a regular ground feeder has wide tracks. And don't just look. Listen, feel, taste, smell—engage your senses as you constantly shift your focus from the ecosystem to the track pattern to an individual track and then back to the big picture again.

Consider the tracks of bald eagles and great blue herons. Even though simple measurements of the length and width of their tracks may be similar, there are obvious differences in the structure, or morphology, of their feet. The massive claws and developed toe pads of the eagle aid in gripping slippery prey, whereas the slender, wide, flat feet of the heron are ideal for supporting the bird during long waits in muddy marshes and as it's stalking. Further, webbing is an obvious sign of swimming. And feathers on feet are an adaptation to conserve heat.

The master tracker goes well beyond identification; he or she is able to accurately read behaviors, the health of a creature, and the moods of an individual from a mere scuff in the ground. At each deepening level of focus, from environment to track, slow down and ask yourself questions that will help you visualize what happened in that moment and thus further understand the bird in relation to its habitat. Use the questions, visualization exercises, and perspectives offered in this book to deepen your enjoyment of track identification. Take the time to immerse yourself in the relationship between individual tracks and trails and the environment.

Foot Morphology

The structure of the foot, related to the arrangement of the toes, aids ornithologists in the organization and classification of bird species. Using a simplification of the ornithologists' categories, we are also able to create track groups, based on how foot structure helps determine the size and shape of the track. These categories not only provide information about how a bird exists—its feeding and nesting habits, and so on—but also allow us to present bird tracks in easy-to-learn groups based on common characteristics, for easier identification in the field.

In most bird species, the feet are four-toed; a few are three-toed. The toes of birds all converge in a central area of the foot called the *metatarsal pad,* or sole. Birds don't place all the bones in their feet equally on the ground. Rather, they usually move with most of their weight supported by their toes, and therefore create deeper impressions at the outer edges than in the center of their tracks. This method of movement is called *digitigrade.* Humans place the entire foot on the ground when walking, known as *plantigrade.*

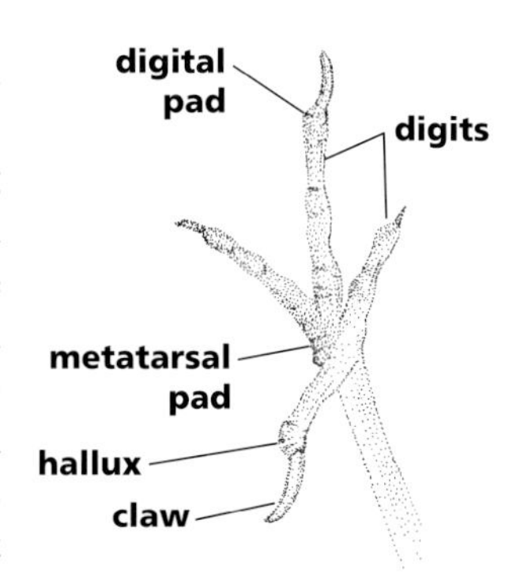

The basic anatomy of a bird foot.

Bird feet vary tremendously, depending upon how the toes, or digits, are ranged around the metatarsal pad. The digital pads, which line each toe, also vary in form and function from bird to bird. The most common toe arrangement is *anisodactyl,* with three toes pointed forward and one backward. The toe that points back is commonly referred to as the *hallux.* It varies in length tremendously across species and is absent altogether in some. The second most common toe arrangement is *zygodactyl,* with two toes pointed backward and two forward.

Ornithologists also number the toes of each foot, from 1 to 4, in a systematic fashion, called the *digital formula*. Toe 1 is always the hallux, whether it is present or not, and the other toes are numbered in sequence, beginning with the inside of the foot and circling out. Thus left and right tracks are circled in different directions. Study the illustrations to better visualize this point.

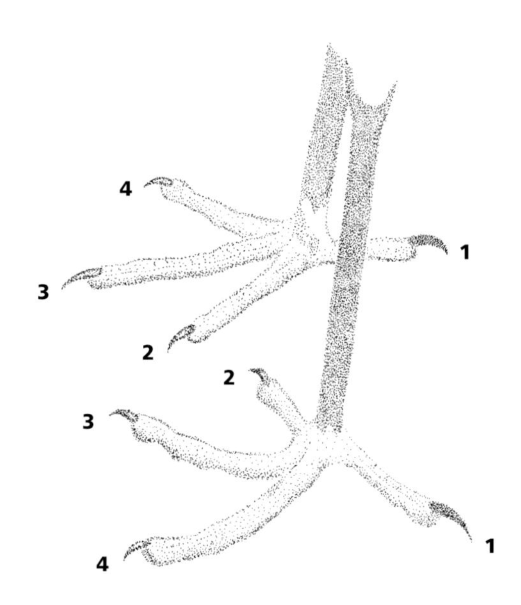

Digital formula: anisodactyl feet **(above);** ***zygodactyl feet*** **(below).**

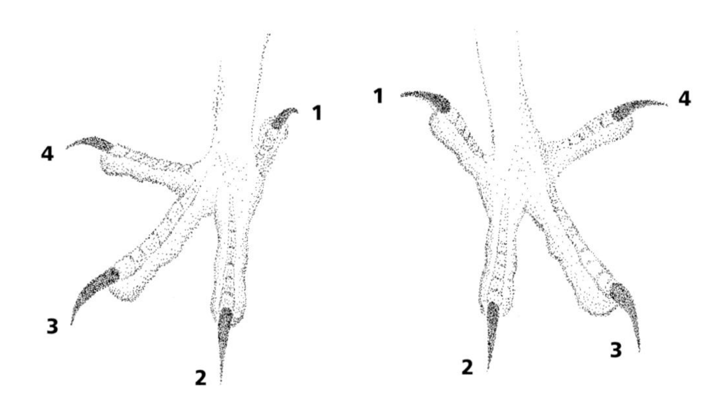

For anisodactyl feet, toe 1 points backward and toes 2, 3, and 4 point forward. For zygodactyl feet, toes 1 and 4 point backward and toes 2 and 3 point forward. This system was adopted to help define bird tracks by Bang and Dahlstrom (1974); Brown et al. (1987) and, more recently, Halfpenny (1998) followed suit. We, too, shall incorporate this system in our track analysis. As we present bird tracks, we will also incorporate information on symmetry, size, webbing, and substrate. These are key identification concepts in bird tracking, just as bill and tail shapes are key identification concepts in bird watching.

Symmetry

Symmetry compares the shapes of the left and right sides of the track. If you were to cut a track in half lengthwise, most easily done by cutting right down the middle of toe 3 in anisodactyl tracks, do the right and left sides match perfectly? Is a mirror image produced, or are the sides misaligned and different sizes and shapes? Bird tracks are rarely perfectly symmetrical, but the degree of asymmetry is useful to note and compare across species.

Size

Generally, the size of a track provides useful information for guessing or extrapolating the probable size of a typical member of the particular bird species. Sometimes simply looking at track size to estimate the overall size of a bird can be misleading, however. Certain species that spend a great deal of time on the ground have developed larger feet in proportion to the size of the body. In addition, different species have developed larger or smaller feet for various reasons, such as for swimming, gripping prey, or hanging upside down.

Webbing

Visible webbing in tracks varies tremendously across species. In ducks and gulls, webs are large and clearly visible. In herons, grouse, and certain plovers, smaller webs are present between two or more toes. Because partial webbing does not always show in the track, species with partial webbing are not grouped in this chapter with species that display obvious webbed tracks, but with those whose tracks do not show webbing.

Even the fully webbed feet of geese and gulls cannot be relied upon to always leave clear webbing marks in their tracks. The depth and hardness of the substrate in which a bird stands affects a track significantly. In cases where webbing is not clearly visible, it can be deduced by noting whether the outer toes are curved. On webbed feet, toes 2 and 4 always curve in toward toe 3.

Substrate

Substrate refers to what the bird has stepped in, be it sand, mud, snow, or grass. The depth of the substrate, which is reflected in the depth of the track, has an enormous influence on the appearance, size, and shape of the track, as well as on how the bird moves. In shallow substrate, the bird can move easily and therefore tends toward its natural gait—walking or hopping—but in deep or slippery substrate, the bird is likely to move differently. Note substrate characteristics as you look at tracks and trails so you can begin to understand and predict the influences of various substrates.

The variety of substrates in which a bird may step are infinite and thus create great challenges for the tracker. If a beginner relies rigidly on memorizing each kind of track, like a file or archive, track identification is made more difficult. Tracks left by the same species in wet sand and in drier conditions, while the bird was moving at different speeds, will look totally different. This is why it's important, when encountering a track, to avoid attempting to quickly label

Compare the perfect tracks of a ruffed grouse in mud . . .

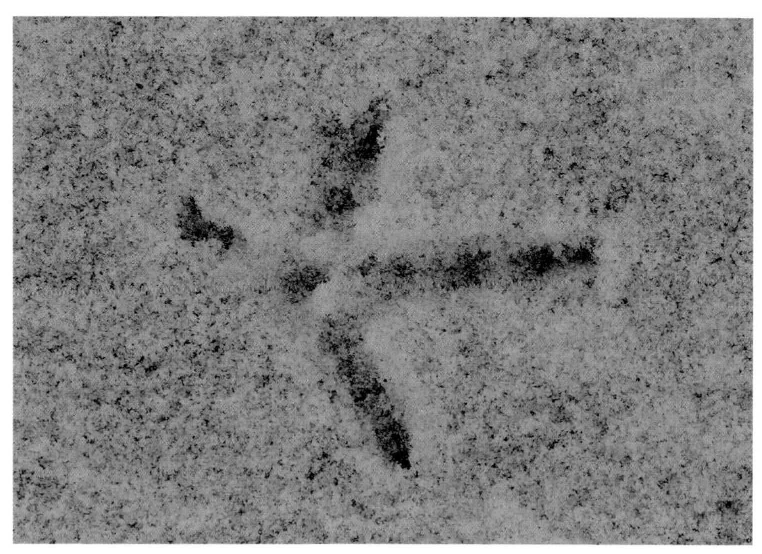

. . . and in snow. (NH)

it and to first ask diagnostic questions. These questions—"Why is this a bobwhite?" Not "is this a bobwhite?"—will help you understand the subtleties of why the track was made by one particular species and not another. It is these subtle defining characteristics or commonalities that are the most important, as they remain consistent across the differences in substrates and speeds which cause variation in track appearance or outline.

Describing all the subtleties of the effects of substrate on track and trail characteristics is beyond the scope of this book. With time in the field, however, you will absorb these nuances without even realizing it. To begin to understand the effects of substrate on tracks, start by examining your own footprints. Compare your barefoot track on firm ground with one made in several inches of mud. Likely, the mud track will look larger and your toes will have spread farther apart than when on firm ground. Trackers often use the word *splay* to indicate track spread due to such factors as depth of substrate, speed, shifts in weight, and change of direction. The deeper the substrate, the larger the track—up to a certain point, at least. In very deep substrates, the track will appear smaller, as the foot has slid into and under a layer of substrate, or the top layers may be so soft that they fold in on the foot as it sinks deeper. What is considered deep substrate is relative to the size and weight of the creature; what's deep for a junco and for

Compare the trails of a spotted sandpiper walking in deep mud (L) and in shallow mud (R).

a mute swan differs. Now try stepping in dry sand, wet sand, and so on. Examine how your tracks in the various substrates differ. If possible, also compare the trail you leave while walking in an inch of snow with that made in 2 feet of snow. It's likely that the length between tracks decreases and the width of the entire trail pattern increases with the depth of the snow.

When looking at trail characteristics in relation to depth of substrate, it may be more accurate to discuss energy output rather than speed. If you were to use the same amount of energy in each of the two trails in snow, you would move more slowly in the deeper conditions. If you wanted to move at the same speed in both, it would take a higher degree of energy output to maintain that speed in the deeper snow. The effects of depth of substrate on a bird or animal trail are dramatic.

A weathered old ruffed grouse trail looks like a line running through the woods. (MA)

Substrates also change over time, due to weather. This can be appreciated only in the field. Watch and be aware of how things change over time. Learning to identify fresh tracks is a start, but begin to challenge yourself with older and older tracks and trails.

Track Categories

Bird tracks tend to fall into one of five categories, depending on their visual characteristics. A few birds fall outside these categories (which, in fact, makes their identification easier). These include swifts, rarely encountered, and three-toed woodpeckers. Once you've placed a track in a category, further narrow down the possibilities for identification by considering the environment, the track's size, and how the bird was moving.

1. Classic Bird Tracks

The classic bird tracks are those technically termed anisodactyl. This is the bird track most people would draw: three toes pointing forward, and one pointing backward. It's toe 1 that points backward,

Classic bird tracks: American pipit (L); trail: walking in mud (R).

and toes 2, 3, and 4 point forward. Many bird species share this track category. Although kingfishers are syndactyl, with toes 2 and 3 fused for a portion of their lengths, they still leave classic bird tracks.

Measuring classic bird tracks is quite simple. Always include claws. The track length is taken from the tip of the claw of toe 1 to that of toe 3. The track width is taken from the tip of the claw of toe 2 to that of toe 4, which is the widest point within the track. Refer to the illustration.

Bird families that fall into this category include bitterns, herons, egrets, ibises, vultures, eagles, hawks, falcons, moorhens, doves, kingfishers, flycatchers, shrikes, vireos, jays, nutcrackers, magpies, crows, ravens, larks, swallows, titmice, creepers, nuthatches, wrens, dippers, kinglets, gnatcatchers, thrushes, mimics, thrashers, starlings, pipits, waxwings, warblers, tanagers, towhees, sparrows, longspurs, buntings, cardinals, bobolinks, meadowlarks, blackbirds, grackles, cowbirds, orioles, and finches.

2. Game Bird Tracks

In game bird tracks (illustrated on page 26), toe 1, the hallux, or the toe pointing backward, is greatly reduced or absent altogether. In many species, toe 1 is raised above the level of the metatarsal and other toes. This foot structure is termed *incumbent*. Technically, game bird tracks are still anisodactyl, as the skeletal nub of toe 1 points backward and toes 2, 3, and 4 point forward. Besides game birds, shorebirds also fall in this category, as well as birds whose feet are only partially webbed, termed *semipalmate*.

For better comparison across species, along with Brown et al. (1987) and Rezendes (1999), we measure the length of game bird tracks from the tip of the claw of toe 3 to the far end of the metatarsal pad. Because toe 1 does not reliably show up in tracks of game birds and shorebirds, we don't include it in measurements. This allows us to have reliable comparative measurements across species, as some don't show a hallux at all, others leave only hints of toe 1 erratically, and still others register toe 1 fairly consistently. The track width is measured as for the classic bird track, from the tip of the claw of toe 2 to that of toe 4.

Families in this category include pheasants, turkeys, grouse, ptarmigan, quail, rails, coots, cranes, plovers, oystercatchers, stilts, sandpipers, and phalaropes.

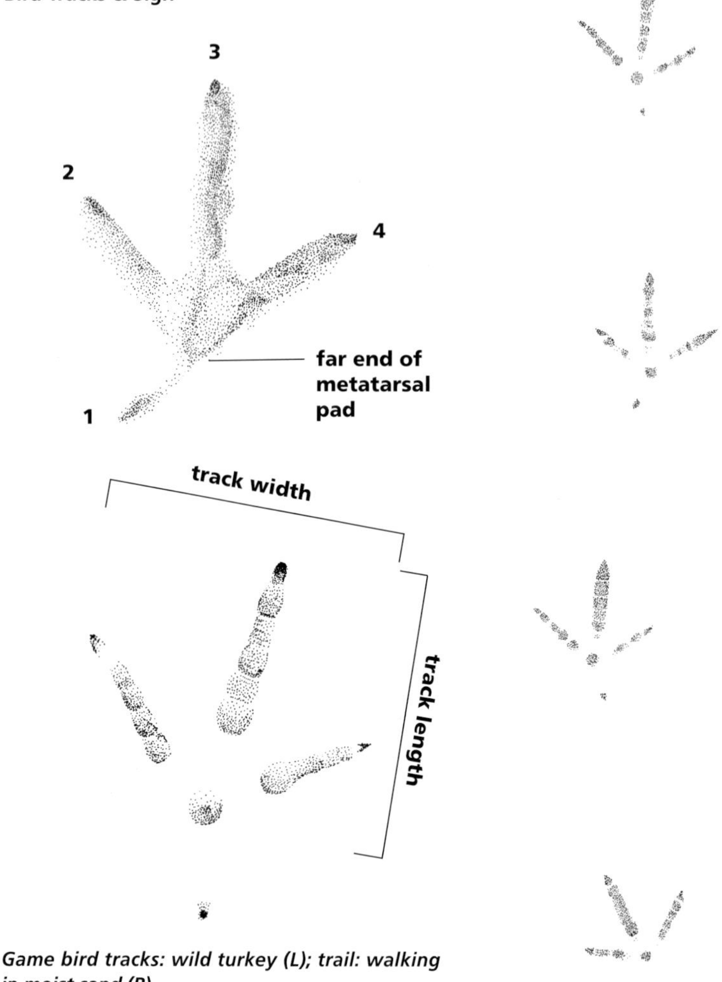

Game bird tracks: wild turkey (L); trail: walking in moist sand (R).

3. Webbed, or Palmate, Tracks

Webbed tracks (illustrated on page 27) are similar to game bird tracks in that toe 1 is greatly reduced or absent altogether, but they are distinct with their full webbing. Like those of game birds, webbed tracks are anisodactyl, and they are also *palmate,* which refers to the distal (full) webbing between toes 2 and 3 and between 3 and 4.

Measurements are taken exactly as for game bird tracks; toe 1 is not included in the length, as it shows unreliably in the tracks of some species. Families include loons, swans, geese, ducks, avocets, gulls, and terns.

4. Totipalmate Tracks

Totipalmate tracks (illustrated on page 28) are also webbed but have additional webbing found between toes 1 and 2; every toe is included in the webbing. Should the webbing be unclear, totipalmate tracks are also distinguished by toe 4 being clearly the longest toe in the track, significantly longer than toe 2 and noticeably longer than toe 3. In webbed tracks, toes 2 and 4 are relatively similar in length, and toe 3 is the longest toe in the track.

Measure the full length of totipalmate tracks from the tip of the claw of toe 1 to that of toe 4. Measure the width of the track at the widest point, which is often along toe 2. Families include boobies, gannets, pelicans, and cormorants.

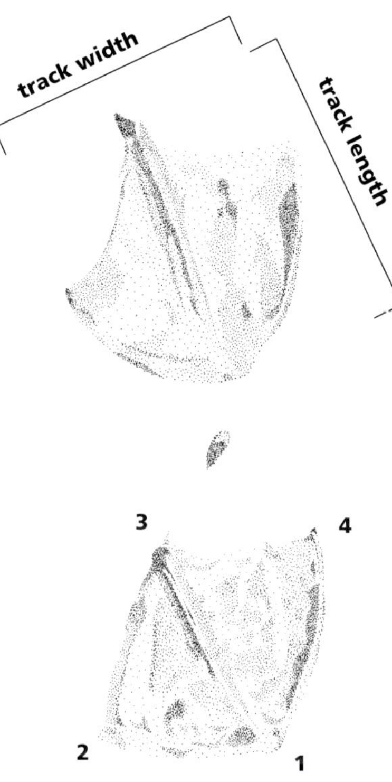

Webbed tracks: green-winged teal.

5. Zygodactyl Tracks

The zygodactyl foot pattern (illustrated on pages 29 and 30), with two toes forward and two backward, is the second most common among birds. Toes 1 and 4 point backward, and toes 2 and 3 point forward. Woodpeckers, cuckoos (which include roadrunners), and parrots consistently show this pattern clearly. Owls also fall into this group, but because they can rotate toe 4 into different positions, their tracks can potentially show great variability. However, owl tracks in the field show a consistent shape. Rather than two toes forward and two backward, owl tracks often have two toes forward, one sideways, and one backward. Three-toed woodpeckers are also included in this category for convenience.

Measure the length of zygodactyl tracks at their longest potential. Do not measure angles, which increases the measurement, but straight up the length and across the track. For owls, measure along the inside of the track, from the tip of the claw of toe 1 to that of toe 2. Measure track width for all species at the widest point. Families include ospreys, parrots, cuckoos, owls, and woodpeckers.

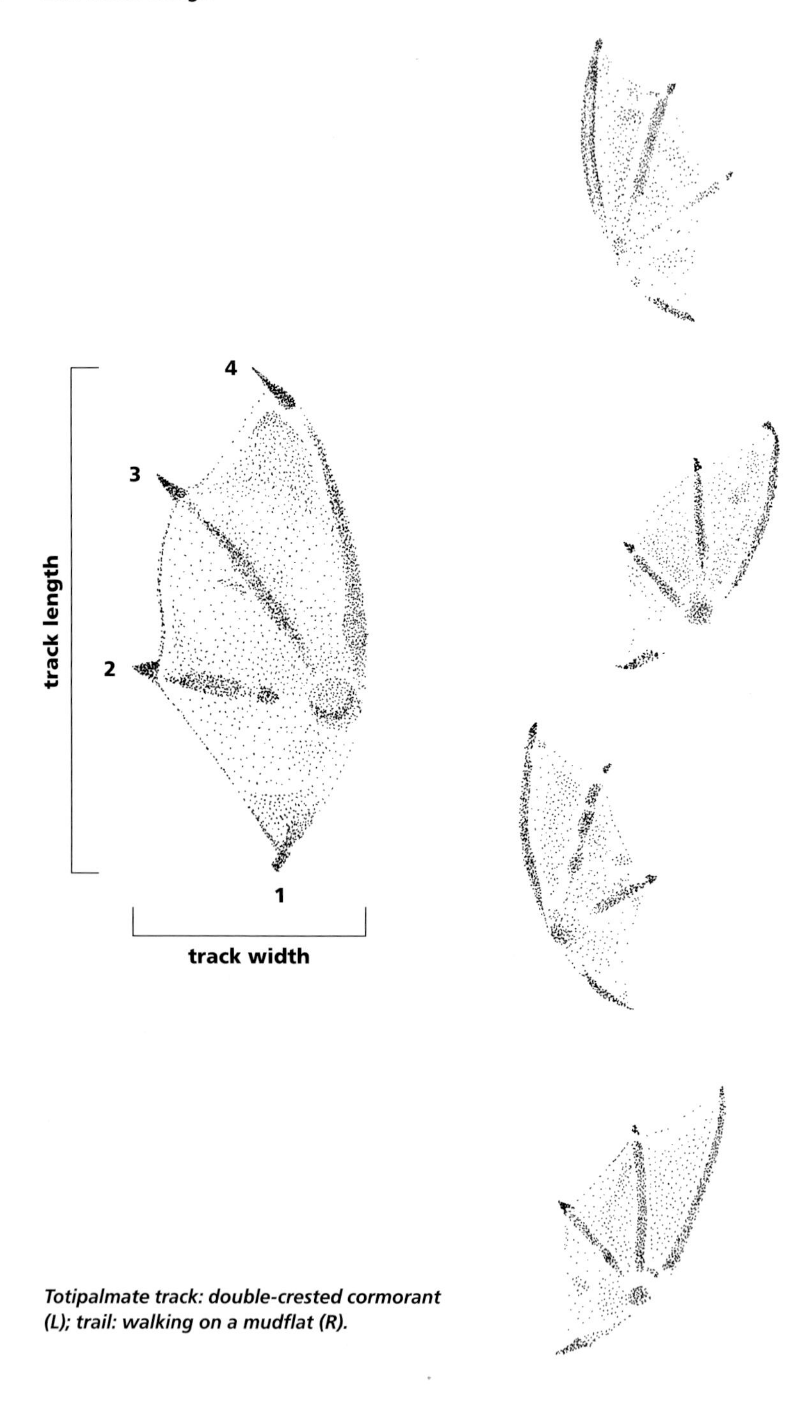

Totipalmate track: double-crested cormorant (L); trail: walking on a mudflat (R).

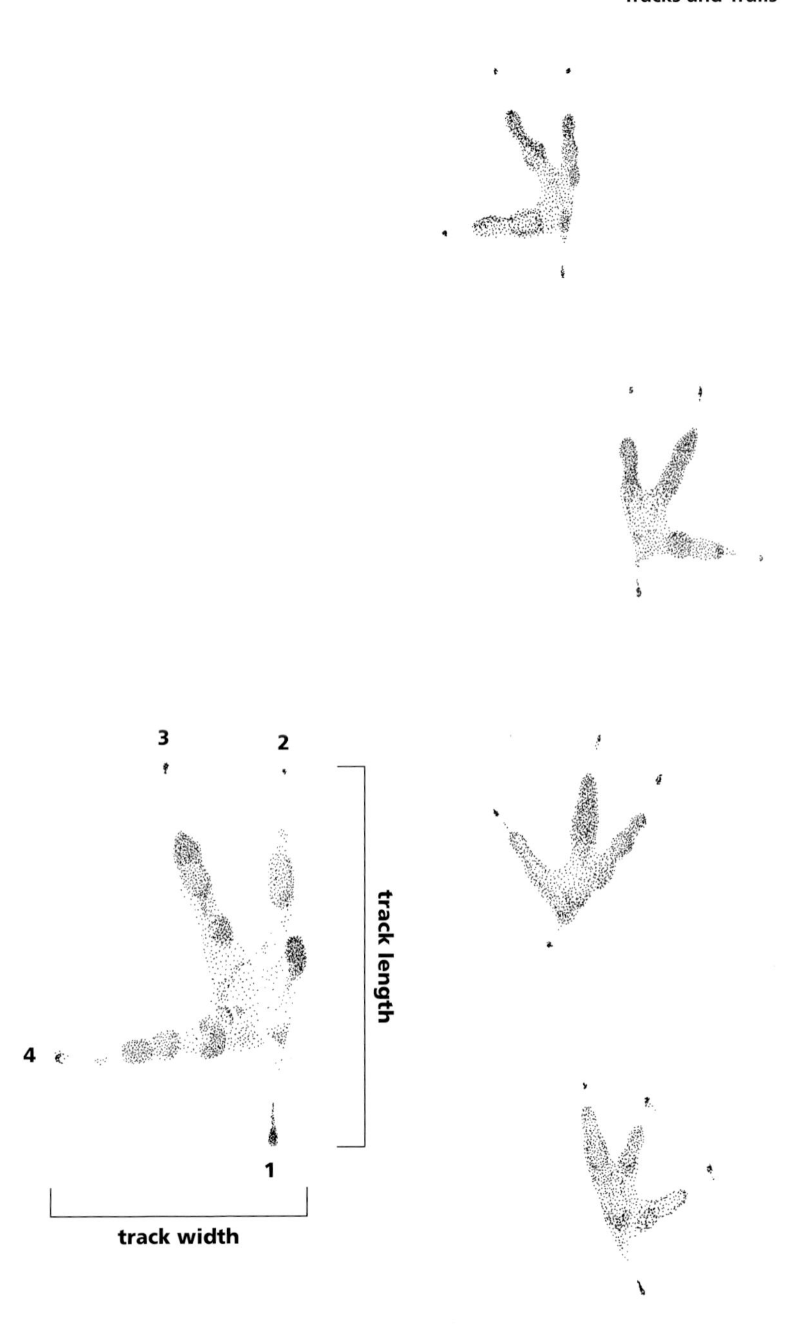

Zygodactyl track: great horned owl (L); trail: walking in mud (R).

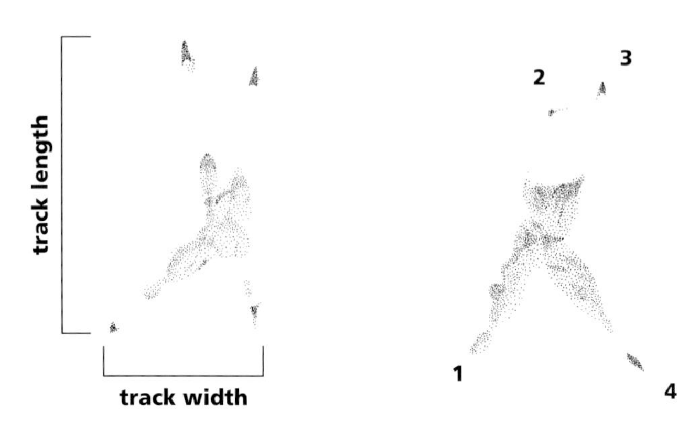

Zygodactyl tracks: pileated woodpecker.

Gaits and Track Patterns

As we dropped down into Death Valley, I just had to pull over and observe the tracks in the beautiful washes in the last few hours of daylight. I stepped out of the car onto a kit fox trail and jumped back, angry at myself, only to land on a jackrabbit trail. "Oh well," I thought, and wandered up a wash. I hadn't gone far when I encountered my first bird trail—that of a sparrow that ran from the cover of bush to bush, stopping to peck at seeds collected in small swales by the wind. I reached for my National Geographic field guide and flipped to sparrows. The sage sparrow was illustrated in midstride, and under the description were these words: "Runs on ground with tail cocked." Hmmm, I thought, a running sparrow. These trails were all over, and as the sun began to set, the birds appeared around us. Identification confirmed my tracking and was also made easier by having read the ground ahead of time.

Gait is the method of locomotion and the distinct way the body moves to propel a bird forward on the ground. There is a relationship between the gait of the bird and the track pattern left behind. Most bird track patterns show one of four distinct gaits: walking, running, hopping, or skipping. Individual bird species, like mammals, tend to use one particular gait more often than the others; this is their *natural rhythm* within an environment. However, a species may use a second and sometimes even a third gait as well. If you see a bird moving in a gait that's not its natural rhythm, ask yourself why. This will help you learn to interpret bird behavior from their trails.

Walking

Walking (illustrated on page 32) is the preferred gait for many birds that normally spend a great deal of time on the ground. All birds with a reduced or absent toe 1 walk. These include ducks, many wading birds, gulls, grouse, and quail. Most raptors walk when they're on the ground, as do crows, ravens, pipits, and magpies.

Walking is a slow gait in which one foot is placed in front of the other, and so on, carrying the bird forward in exactly the same way as we walk around. At least one foot is on the ground at all times, and once in each cycle, both feet are on the ground. The remaining track pattern is of single tracks at regular intervals, in a zigzag fashion or straight line, depending on the width of the trail, or *straddle,* which can vary. In some species, such as quail, the straddle is narrow, as they place one foot almost directly in front of the other. In other species, such as doves or ducks, the straddle is wider, as they do not place each foot in front of the other. Watch birds as often as possible, observing how they walk and how their tails swing from side to side. Watching birds move will help you make the leap from identifying the gait of a bird from a track pattern to fully visualizing the bird moving in front of you.

Measure a walking trail stride from the tip of toe 3 of one track to the tip of toe 3 of the next track—in other words, the distance between two tracks added to the length of the track in front.

Running

Running (illustrated on page 33) is another gait used by birds that tend to spend a lot of time on the ground. Just as in walking, one foot is placed in front of the other, but in rapid succession. There is also a moment in each cycle where the bird is airborne and there is no foot contact with the ground at all. Running can be observed almost anywhere, especially in backyards with American robins or along shores with sanderlings. Similar to walking trails, running trails are created by single tracks left at regular intervals.

Except with shorebirds and waders, running strides are generally two to five times the length of a single track. Long-legged birds, such as waders, often have walking strides of three to five times their individual track lengths. Because of the incredible diversity of bird leg lengths across species, it's not possible to place an exact ratio between running strides and track length to help you determine for certain that a bird is running and not walking. Towhees run at two and a half

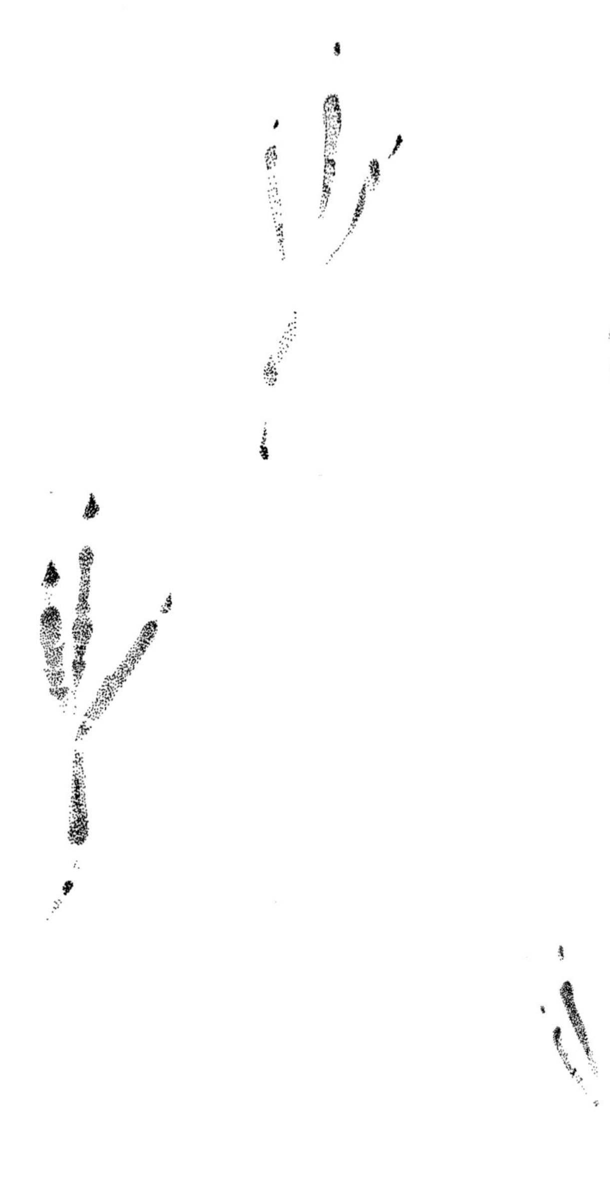

Red-winged blackbird tracks (L); trail: walking in sand (R).

running stride length

Greater roadrunner tracks (L); trail: running in dust (R).

to three times their track length. Robins are similar. Magpies run at four to six times, and thrashers three to five times. You must learn to recognize runs within a particular species, which does not take too long in the field.

Additional clues can be used to help identify a particular track pattern as a running gait. Are the tracks fully registering? Or did the bird shift its weight forward? Is there a greater disturbance or debris thrown back around the track? Has the track become blurred by the force and speed with which it was created? Has the track splayed more than usual? Has the trail width narrowed or become more pigeon-toed? These are all good things to look for.

Running trails are measured just like walking trails. Measure a stride from the tip of the claw of toe 3 in one track to that of toe 3 of the next track.

Hopping

Both hopping and skipping involve trails where tracks are paired, rather than falling independently at regular intervals, as in the above gaits. In hopping trails, paired tracks appear right next to each other, or nearly so. This pattern is possible because both feet hit the ground at the same time, or nearly so. Juncos and finches provide wonderful examples of hopping. When you see birds hopping, note the vertical component of the movement, which creates an arcing line of travel between the moments when the bird is on the ground.

The strides of hopping trails are measured from one pair to the next. Measure from the tip of the claw of toe 3 in the leading track in a pair to that of toe 3 of the lead track in the next pair. Measure trail width, or straddle, within a track pair across the widest point, which is generally from the tip of the claw of toe 4 of the left track to that of toe 4 of the right track. Trail width is a useful measurement to help distinguish species.

Skipping

We are distinguishing hopping from skipping because the trail pattern is markedly different and relates to an important difference in the body mechanics of movement. In skipping trails (illustrated on page 37), tracks are also paired, but each foot lands independently of the other. A trail pattern is a skip rather than a hop when one foot registers completely in front of the other. Birds that skip tend to be small perching birds that nevertheless spend a great deal of time on the ground, such as snow buntings and song sparrows. One theory

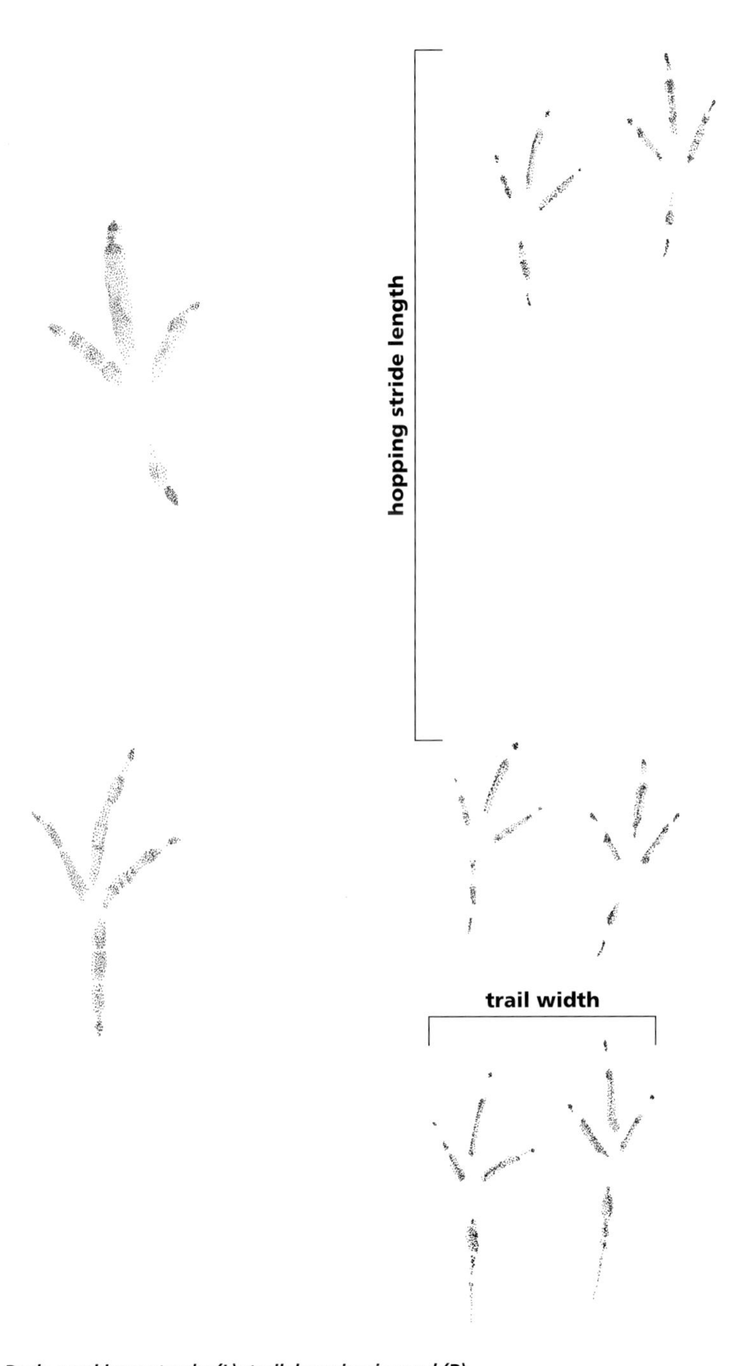

Dark-eyed junco tracks (L); trail: hopping in mud (R).

suggests that placing the feet in this way, rather than right next to each other, increases maneuverability.

When you see a bird skipping, it stays very low to the ground, which can give the impression of running. The head of the bird remains nearly level, and all the rotation comes from the lower body and legs. Both feet rotate forward, one foot strikes down, and as the body moves forward over this foot, the second foot touches down. The momentum continues propelling the body forward over the second foot, while the first foot lifts up behind the bird. Continuing forward, the second foot joins the first behind the bird, lifts off, and together they rotate forward for another cycle. Momentum is horizontal, and little energy is wasted with vertical rise.

Strides in skip trails are measured exactly like those of hopping trails, from the tip of the leading track in a pair to the tip of the leading track of the next pair. Trail width is measured straight across the pair and will be noticeably narrower than that of hop trails. Do not measure at an angle, as this will greatly increase and distort the measurements. As paired tracks spread farther apart in skipping trails, they could be mistaken for a run at a glance. Measurements will make this clear.

Track Data

The following pages of track data are first organized by track category. Once you've identified the category your bird track falls into, measure the length of the track in the manner appropriate to the track category. In general, tracks are organized within each category from the shortest length to the longest. Begin tracking by using measurements to help in your analysis. It will help improve your visual and intuitive skills, as well as provide scientific evidence or documentation when needed. Later, if you so desire, you can leave your ruler behind or keep it in your pack rather than in your pocket.

Tracks and gaits are illustrated here for nearly all the species researched. Gaits illustrated are the natural rhythms of the species, those being the trail patterns you're most likely to encounter afield. Combining track measurements and characteristics with the track patterns provides a powerful tool in species identification.

Drawings and photos are given for many species, along with descriptive captions. The illustrations were all rendered accurately from photographs taken in the field. You can answer questions such

Snow bunting tracks (L); trail: skipping in sand (R).

as whether the metatarsal registers or what the proportion of toe 1 is to the entire track by studying the track illustrations. Gait illustrations realistically reflect the effects of the substrate in which they were photographed and provide a large range of examples of track variations and effects due to substrate. Also note the trail width (TW) and the degree to which a bird is pigeon-toed.

If the bird track is a common one, you will likely find it within these pages, but even if not, you can still decipher the species if you think ecologically. One representative of most families is illustrated, and measurement comparisons for other members of the family are often given. In some cases, entire families have similar tracks. Ask yourself as many questions as possible to aid in identification. Which birds are common in the area you're tracking? What was the behavior of the bird whose track you found? Then try to determine which photo or illustration here is most similar to the track you're trying to identify.

Classic Bird Tracks

Barn swallow tracks on a riverbank. (WY)

Barn Swallow *(Hirundo rustica)*

Track: 11/16–1 in. (1.7–2.5 cm) L x 1/4–7/16 in. (.6–1.1 cm) W

Classic bird track. Very small. Anisodactyl. Metatarsal weakly registers or is absent altogether.

Notes: Common only in muddy areas during the nest-building season; otherwise, rarely encountered.

Redpolls *(Carduelis* sp.)

Track: 7/8–1 in. (2.2–2.5 cm) L x 5/16–3/8 in. (.8–1 cm) W

Classic bird track. Very small. Anisodactyl. Metatarsal weakly registers or is absent altogether. Toes 1 and 3 often curve in toward the center of the trail, which is typical for finch species.

Similar species: Kinglet tracks are very similar. Other finches share the same curved track characteristics and should be distinguished by size and behaviors. Barn swallows should be easily distinguished using ecological clues.

Trail: Hop TW 1–1 9/16 in. (2.5–4 cm)
Strides 2–10 in. (5.1–25.4 cm)

Notes: Redpolls tend to feed in flocks, and their combined tracks carpet a feeding area. They will also forage on the ground with other species, most notably grosbeaks and crossbills.

Redpoll tracks (L); trail: hopping in a dusting of snow (R).

Black-capped Chickadee *(Poecile atricapillus)*

Track: $\frac{15}{16}$–$1\frac{1}{8}$ in. (2.4–2.8 cm) L x $\frac{1}{8}$–$\frac{1}{4}$ in. (.3–.6 cm) W

Classic bird track. Very small. Anisodactyl. Metatarsal weakly registers or is absent altogether. Toes 1 and 2 form a straight line on the inside of the track, which is unique.

Similar species: Redpolls and other finches have curved tracks, as toes 1 and 3 point in toward the center of the trail. Finches, swallows, and warblers also tend to splay their toes farther.

Trail: Hop TW $1\frac{1}{8}$–$1\frac{7}{16}$ in. (2.8–3.7 cm)
Strides 1–5 in. (2.5–12.7 cm)

Black-capped chickadee tracks (L); trail: hopping in mud (R).

Common yellowthroat tracks on a streambank. (CA)

Common Yellowthroat
(Geothlypis trichas)

Track: 15/16–1 1/8 in. (2.4–2.8 cm) L x 5/16–9/16 in. (.8–1.4 cm) W

Classic bird track. Very small. Anisodactyl. Metatarsal weakly registers or is absent altogether.

Similar species: Redpolls and other finches have curved tracks, as toes 1 and 3 point in toward the center of the trail. Chickadees have very narrow tracks. Swallows alight and collect mud, pick up an injured insect, or bathe, but don't hunt and explore, which would leave long trails on the ground.

Trail: Hop TW 7/8–1 7/16 in. (2.2–3.7 cm)
Strides 1–5 in. (2.5–12.7 cm)

(Parameters created from small data pool.)

American Goldfinch *(Carduelis tristis)*

Track: 1–1 1/8 in. (2.5–2.8 cm) L x 1/4–3/8 in. (.6–1 cm) W

Classic bird track. Very small. Anisodactyl. Metatarsal weakly registers or is absent altogether. Toes 1 and 3 often curve in toward the center of the trail, which is typical for finch species.

Similar species: Other finches share the same curved track characteristics and should be distinguished by size and behaviors.

Trail: Hop TW 1 1/16–1 7/16 in. (2.7–3.7 cm)
Strides 1–6 in. (2.5–15.2 cm)

American goldfinch tracks.

Louisiana waterthrush track (L); trail: walking in soft mud (R).

Louisiana Waterthrush *(Seiurus motacilla)*

Track: 1–1 3/16 in. (2.5–3 cm) L x 3/8–1/2 in. (1–1.2 cm) W
Classic bird track. Very small. Anisodactyl. Metatarsal weakly registers or is absent altogether.

Similar species: Finches, other warblers, and chickadees all hop rather than walk.

Trail: Walk Strides 1 1/8–1 5/8 in. (2.8–4.1 cm)

Notes: Tracks are fairly common along riverbanks where the birds reside. Trails are a combination of short and long strides, with frequent pauses, which will appear as a paired set of tracks.

Belted kingfisher tracks.

Belted Kingfisher *(Ceryle alcyon)*

Track: 1 1/8–1 1/4 in. (2.8–3.1 cm) L x 3/8 in. (1 cm) W
Classic bird track. Very small. Syndactyl. Metatarsal weakly registers or is absent altogether.

(Parameters created from small data pool.)

Brewer's Sparrow
(Spizella breweri)

Track: 1 1/8–1 1/4 in. (2.8–3.1 cm) L x 3/8–9/16 in. (1–1.4 cm) W

Classic bird track. Very small. Anisodactyl. Metatarsal weakly registers or is absent altogether.

Similar species: Hard to distinguish from chipping sparrow tracks. Other sparrows and juncos have larger tracks. Finches and winter wrens have curved tracks.

Trail: Hop TW 1 5/8–1 7/8 in. (4.1–4.7 cm)
Strides 2–6 in. (5.1–15.2 cm), while feeding 1/2–2 1/2 in. (1.2–6.3 cm)

Notes: Brewer's sparrows tend to feed in flocks during the winter months.

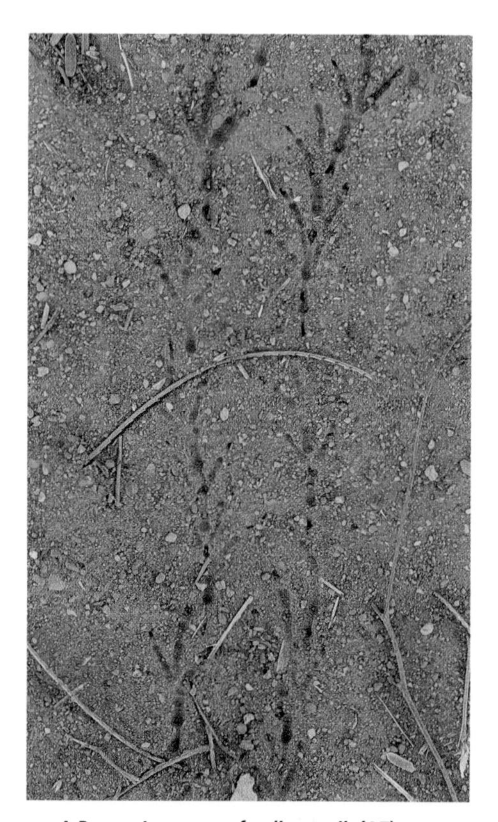

A Brewer's sparrow feeding trail. (AZ)

Brewer's sparrow trail: hopping in dust.

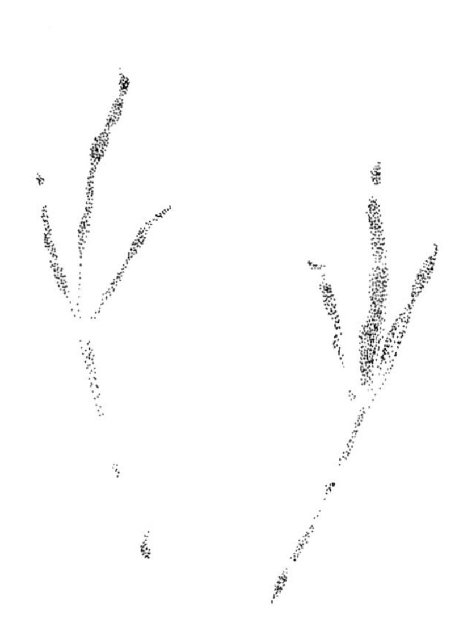

Winter wren tracks (L); trail: hopping in mud (R).

Winter Wren *(Troglodytes troglodytes)*

Track: 1 1/8–1 1/4 in. (2.8–3.1 cm) L x 5/16–7/16 in. (.8–1.1 cm) W

Classic bird track. Very small. Anisodactyl. Metatarsal weakly registers or is absent altogether. Toe 1 is very long, and the tracks appear curved, as toes 1 and 3 point inward toward the center of the trail.

Similar species: Sparrows and warblers have a shorter toe 1 in proportion to the length of the track.

Trail: Hop TW 1 1/8–1 3/8 in. (2.8–3.5 cm)
Strides 1 1/2–4 in. (3.8–10.1 cm)

Notes: Tracks are fairly common in dusty or muddy patches under logs in northern forests.

(Parameters created from small data pool.)

Common Ground-dove
(Columbina passerina)

Track: 1 1/8–1 1/4 in. (2.8–3.1 cm) L x 11/16–7/8 in. (1.8–2.2 cm) W

Classic bird track. Very small. Anisodactyl. Metatarsal weakly registers or is absent altogether.

Similar species: Other dove tracks are larger.

Trail: Walk Strides 1 1/8–2 3/8 in. (2.8–6 cm) Short, wide trail.

Notes: Drag marks are often present in soft substrates.

Common ground-dove tracks (L); trail: walking in deep dust (R).

House Sparrow
(Passer domesticus)

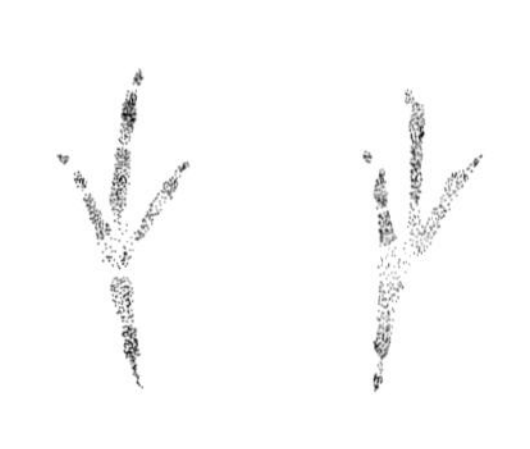

Track: 1 1/8–1 5/16 in. (2.8–3.3 cm) L x 3/8–9/16 in. (1–1.4 cm) W

Classic bird track. Very small. Anisodactyl. Metatarsal weakly registers or is absent altogether.

Similar species: Other small sparrows.

Notes: Common around human habitations and urban centers.

House sparrow trail: hopping in mud.

Black-throated Sparrow
(Amphispiza bilineata)

Track: 1 1/4–1 1/2 in. (3.2–3.8 cm) L x 7/16–5/8 in. (1.1–1.6 cm) W

Classic bird track. Very small. Anisodactyl. Metatarsal weakly registers or is absent altogether.

Similar species: Other sparrows and juncos have larger tracks. Finch tracks are curved.

Trail: Hop TW 1 1/4–1 11/16 in. (3.2–4.3 cm)

(Parameters created from small data pool.)

Other Small Animal Tracks

Certain other creatures leave tracks that look very much like those of small birds. The secret to avoiding such misidentification is to patiently and realistically be conscious that such mistakes are not only possible, but truly inevitable over time. The tracks of frogs and jumping mice, especially when partially registering, can look like very small passerines. The feet of the mice, which are unique, leave impressions that are especially likely to mimic bird tracks. The three center toes of jumping mice hind feet are long and thin, and are often all that clearly registers in their trails. When their feet land parallel, you may misread the incomplete registration of their feet as being toes 2, 3, and 4 of a small perching bird that is hopping along.

Also look closely in deep substrates, especially snow, where feeding redpoll and junco trails may resemble those of bounding mice or trotting voles. Follow obscure snow trails when in doubt, looking for flight or wing impressions that might make things clearer—or to see whether the trail disappears into a woodpile or down a burrow.

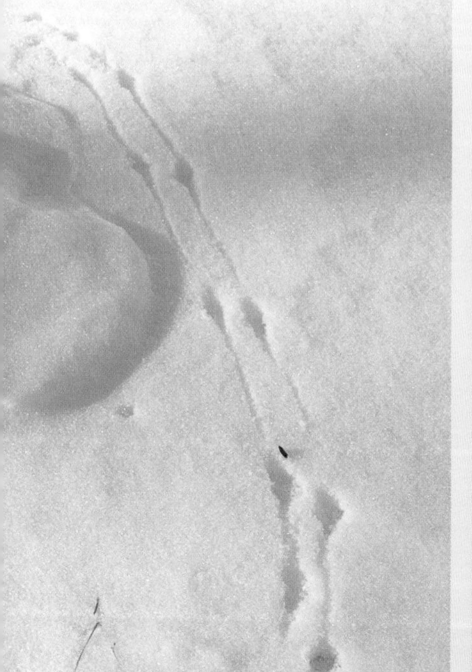

Left: *A dark-eyed junco trail in snow. (NH)* **Right:** *For comparison, look at these mice trails in snow. (UT)*

A sage sparrow running trail. (CA)

Sage Sparrow *(Amphispiza belli)*

Track: 1¼–1½ in. (3.2–3.8 cm) L x 7/16–¾ in. (1.1–1.9 cm) W

Classic bird track. Small. Anisodactyl. Metatarsal weakly registers or is absent altogether.

Similar species: Other sparrows.

Trail: Run Strides 3½–5½ in. (8.9–14 cm)
Skip Strides 5–8 in. (12.7–20.3 cm)

Notes: Claw of toe 1 often drags in run, which gives the impression of an elongated track.

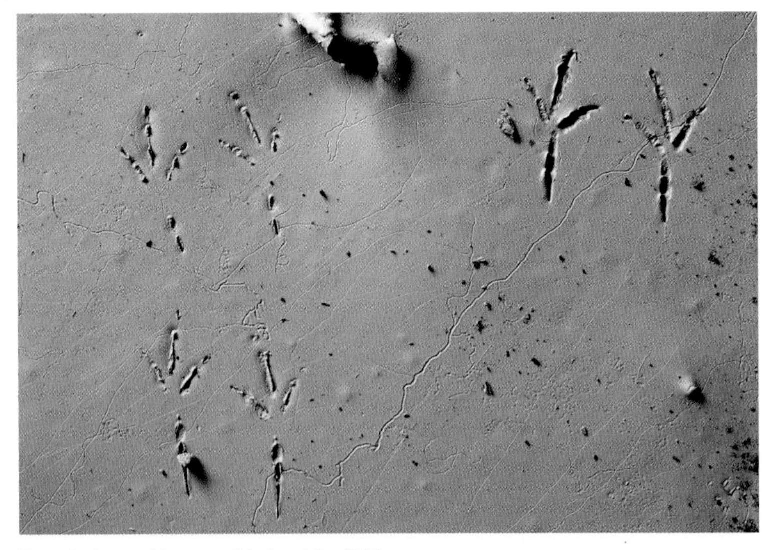

Two dark-eyed juncos, side by side. (NH)

Dark-eyed junco tracks (L); trail: hopping in mud (R).

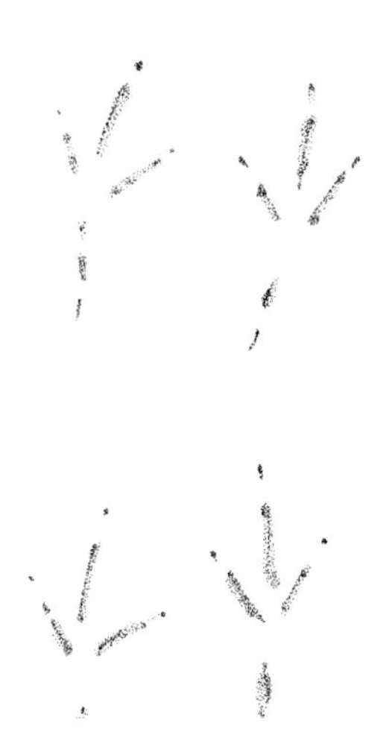

Dark-eyed Junco
(Junco hyemalis)

Track: 1¼–1½ in. (3.2–3.8 cm) L
x 7/16–5/8 in. (1.1–1.6 cm) W

Classic bird track. Very small. Anisodactyl. Metatarsal weakly registers or is absent altogether.

Similar species: Most sparrows have a longer toe 1 in proportion to the overall track length. Song sparrows skip more often than hop. Finch tracks are curved.

Trail: Hop TW 1⅜–2⅛ in. (3.5–5.3 cm)
Strides 1–6 in. (2.5–15.2 cm)

Notes: This is a common track in mud and snow, and an important one to know. Juncos will also skip from time to time.

Northern Cardinal
(Cardinalis cardinalis)

Track: 1 1/4–1 9/16 in. (3.2–4 cm) L x 7/16–11/16 in. (1.1–1.7 cm) W

Classic bird track. Very small to small. Anisodactyl. Metatarsal weakly registers or is absent altogether.

Similar species: Sparrow tracks are the same size, but have a different foot morphology.

Trail: Hop TW 1 7/8–2 1/4 in. (4.7–5.4 cm)
Strides 3 1/2–12 in. (8.9–30.5 cm)

(Parameters created from small data pool.)

Cardinal tracks. (MA)

Northern cardinal trail: hopping in mud.

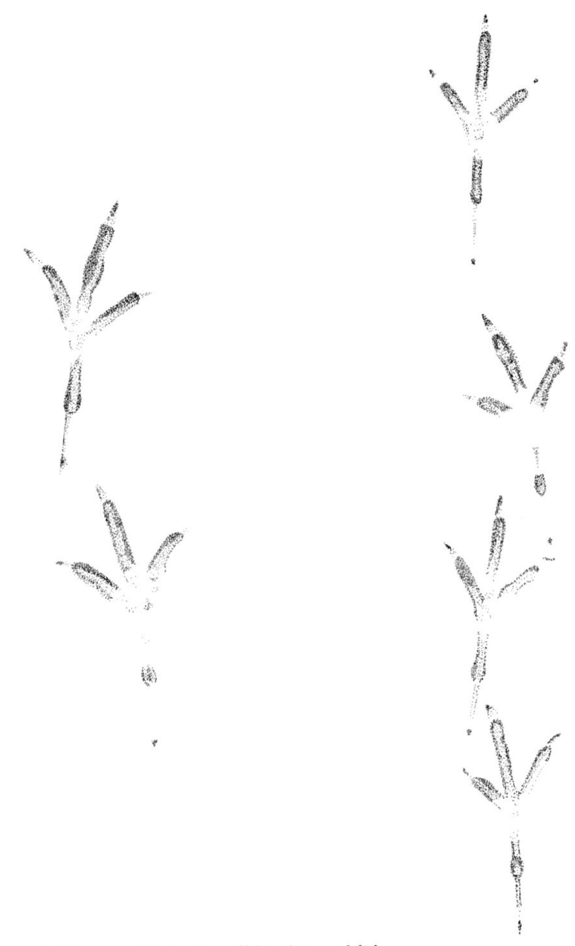

American pipit tracks (L); trail: walking in mud (R).

American Pipit *(Anthus rubescens)*

Track: 1 3/8–1 1/2 in. (3.5–3.8 cm) L x 5/8–7/8 in. (1.6–2.2 cm) W

Classic bird track. Small. Anisodactyl. Metatarsal weakly registers or is absent altogether. The long claw on toe 1 is an important characteristic. Long claw on toe 1 measures 1/4–5/16 in. (.6–.7 cm) L.

Similar species: The trail resembles that of a common ground-dove, but can be distinguished by the habitat and long claw of toe 1. Horned lark tracks are also very similar but should be differentiated by habitat and the tendency for horned larks to run unless feeding.

Trail: Walk Strides 1 1/2–3 1/4 in. (3.8–8.2 cm)

Notes: Pipits are slightly pigeon-toed. They walk unless pushed by a predator, when they may run a short distance or fly to safety.

Horned Lark *(Eremophilia alpestris)*

Track: 1 3/8–1 5/8 in. (3.5–4.1 cm) L x 1/2–1 in. (1.2–2.5 cm) W

Classic bird track. Small. Anisodactyl. Metatarsal weakly registers or is absent altogether. Long claw on toe 1. Toe 3 often curves inward towards toe 2.

Similar species: Pipits tend to walk; snow buntings and longspurs tend to skip.

Trail: Run Strides 3 1/4–7 in. (8.2–17.8 cm)

Notes: Horned larks are very pigeon-toed. The classic trail of the horned lark is the run, although it will walk while feeding; the resulting trail may be confused with a pipit trail. Use habitat and behavioral clues to aid in identification.

Horned lark track (L); trail: running in moist sand (R).

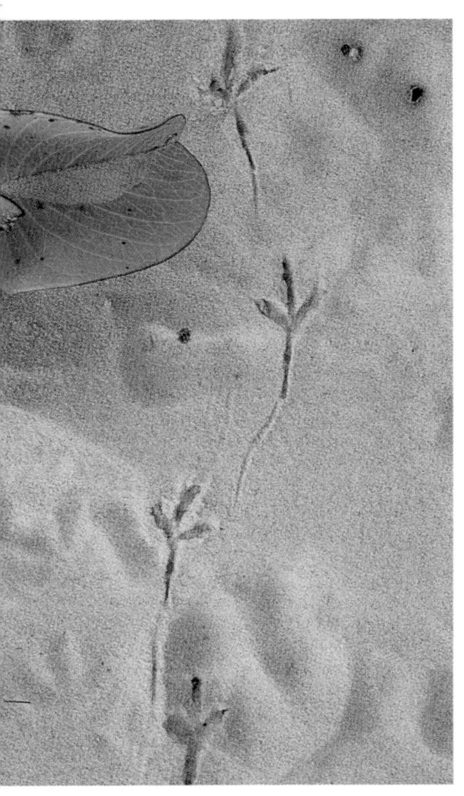

The walking trail of a horned lark. (TX)

White-throated Sparrow *(Zonotrichia albicollis)*

Track: $1\frac{5}{16}$–$1\frac{1}{2}$ in. (3.3–3.8 cm) L x $\frac{1}{2}$–$\frac{3}{4}$ in. (1.2–1.9 cm) W

Classic bird track. Small. Anisodactyl. Metatarsal weakly registers or is absent altogether.

Similar species: Unlike most other sparrows, the proportion of a white-throated sparrow's hallux to the overall track length is similar to that of a junco.

Trail: Hop — TW $1\frac{5}{8}$–$2\frac{1}{8}$ in. (4.1–5.3 cm)

Strides $2\frac{1}{2}$–6 in. (6.3–15.2 cm)

White-throated sparrow tracks. (NH)

The hopping trail of a vesper sparrow. (AZ)

Vesper Sparrow *(Pooecetes gramineus)*

Track: 1 3/8–1 5/8 in. (3.5–4.1 cm) L x 1/2–5/8 in. (1.2–1.6 cm) W

Classic bird track. Small. Anisodactyl. Metatarsal weakly registers or is absent altogether.

Similar species: Other sparrows.

Trail: Hop TW 1 1/2–2 1/4 in. (3.8–5.7 cm)
Strides 2 1/2–8 in. (6.3–20.3 cm)

Skips at speed. Strides up to 15 in. (38.1 cm)

Notes: Feeds in flocks during the winter months.

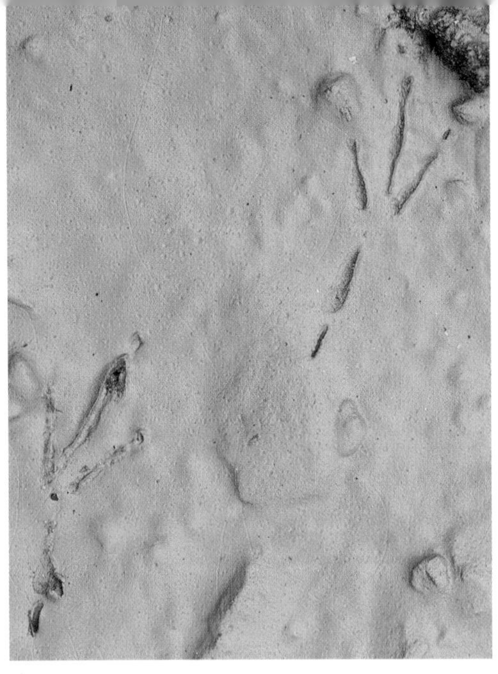

Song sparrow tracks, within a skipping trail. (NH)

Song sparrow tracks (L); trail: skipping in dust (R).

Song Sparrow *(Melospiza melodia)*

Track: 1 3/8–1 5/8 in. (3.5–4.1 cm) L x 1/2–5/8 in. (1.2–1.6 cm) W

Classic bird track. Small. Anisodactyl. Metatarsal weakly registers or is absent altogether.

Similar species: Other sparrows have similar tracks. But few other sparrows skip as often as song sparrows. Junco tracks are more robust and proportionately have a shorter halllux.

Trail: Skip TW 1 1/4–2 in. (3.2–5.1 cm)
Strides 3 3/4–15 in. (9.5–38.1 cm)

Notes: A common track, as this sparrow is very comfortable on the ground.

Savannah Sparrow
(Passerculus sandwichensis)

Track: $1^3/_8$–$1^5/_8$ in. (3.5–4.1 cm) L x $^1/_2$–$^5/_8$ in. (1.2–1.6 cm) W

Classic bird track. Small. Anisodactyl. Metatarsal weakly registers or is absent altogether.

Similar species: Other sparrows, but use track patterns to help differentiate. On occasion, song sparrows will run for several steps before switching into a skip. Follow the trail a bit to feel more comfortable in your assessment.

Trail:	Run	Strides $3^1/_2$–$5^1/_2$ in. (8.9–14 cm)
	Skip	Strides 5–8 in. (12.7–20.3 cm)

Savannah sparrow trail: running in sand.

White-crowned Sparrow
(Zonotrichia leucophrys)

Track: $1^7/_{16}$–$1^5/_8$ in. (3.6–4.1 cm) L x $^9/_{16}$–$^3/_4$ in. (1.4–1.9 cm) W

Classic bird track. Small. Anisodactyl. Metatarsal weakly registers or is absent altogether.

White-breasted nuthatch track (L); trail: hopping in mud (R).

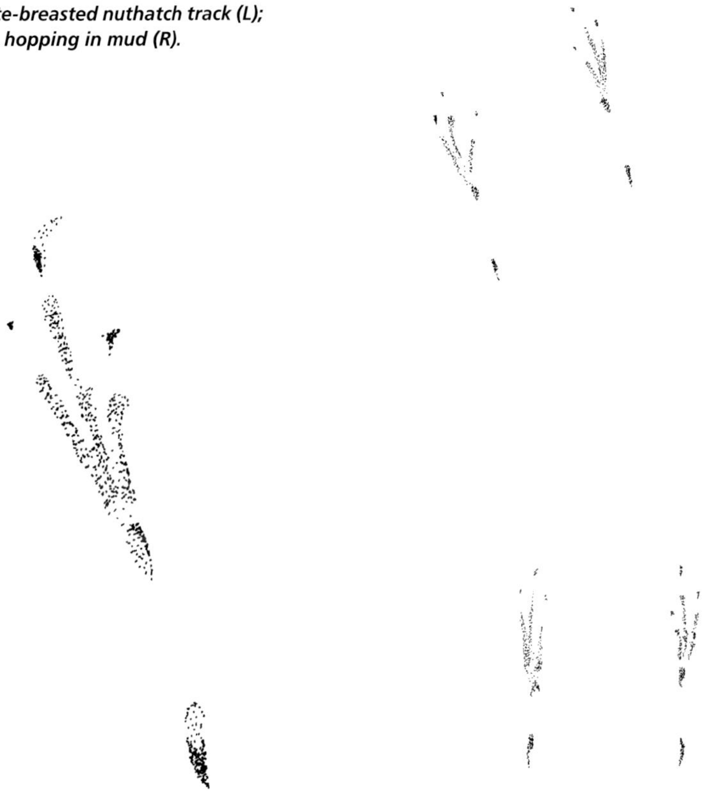

White-breasted Nuthatch
(Sitta carolinensis)

Track: $1\frac{1}{2}$–$1\frac{5}{8}$ in. (3.8–4.1 cm) L x $\frac{3}{16}$–$\frac{3}{8}$ in. (.5–1 cm) W

Classic bird track. Small. Anisodactyl. Metatarsal weakly registers or is absent altogether. Long claw on toe 1 measures $\frac{1}{2}$–$\frac{5}{8}$ in. (1.2–1.6 cm) L.

Similar species: The narrow track and incredible claw of toe 1 make this track hard to confuse with those of other species. Other nuthatches and brown creepers have smaller tracks. Woodpeckers and sapsuckers are zygodactyl.

Trail: Hop TW $1\frac{3}{8}$–$1\frac{7}{8}$ in. (3.5–4.7 cm)
Strides 2–6 in. (5.1–15.2 cm)

Evening Grosbeak
(Coccothraustes vespertinus)

Track: $1\frac{3}{8}$–$1\frac{5}{8}$ in. (3.5–4.1 cm) L x $\frac{3}{8}$–$\frac{5}{8}$ in. (1–1.6) W

Classic bird track. Small. Anisodactyl. Metatarsal weakly registers or is absent altogether. Track is curved and narrow, typical of the finch family.

Similar species: Other grosbeaks. Other finch tracks are smaller.

Trail: Hop TW $1\frac{5}{8}$–$2\frac{1}{2}$ in. (4.1–6.3 cm)
Strides $3\frac{1}{2}$–8 in. (8.9–20.3 cm)

Notes: Often feed in flocks, including mixed flocks with other species, such as redpolls and crossbills. Males are significantly larger than females.

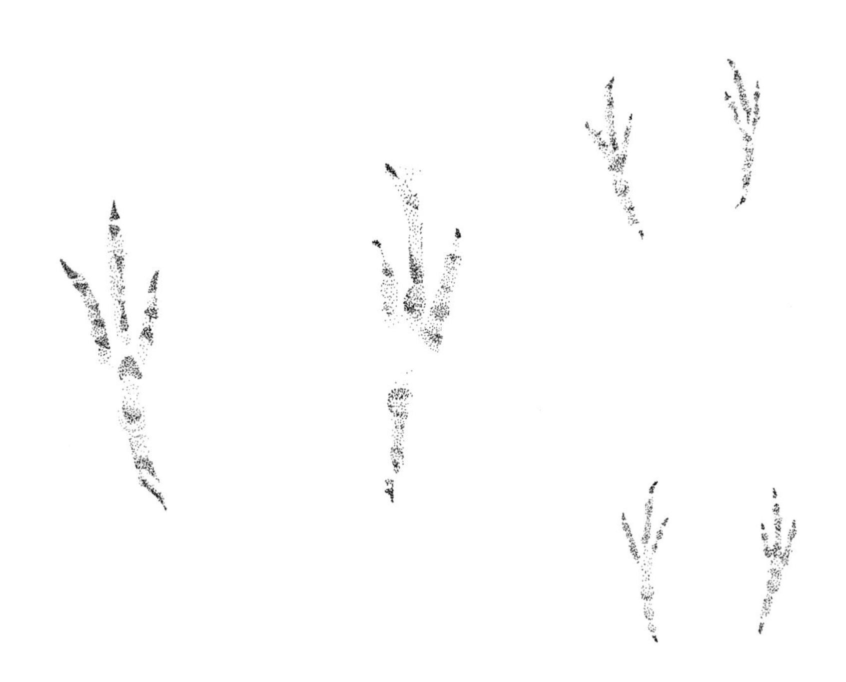

Evening grosbeak tracks (L); trail: hopping in shallow snow (R).

Snow Bunting *(Plectrophenax nivalis)*

Track: $1\frac{3}{8}$–$1\frac{5}{8}$ in. (3.5–4.1 cm) L x $\frac{5}{8}$–$\frac{3}{4}$ in. (1.6–1.9 cm) W

Classic bird track. Small. Anisodactyl. Metatarsal weakly registers or is absent altogether.

Similar species: Longspurs have similar tracks and gaits. Horned larks, which often share the same coastal habitat in winter months, run rather than skip.

Trail:	Skip	TW $1\frac{3}{8}$–$2\frac{1}{4}$ in. (3.5–5.7 cm)
		Strides 4–10 in. (10.1–25.4 cm)
	Hop slanted	Strides 2–5 in. (5.1–12.7 cm)

Notes: Often found in flocks and will readily mix with other species, such as horned larks and longspurs.

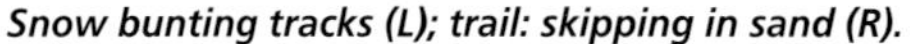

Snow bunting tracks (L); trail: skipping in sand (R).

Spotted Towhee *(Pipilo maculatus)*

Track: 1 5/8–1 7/8 in. (4.1–4.7 cm) L x 3/4–1 in. (1.9–2.5 cm) W

Classic bird track. Small. Anisodactyl. Metatarsal weakly registers or is absent altogether.

Similar species: Other towhees, especially the eastern towhee. Canyon towhees tend to run, and spotted towhees far more often hop. Toes 2 and 4 appear longer than those of robins and mimic thrushes.

Trail:		
	Hop	TW 1 7/8–2 1/8 in. (4.7–5.4 cm)
		Strides 3–8 in. (7.6–20.3 cm)
	Run	Strides 3 1/4–5 1/2 in. (8.2–14 cm)

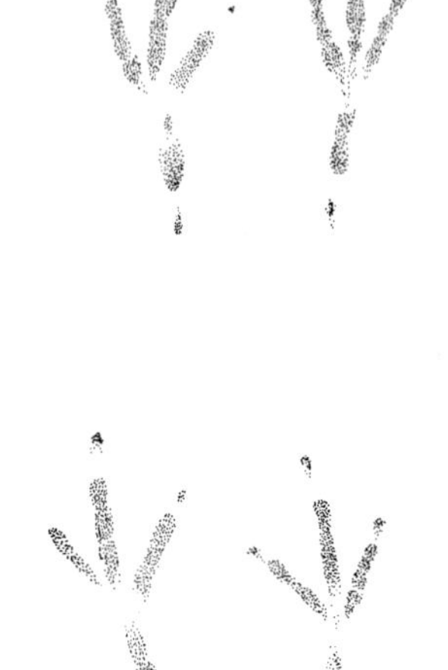

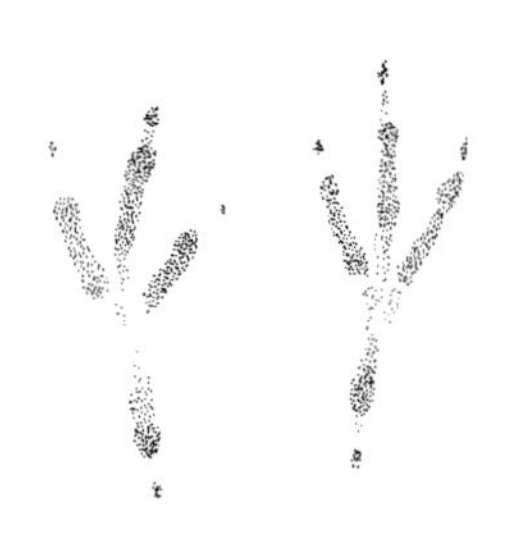

Spotted towhee trail: hopping in dust.

Spotted towhee tracks. (TX)

Cactus Wren *(Campylorhynchus brunneicapillus)*

Track: 1 1/2–1 3/4 in. (3.8–4.4 cm) L x 9/16–3/4 in. (1.4–1.9 cm) W

Classic bird track. Small. Anisodactyl. Metatarsal weakly registers or is absent altogether.

Notes: Runs and hops.

(Parameters created from small data pool.)

Strong Perching Birds

In the tracks of birds that spend little time on the ground, you'll note a greater development of the toe pad on the hallux, just behind the metatarsal. This wide, flat pad is a good indicator of how much time a species spends on the ground. The lack of evidence of this pad does not mean that the bird spends considerable time on the ground, however. Possibly it did not register well, or the substrate may have affected the track. Or consider the tracks of nuthatches, which spend little time on the ground; here, only the claw of the hallux typically shows.

Green-tailed Towhee
(Pipilo chlorurus)

Track: 1 9/16–1 3/4 in. (4–4.4 cm) L x 11/16–13/16 in. (1.7–2 cm) W

Classic bird track. Small. Anisodactyl. Metatarsal weakly registers or is absent altogether.

Similar species: Toes and tracks are more robust than those of sparrows. Study the gait for positive identification. Other towhee tracks are larger.

Trail:	Run	Strides 3–6 1/2 in. (7.6–16.5 cm)
	Hop	TW 2 1/4–2 5/8 in. (5.7–6.6 cm)
		Strides 5–8 in. (12.7–20.3 cm)

Notes: The green-tailed towhee is slightly pigeon-toed. It skips too.

(Parameters created from small data pool.)

Green-tailed towhee trail: running in dust.

Canyon Towhee *(Pipilo fuscus)*

Track: 1 5/8–1 7/8 in. (4.1–4.7 cm) L x 3/4–1 in. (1.9–2.5 cm) W

Classic bird track. Small. Anisodactyl. Metatarsal weakly registers or is absent altogether.

Similar species: Other towhees. Spotted towhees tend to hop.

Trail:	Run	Strides 4 1/4–7 1/4 in. (10.8–18.4 cm)

Gray Catbird *(Dumetella carolinesis)*

Track: 1¾–1$^{15}/_{16}$ in. (4.4–4.9 cm) L x ¾–$^{15}/_{16}$ in. (1.9–2.3 cm) W

Classic bird track. Medium. Anisodactyl. Metatarsal weakly registers or is absent altogether. Slim toes of even width.

Similar species: Robins and mockingbirds leave similar trails but have curved toes.

Trail: Run Strides 4–6 in. (10.2–15.3 cm)

(Parameters created from small data pool.)

Gray catbird track (L); trail: running in mud (R).

California Towhee *(Pipilo crissalis)*

Track: 1¾–2 in. (4.4–5.1 cm) L x $^{9}/_{16}$–⅞ in. (1.4–2.2 cm) W

Classic bird track. Medium. Anisodactyl. Metatarsal weakly registers or is absent altogether.

Similar species: Jays and nutcrackers have very narrow tracks. Other towhees may be considered, depending on habitat and size.

Trail: Run Strides 4–6½ in. (10.2–16.5 cm)

Mourning Dove *(Zenaida macroura)*

Track: 1 5/8–1 7/8 in. (4.1–4.7 cm) L x 1–1 1/4 in. (2.5–3.2 cm) W

Classic bird track. Small. Anisodactyl. Metatarsal weakly registers or is absent altogether.

Similar species: When toe 1 registers lightly, track appears like that of a small quail. Pigeon and white-winged dove tracks are larger, and other dove tracks are smaller.

Trail: Walk Strides 2–4 1/2 in. (5.1–11.5 cm)

Notes: Mourning dove trails are very common. This is an important bird track to learn to identify confidently.

Mourning dove tracks (L); trail: walking in mud (R).

The walking trail of a mourning dove at sunup. (TX)

Gray Jay *(Perisoreus canadensis)*

Track: 1 3/4–2 in. (4.4–5.1 cm) L x 1/2–3/4 in. (1.2–1.9 cm) W

Classic bird track. Medium. Anisodactyl. Metatarsal weakly registers or is absent altogether.

Similar species: Other jays and nutcrackers have larger tracks. Towhees splay their feet more.

Trail: Hop TW 1 3/4–2 7/8 in. (4.4–7.3 cm)
Strides 6–18 in. (15.3–45.8 cm)

American Dipper *(Cinclus mexicanus)*

Track: 1 3/4–1 15/16 in. (4.4–4.9 cm) L x 7/8–1 in. (2.2–2.5 cm) W

Classic bird track. Medium. Anisodactyl. Metatarsal weakly registers or is absent altogether. Slim toes or even width.

Similar species: Robins tend to run, rather than walk, and their toe pads are very bulbous. They often do not register their metatarsal area or complete toe 1 pads, which show far more often in dipper tracks. Blackbirds have narrower and larger tracks, in general, and in blackbird tracks, toe 3 hugs toe 4 near the metatarsal pad.

Trail: Walk Strides 2–4 1/2 in. (5.1–11.5 cm)

American dipper track (L); trail: walking in mud (R).

American dipper tracks. (CO)

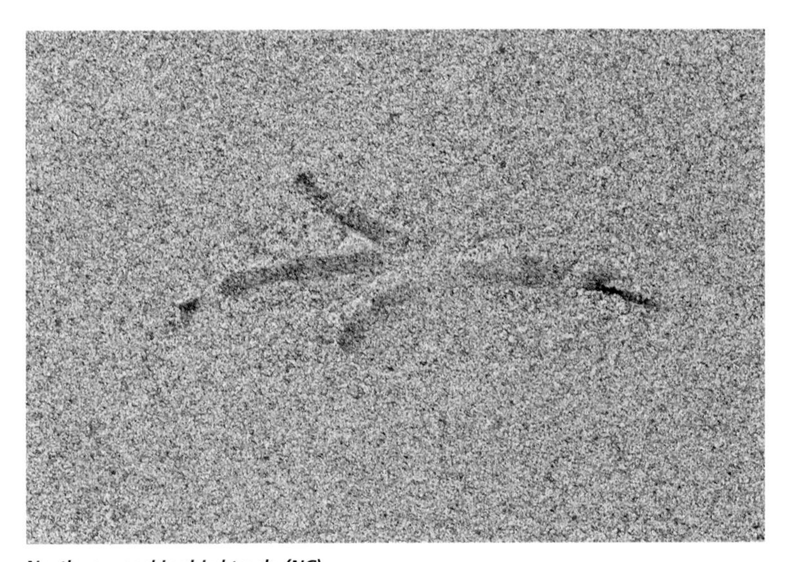

Northern mockingbird track. (NC)

Northern Mockingbird *(Mimus polyglottos)*

Track: 1 3/4–2 in. (4.4–5.1 cm) L x 3/4–1 in. (1.9–2.5 cm) W

Classic bird track. Medium. Anisodactyl. Metatarsal weakly registers or is absent altogether.

Similar species: Other jays and nutcrackers have larger tracks. Catbird has straighter toes.

Trail: Run Strides 4–6 in. (10.2–15.3 cm)

Red-winged Blackbird
(Agelaius phoeniceus)

Track: 1 11/16–2 1/16 in. (4.3–5.2 cm) L x 1/2–7/8 in. (1.2–2.2 cm) W

Classic bird track. Medium. Anisodactyl. Metatarsal weakly registers or is absent altogether. Slim toes. Toe 3 hugs toe 4 near the metatarsal pad, which is a blackbird and grackle characteristic.

Similar species: Jays have a very narrow track and hop. Dippers have wider tracks and are more habitat specific. Robin tracks are often shorter, and toes 2 and 4 curve out and down. Robins also run. Grackle tracks are larger.

Trail: Walk Strides 2–4 in. (5.1–10.2 cm)

Red-winged blackbird tracks (L); trail: walking in sand (R).

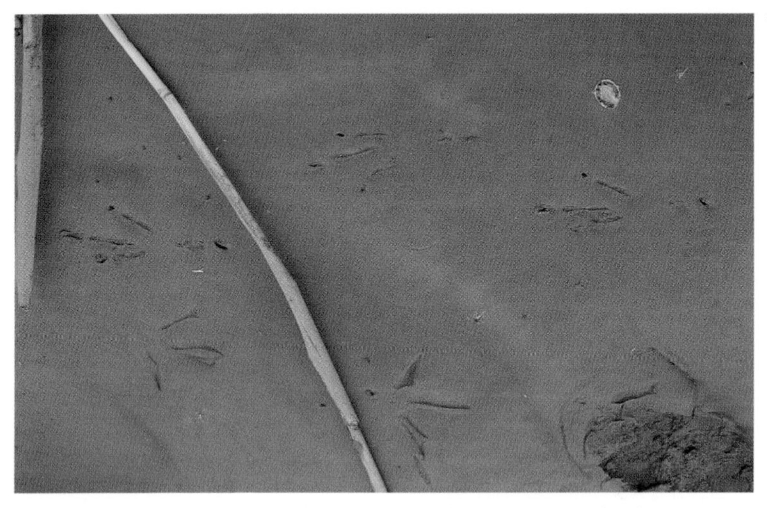

A red-winged blackbird and spotted sandpiper share the banks of a clay wash in a rock quarry. (NH)

Curve-billed thrasher tracks. (AZ)

Curve-billed Thrasher *(Toxostoma curvirostre)*

Track: 1 7/8–2 1/8 in. (4.7–5.4 cm) L x 3/4–1 1/8 in. (1.9–2.8 cm) W

Classic bird track. Medium. Anisodactyl. Metatarsal weakly registers or is absent altogether.

Similar species: Other thrashers. Robin tracks are a bit smaller and have bulbous toe pads.

Trail: Hop TW 2–2 1/2 in. (5.1–6.3 cm)
Strides 3–12 1/2 in. (7.6–31.7 cm)

American Robin *(Turdus migratorius)*

Track: 1¾–2⅛ in. (4.4–5.4 cm) L x ¾–1 in. (1.9–2.5 cm) W

Classic bird track. Medium. Anisodactyl. Metatarsal rarely registers at all. Toe pads bulbous, and overall track very curved. Toes 2 and 4 often point out and back.

Similar species: Dippers have slim, even toes and often register the metatarsal pad, and they walk, rather than run. Blackbirds have slim tracks and walk. Thrashers don't have the same bulbous toe pads, and their toes 2 and 4 don't splay in the same way.

Trail: Run Strides 4–6½ in. (10.2–16.5 cm)
Skip Strides 8–15 in. (20.3–38.1 cm)

Notes: This is a common track in mud puddles, where robins hunt. Also, look for mud collection at the appropriate stage of nest-building season.

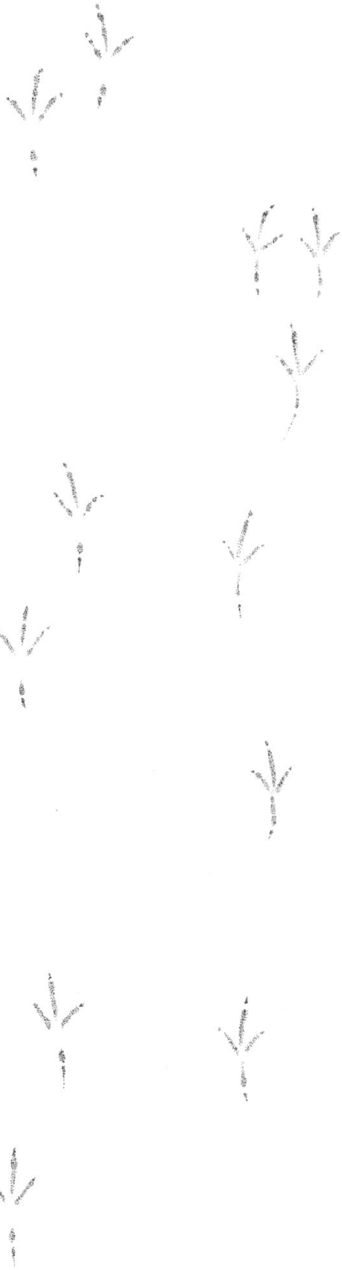

American robin tracks (L); trails: skipping in mud (M), running in mud (R).

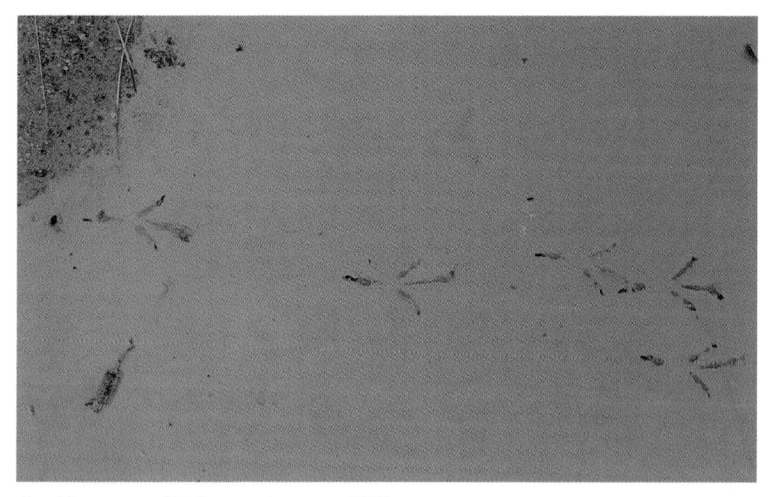

A robin runs—with frequent pauses. (NH)

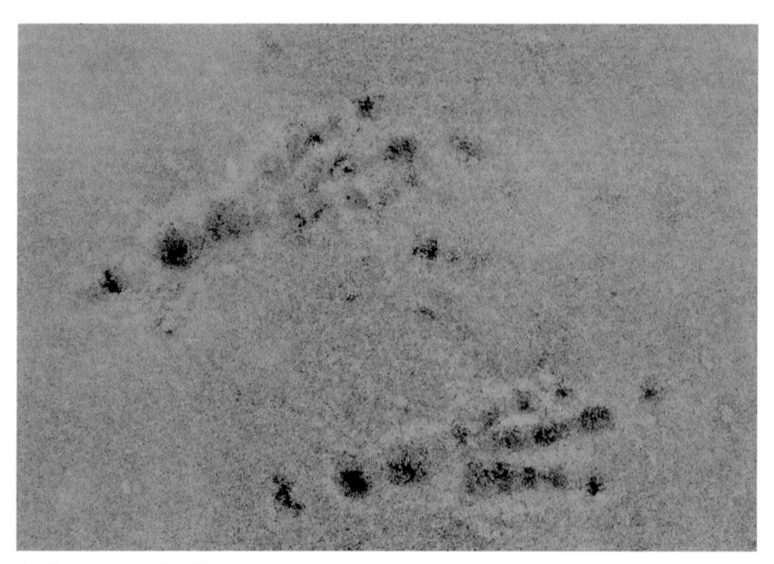

Steller's jay tracks. (CO)

Steller's Jay *(Cyanocitta stelleri)*

Track: 1 7/8–2 3/8 in. (4.7–6 cm) L x 9/16–3/4 in. (1.4–1.9 cm) W

Classic bird track. Medium. Anisodactyl. Metatarsal weakly registers or is absent altogether.

Similar species: Other jays and Clark's nutcracker. Jay tracks are very narrow in comparison with other species with overlapping track lengths. Pinyon jays are unique in that they walk, rather than hop. Black-billed magpies have stout, robust toes.

Trail: Hop TW 2–3 1/2 in. (5.1–8.9 cm)
Strides 5–18 in. (12.7–45.8 cm) and up to 3 ft. (91.5 cm)

European starling track (L); trail: walking in snow (R).

European Starling *(Sturnus vulgaris)*

Track: 1 13/16–2 1/8 in. (4.6–5.4 cm) L x 7/8–1 1/8 in. (2.2–2.8 cm) W

Classic bird track. Medium. Anisodactyl. Metatarsal rarely registers at all.

Similar species: Habitat and track patterns can usually eliminate the possibilities of other species, along with the fact that starlings are often found in flocks.

Trail: Walk Strides 2 3/4–4 3/4 in. (7–12.1 cm)

Le Conte's Thrasher
(Toxostoma lecontei)

Track: $1\frac{7}{8}$–$2\frac{1}{8}$ in. (4.7–5.4 cm) L x $\frac{3}{4}$–$1\frac{1}{8}$ in. (1.9–2.8 cm) W

Classic bird track. Medium. Anisodactyl. Metatarsal weakly registers or is absent altogether.

Similar species: Other thrashers. Robin tracks are a bit smaller and have bulbous toc pads.

Trail: Run Strides $5\frac{1}{2}$–11 in. (14–28 cm)

Notes: This thrasher is comfortable running across great open areas, where other thrashers would fly.

Le Conte's thrasher trail: running in loose sand.

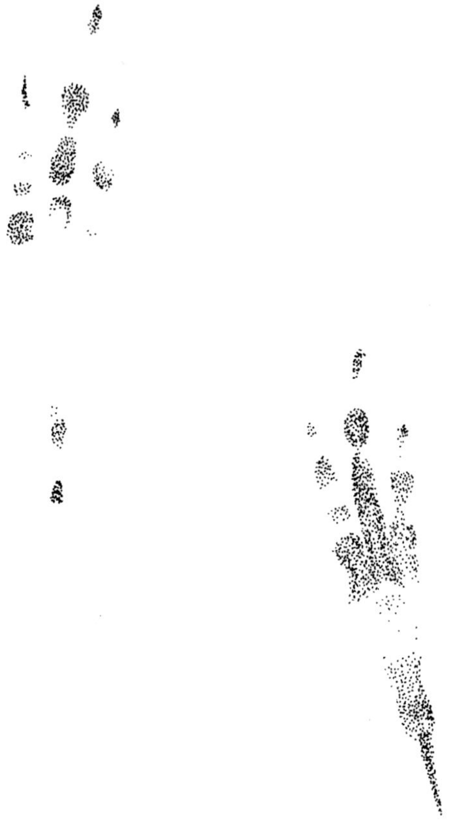

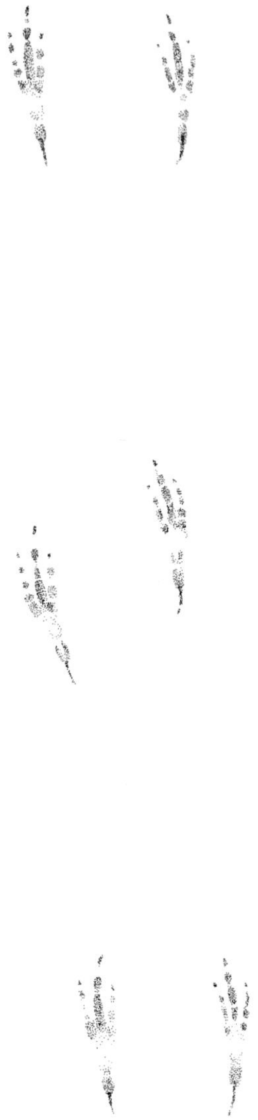

Blue jay tracks (L); trail: hopping in mud (R).

Blue Jay *(Cyanocitta cristata)*

Track: 1¾–2¼ in. (4.4–5.7 cm) L x ⅜–⅝ in. (1–1.6 cm) W

Classic bird track. Medium. Anisodactyl. Metatarsal weakly registers or is absent altogether.

Similar species: Other jays and Clark's nutcracker. Jay tracks are very narrow in comparison with other species with overlapping track lengths. Pinyon jays are unique in that they walk, rather than hop. Black-billed magpies have stout, robust toes.

Trail: Hop TW 1⅞–3 in. (4.7–7.6 cm)
Strides 5–18 in. (12.7–45.8 cm) and up to 3 ft. (91.5 cm)

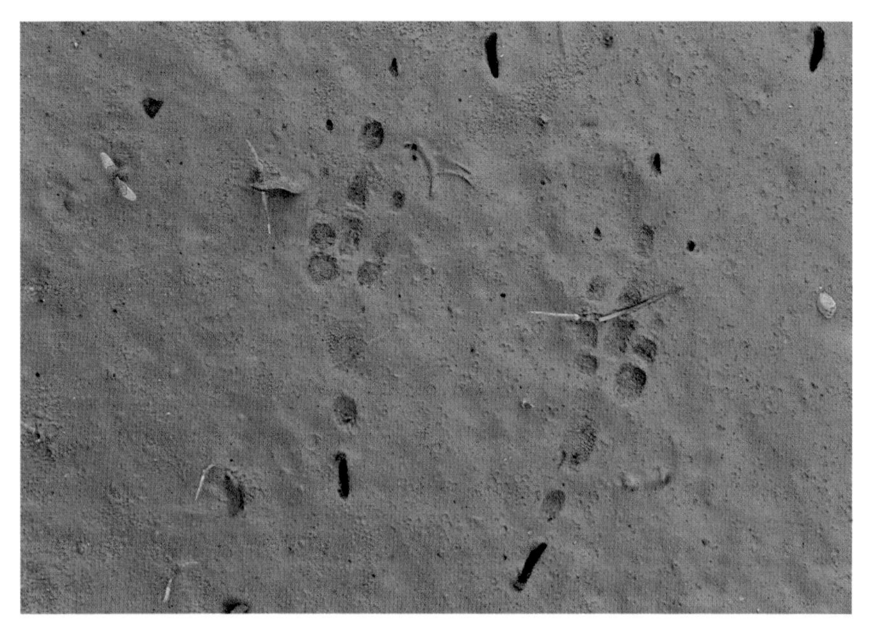

Blue jay tracks in a mud puddle. (NH)

A blue jay trail—note the skinny feet. (NH)

American Kestrel *(Falco sparverius)*

Track: $1\frac{3}{4}$–$2\frac{1}{4}$ in. (4.4–5.7 cm) L x $\frac{3}{4}$–1 in. (1.9–2.5 cm) W

Classic bird track. Medium. Anisodactyl. Metatarsal weakly registers or is absent altogether.

Similar species: Jay tracks are narrower. Black-billed magpies have stout, robust toes. Merlins have larger tracks and a much longer toe 3.

(Parameters created from small data pool.)

Mexican Jay *(Aphelocoma ultramarina)*

Track: $1\frac{7}{8}$–$2\frac{3}{8}$ in. (4.7–6 cm) L x $\frac{9}{16}$–$\frac{3}{4}$ in. (1.4–1.9 cm) W

Classic bird track. Medium. Anisodactyl. Metatarsal weakly registers or is absent altogether.

Similar species: Other jays and Clark's nutcracker. Jay tracks are very narrow in comparison with other species with overlapping track lengths. Pinyon jays are unique in that they walk, rather than hop. Black-billed magpies have stout, robust toes.

Trail: Hop TW 2–$3\frac{1}{8}$ in. (5.1–8 cm)
Strides 5–18 in. (12.7–45.8 cm) and up to 3 ft. (91.5 cm)

Clark's nutcracker tracks. (CO)

Clark's Nutcracker *(Nucifraga columbiana)*

Track: 2–$2\frac{3}{8}$ in. (5.1–6 cm) L x $\frac{15}{16}$–$1\frac{1}{4}$ in. (2.4–3.2 cm) W

Classic bird track. Medium. Anisodactyl. Metatarsal weakly registers or is absent altogether.

Similar species: Jay tracks are similar. Black-billed magpies have stout, robust toes.

Rock Dove *(Columba livia)*

Track: 2–2 1/2 in. (5.1–6.3 cm) L x 1 3/4–2 1/8 in. (4.4–5.4 cm) W
Classic bird track. Medium. Anisodactyl. Metatarsal weakly registers or is absent altogether.

Similar species: Other dove tracks are smaller. Jays and nutcrackers hop, except the pinyon jay. Blackbirds, robins, magpies, and grackles have considerably narrower tracks. Crow tracks are longer.

Trail: Walk Strides 2 1/4–6 1/4 in. (5.7–15.9 cm)

Notes: Where you find the trail of one rock dove, you often find another.

Rock dove (pigeon) tracks (L); trail: walking in mud (R).

Black-billed Magpie *(Pica pica)*

Track: 2–$2\frac{3}{8}$ in. (5.1–6 cm) L x $\frac{7}{8}$–$1\frac{1}{4}$ in. (2.2–3.2 cm) W

Classic bird track. Medium. Anisodactyl. Metatarsal weakly registers or is absent altogether.

Similar species: Jays and nutcrackers hop, except the pinyon jay. Crow tracks are larger. Meadowlark tracks are similar.

Trail:	Walk	Strides $3\frac{1}{2}$–7 in. (8.9–17.8 cm)
	Run	Strides 7–12 in. (17.8–30 .5 cm)
	Hop/Skip	TW $1\frac{3}{4}$–$3\frac{1}{2}$ in. (4.5–8.9 cm)
		Strides 9–$34\frac{1}{2}$ in. (22.9–90.6 cm)

Notes: The magpie has very small, weak feet for such a large bird with a bold nature.

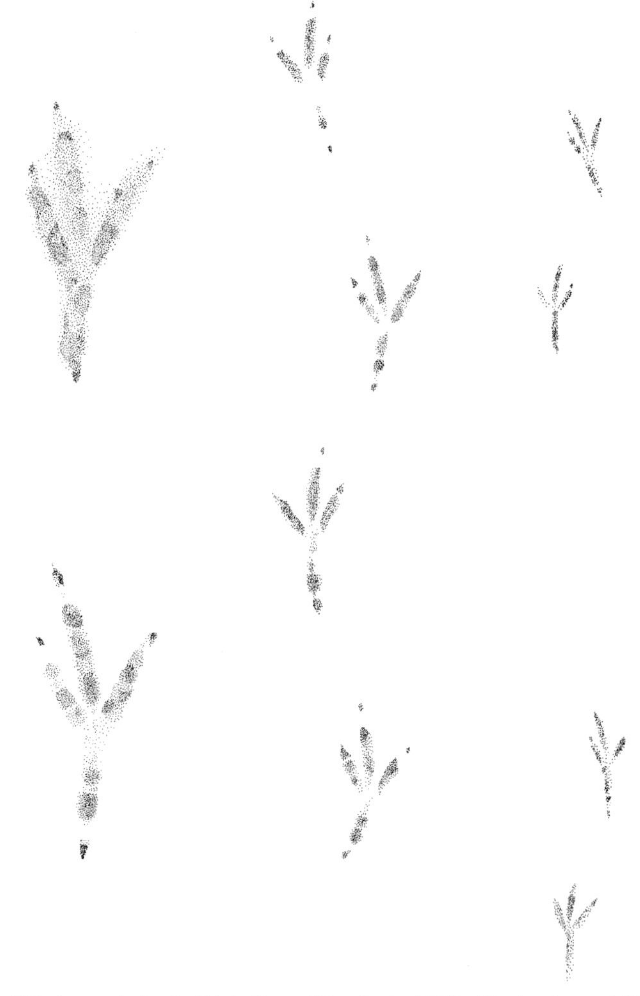

Black-billed magpie tracks (L); trails: walking in dust (M), skipping in sand (R).

A black-billed magpie runs toward the camera, while another skips away. (CO)

Western Meadowlark *(Sturnella neglecta)*

Track: 2¼–2⅝ in. (5.7–6.7 cm) L x 1–1⅛ in. (2.5–2.8 cm) W

Classic bird track. Medium. Anisodactyl. Metatarsal weakly registers or is absent altogether.

Similar species: Jays and nutcrackers hop, except the pinyon jay. Magpies and grackles leave similar tracks.

Trail: Walk Strides 2–6 in. (5.1–15.3 cm)

(Parameters created from small data pool.)

Western meadowlark trail: walking in mud.

Common Grackle *(Quiscalus quiscula)*

Track: 2 1/8–2 5/8 in. (5.4–6.7 cm) L x 5/8–1 in. (1.6–2.5 cm) W

Classic bird track. Medium. Anisodactyl. Metatarsal weakly registers or is absent altogether.

Similar species: Jays and nutcrackers hop, except the pinyon jay. Crows and other grackles have larger tracks. Blackbird tracks are smaller.

Trail:	Walk	Strides 2 1/2–4 1/2 in. (6.3–11.4 cm)
	Run	Strides 4 1/2–6 in. (11.4–15.3 cm)

Common grackle tracks (L); trail: walking in mud (R).

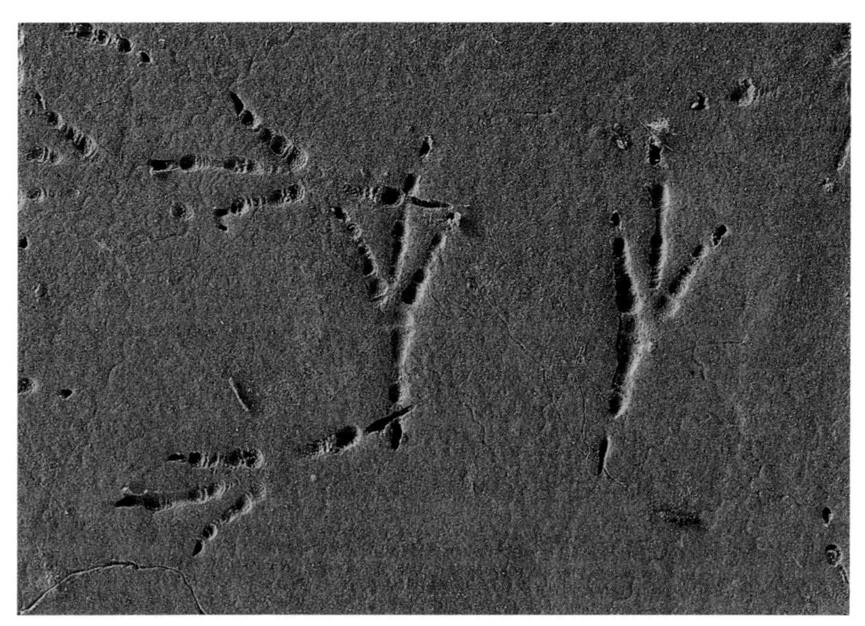

Common grackle tracks under a bridge. (WY)

Merlin *(Falco columbarius)*

Track: 2½–2¾ in. (6.3–7 cm) L x 1¾–1⅞ in. (4.4–4.8 cm) W

Classic bird track. Medium. Anisodactyl. Metatarsal weakly registers or is absent altogether. Toe 3 is significantly longer than toes 2 and 4.

Similar species: Larger foot and longer toe 3 differentiate merlin track from that of kestrel.

(Data taken from one female bird.)

Boat-tailed Grackle *(Quiscalus major)*

Track: 2 3/8–3 1/4 in. (6–8.2 cm) L x 1–1 1/2 in. (2.5–3.8 cm) W

Classic bird track. Medium. Anisodactyl. Metatarsal weakly registers or is absent altogether. Toe 3 hugs toe 4 near the metatarsal.

Similar species: Jays and nutcrackers hop, except the pinyon jay. Blackbird tracks are smaller. Common grackle is smaller. Crow and meadowlark tracks are similar.

Trail: Walk Strides 3 1/2–8 in. (8.9–20.3 cm)

Notes: If track is beyond 2 7/8 in. (7.3 cm) L, the bird is almost certainly a male.

Boat-tailed grackle tracks (L); trail: walking in moist sand (R).

Grackles vs. Crows

There is an overlap in size between male boat-tailed and great-tailed grackles and American and fish crows. Identification seems to be the greatest challenge along the southeast coast, where boat-tailed grackles and fish crows patrol the beaches.

There is a track characteristic that helps in identification. Note the relationship of toe 3 with either toe 2 or 4, especially nearest the metatarsal pad in the center of the foot. In crows and ravens, toe 3 hugs the side of toe 2, but in grackles and blackbirds, it hugs the side of toe 4. Allow for effects of substrate, and be careful of splayed tracks. Always look at several tracks for the characteristics of the overall trail rather than jump to conclusions from a single track.

Great-tailed Grackle
(Quiscalus mexicanus)

Track: 2½–3¼ in. (6.3–8.2 cm) L x 1⅛–1⅝ in. (2.8–4.1 cm) W

Classic bird track. Medium. Anisodactyl. Metatarsal weakly registers or is absent altogether. Toe 3 hugs toe 4 near the metatarsal.

Similar species: Jays and nutcrackers hop, except the pinyon jay. Blackbird tracks are smaller. Common grackle is smaller. Crow and meadowlark tracks are similar.

Trail: Walk Strides 3½–9 in. (8.9–22.9 cm)

Run Strides 9–11 in. (22.9–28 cm)

Notes: If track is beyond 2⅞ in. (7.3 cm) L, the bird is almost certainly a male.

The walking trail of a great-tailed grackle. (TX)

Fish Crow *(Corvus ossifragus)*

Track: $2\frac{15}{16}$–$3\frac{3}{8}$ in. (7.5–8.6 cm) L x $1\frac{1}{4}$–$1\frac{3}{4}$ in. (3.2–4.5 cm) W

Classic bird track. Medium. Anisodactyl. Metatarsal weakly registers or is absent altogether. Well-developed toe pads.

Similar species: Difficult to distinguish from American crow tracks. Grackle tracks are similar.

Trail: Walk Strides $4\frac{1}{2}$–8 in. (11.4–20.3 cm)

Notes: This crow is common along the beaches of the southeast, where the American crow is rare.

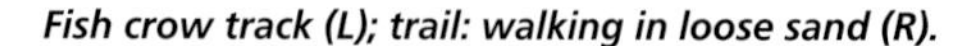

Fish crow track (L); trail: walking in loose sand (R).

Sexual Dimorphism

In most birds, the males are larger and heavier than their female counterparts, but this is not always the case. Certainly, males are notably larger in finches and blackbirds. There is also a notable size difference in turkeys, which have been studied at great lengths, and so have their tracks.

Hunters and naturalists have long been measuring the track length—just as done in this book—to determine the sex of the turkey. Mosby and Hadley (1943) found that eastern turkeys with a track length less than $4^{1}/_{4}$ inches (10.8 cm) were hens, whereas those with greater track lengths were males, or gobblers. Smith (1982) suggests measuring only the center toe, and that the cutoff is $2^{1}/_{2}$ inches (6.3 cm)—below this for hens, above for gobblers. Rezendes (1999) offers slightly different measurements than Mosby and Hadley. He feels that tracks $4^{1}/_{8}$ inches (10.6 cm) and below are hens, and those $4^{1}/_{4}$ inches (10.8 cm) and above are males, or toms. He also adds two alternate measurements. Tracks wider than $4^{3}/_{4}$ inches (12.1 cm) and toe 3 widths greater than $^{1}/_{2}$ inch (1.2 cm) are toms. Take your pick.

In many shorebirds, females are larger than males. This is also true of raptors, or predatory birds, which have been studied at length. Ehrlich, Dobkin, and Wheye (1988), in their compilation work, say that there seems to be a correlation in raptors between the breadth of the size difference between sexes and the speed of the prey the birds hunt. Vultures show little dimorphism (carrion doesn't run), whereas the falcons, which hunt birds, show the greatest size difference. This is useful to note when comparing tracks. If a track falls at the high end of the given range, then it is likely the larger of the sexes. If it is at the low end, it could be the smaller of the sexes or an immature bird of either sex.

There are also many birds in which there is a great deal of overlap between the track measurements of both sexes. Certain species show less overlap than others. Atwater and Schnell (1989) note the validity of sex determination in the ruffed grouse from a measurement of toe 3. There was, indeed, overlap of measurements between sexes, but only 10 percent of forty-four females were greater than $1^{9}/_{16}$ inches (4 cm), and the longest female's toe length was $1^{5}/_{8}$ inches (4.1 cm). Measurements beyond these figures are more likely to be from a male, but not definitely.

American Crow *(Corvus brachyrhynchos)*

Track: 2 15/16–3 5/8 in. (7.5–9.2 cm) L x 1 1/8–1 5/8 in. (2.8–4.1 cm) W

Classic bird track. Medium. Anisodactyl. Metatarsal weakly registers or is absent altogether. Well-developed toe pads.

Similar species: Often hard to distinguish from tracks of other crow species, although they are a bit smaller. Raven tracks are larger.

Trail: Walk Strides 4 1/2–8 in. (11.4–20.3 cm)

Hop/Skip Strides 8–18 in. (20.3–45.9 cm)

American crow tracks (L); trails: walking in mud (M), skipping in loose sand (R).

American crow tracks. (MA)

Northern Goshawk *(Accipiter gentilis)*

Track: 4½–4¾ in. (11.4–12.1 cm) L x 2¾–3 in. (7–7.6 cm) W

Classic bird track. Large. Anisodactyl. Metatarsal weakly registers or is absent altogether. Toe pads bulbous and foot strong, which aids in gripping prey.

Trail: Run Strides 12–15 in. (30.5–38.1 cm)

Notes: Goshawks move quite comfortably on the ground, which is rare for raptors.

(Data taken from one female bird; males would be smaller.)

Green Heron *(Butorides virescens)*

Track: 3 1/8–3 3/4 in. (7.9–9.5 cm) L x 2–2 3/8 in. (5.1–6 cm) W

Classic bird track. Medium. Anisodactyl. Webbing is present between toes 3 and 4. Metatarsal weakly registers or is absent altogether. Slim, even toes.

Similar species: Crows have bulbous toe pads and are narrower. Ravens, egrets, and other herons have larger tracks.

Trail: Walk Strides 3 1/2–9 in. (8.9–22.9 cm)

Green heron tracks (L); trail: walking in mud (R).

A green heron track. (MA)

Red-tailed Hawk *(Buteo jamaicensis)*

Track: 3¾–5⅜ in. (9.5–13.7 cm) L x 2½–4 in. (6.3–10.2 cm) W

Classic bird track. Large. Anisodactyl. Metatarsal weakly registers or is absent altogether. Toe pads bulbous.

Similar species: Raven tracks are narrower. Eagle tracks are longer.

Trail: Walk Strides 2¾–5½ in. (7–14 cm)

(Parameters created from small data pool of females.)

Red-tailed hawk tracks.

Tricolored Heron *(Egretta tricolor)*

Track: $3\frac{3}{4}$–$4\frac{1}{2}$ in. (9.5–11.4 cm) L x 2–$3\frac{1}{8}$ in. (5.1–7.9 cm) W

Classic bird track. Large. Anisodactyl. Webbing is present between toes 3 and 4. Metatarsal weakly registers or is absent altogether. Slim toes.

Similar species: Reddish egret tracks are larger. Snowy and cattle egrets have similar tracks; here, use behaviors to aid in identification.

Trail: Run (more of a "jog") Strides 9–14 in. (22.9–35.6 cm)

Notes: Rather than still-hunt fish, tricolored herons tend to chase and push their target into ideal depths for capture. This behavior is most apparent in the strides of the bird.

(Parameters created from small data pool.)

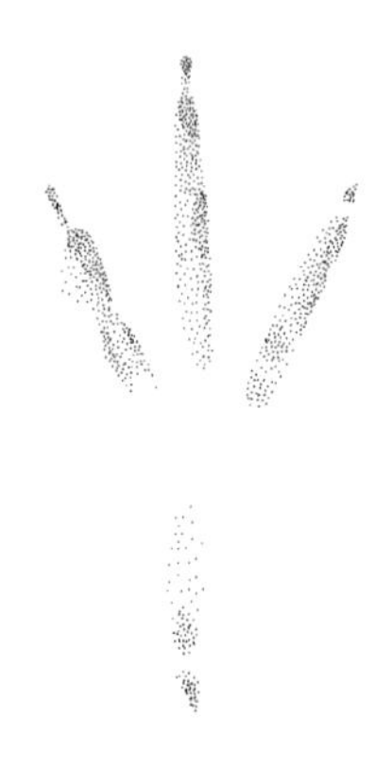

Tricolored heron track (L); trail: "jogging" across a mudflat (R).

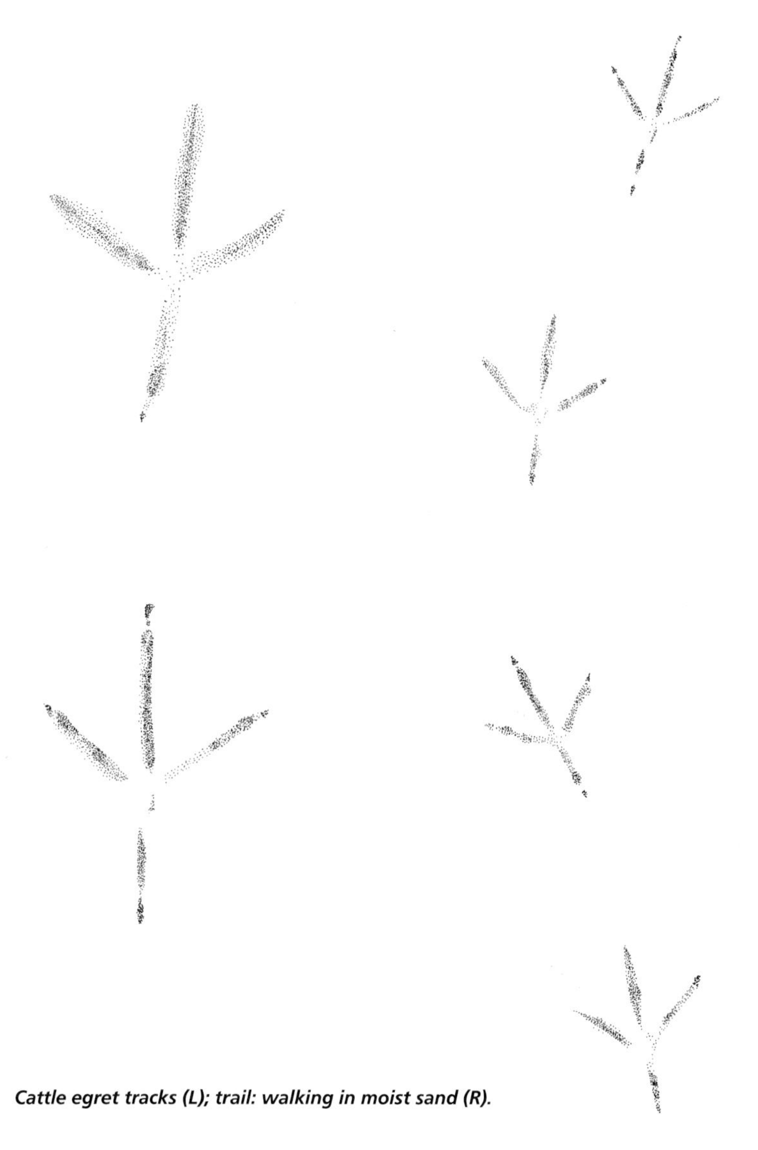

Cattle egret tracks (L); trail: walking in moist sand (R).

Cattle Egret *(Bubulcus ibis)*

Track: 4–4 5/8 in. (10.2–11.7 cm) L x 2 7/8–4 1/8 in. (7.3–10.5 cm) W

Classic bird track. Large. Anisodactyl. Webbing is present between toes 3 and 4. Metatarsal weakly registers or is absent altogether. Slim, even toes.

Similar species: Difficult to distinguish from snowy egret tracks. Ibis tracks are larger.

Trail: Walk Strides 4 1/2–10 in. (11.4–25.4 cm)

Common Raven *(Corvus corax)*

Track: 3¾–4¾ in. (9.5–12.1 cm) L, occasionally to 5¼ in. (13.3 cm) L, x 1⅝–2½ in. (4.1–6.3 cm) W

Classic bird track. Large. Anisodactyl. Metatarsal weakly registers or is absent altogether.

Similar species: The Chihuahuan raven is a bit smaller but overlaps at the low end in size. Crow tracks are smaller. Hawks, which spend far less time on the ground, have more bulbous toe pads, and their toes splay sporadically.

Trail: Walk Strides 5½–9 in. (14–22.9 cm)
Hop TW 4⅞–6 in. (12.4–15.3 cm)
Strides 18–30 in. (45.9–79.2 cm)

Common raven tracks (L); trail: walking in snow (R).

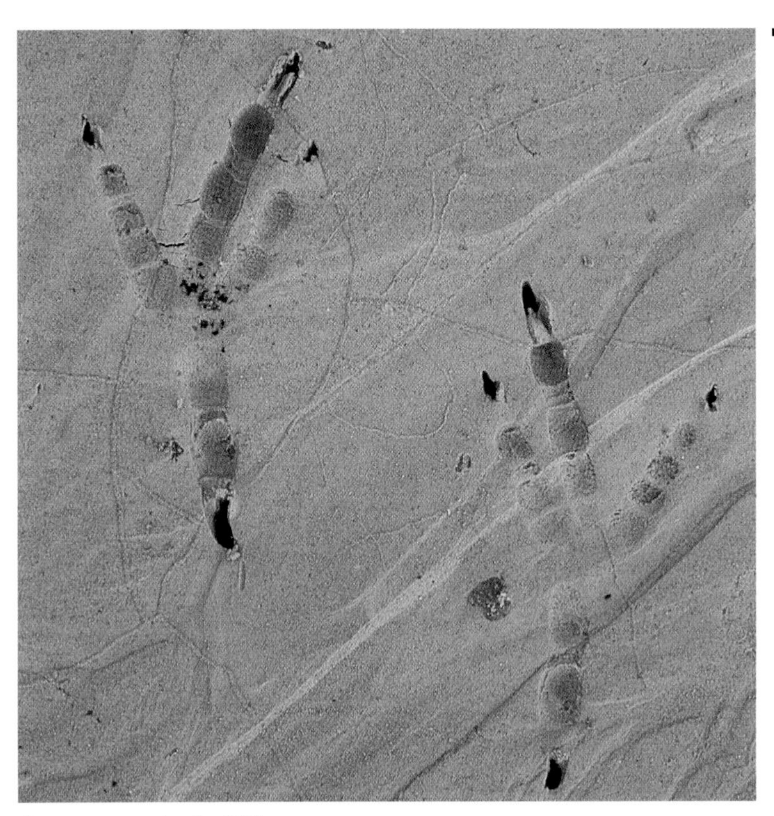

Common raven tracks. (CA)

White-faced Ibis
(Plegadis chihi)

Track: 4¾–5⅜ in. (12.1–13.7 cm) L x 3⅝–4⅞ in. (9.2–12.4 cm) W

Classic bird track. Large to very large. Anisodactyl. Significant webbing is present between toes 3 and 4, and a smaller web is present between 2 and 3. Metatarsal weakly registers or is absent altogether.

Similar species: Difficult to distinguish from other ibis tracks. Egret and heron tracks are narrower.

Trail: Walk Strides 4–11 in. (10.2–28 cm)

White-faced ibis tracks.

Snowy Egret *(Egretta garzetta)*

Track: $3\frac{7}{8}$–$4\frac{5}{8}$ in. (9.8–11.7 cm) L x $2\frac{1}{2}$–$3\frac{3}{4}$ in. (6.3–9.5 cm) W

Classic bird track. Large. Anisodactyl. Webbing is present between toes 3 and 4. Metatarsal weakly registers or is absent altogether. Slim, even toes.

Similar species: Difficult to distinguish from cattle egret tracks. Ibis tracks are larger.

Trail: Walk Strides $4\frac{1}{2}$–12 in. (11.4–30.5 cm)

Snowy egret tracks.

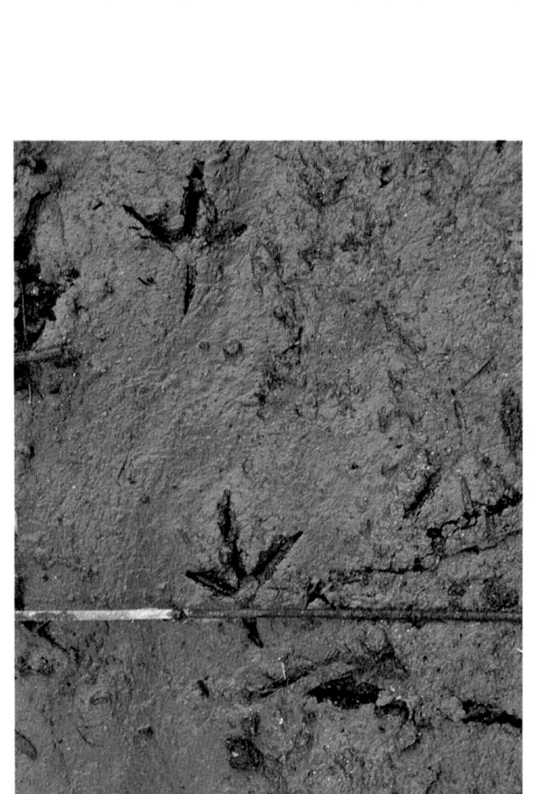

A snowy egret trail. (FL)

Reddish egret tracks (L); trail: walking on hard-packed sand (R).

Reddish Egret *(Egretta rufescens)*

Track: 4½–5¼ in. (11.4–13.3 cm) L x 2⅞–3⅞ in. (7.3–9.8 cm) W

Classic bird track. Large. Anisodactyl. Webbing is present between toes 3 and 4. Metatarsal weakly registers or is absent altogether.

Similar species: Other egrets have smaller tracks, except the great egret, which has larger ones. Ibis tracks are wider. Tricolored herons have smaller tracks and tend to move at a jog rather than a walk.

Trail: Walk Strides 8–15 in. (20.3–38.1 cm)

Turkey vulture walking trail. (TX)

Turkey vulture track (L); trail: hopping in snow (R).

Turkey Vulture *(Cathartes aura)*

Track: 3 3/4–5 1/2 in. (9.5–14 cm) L x 3 1/4–4 in. (8.2–10.2 cm) W

Classic bird track. Large. Anisodactyl. Webbing is present between toes 2 and 3 and toes 3 and 4. Metatarsal weakly registers or is absent altogether. Toe pads less bulbous.

Similar species: Other raptors have very bulbous toe pads and tend to walk.

Trail:	Walk	Strides 5 1/2–11 in. (14–28 cm). Runs up to 16 in. (40.6 cm)
	Hop	Strides 18–24 in. (45.8–61.4 cm)

Turkey vulture track. (MA)

Great Egret *(Ardea alba)*

Track: $6^1/_2$–$7^1/_4$ in. (16.5–18.4 cm) L x 4–$5^1/_2$ in. (10.2–14 cm) W

Classic bird track. Very large. Anisodactyl. Webbing is present between toes 3 and 4. Metatarsal weakly registers or is absent altogether. Toes slim and of an even width for their entire length.

Similar species: Raptor tracks, such as those of eagles, have bulbous pads used for gripping prey. Toes of great blue herons appear more robust in tracks, but are still difficult to distinguish.

Trail: Walk Strides 10–18 in. (25.4–45.8 cm)

Great egret tracks.

White Ibis *(Eudocimus albus)*

Track: 4 3/4–5 3/8 in. (12.1–13.7 cm) L x 3 5/8–4 7/8 in. (9.2–12.4 cm) W

Classic bird track. Large to very large. Significant webbing is present between toes 3 and 4, and a smaller web is present between 2 and 3. Anisodactyl. Metatarsal weakly registers or is absent altogether.

Similar species: Difficult to distinguish from other ibis tracks. Egret and heron tracks are narrower. Consider how the hallux registers.

Trail: Walk Strides 4–12 in. (10.2–30.5 cm)

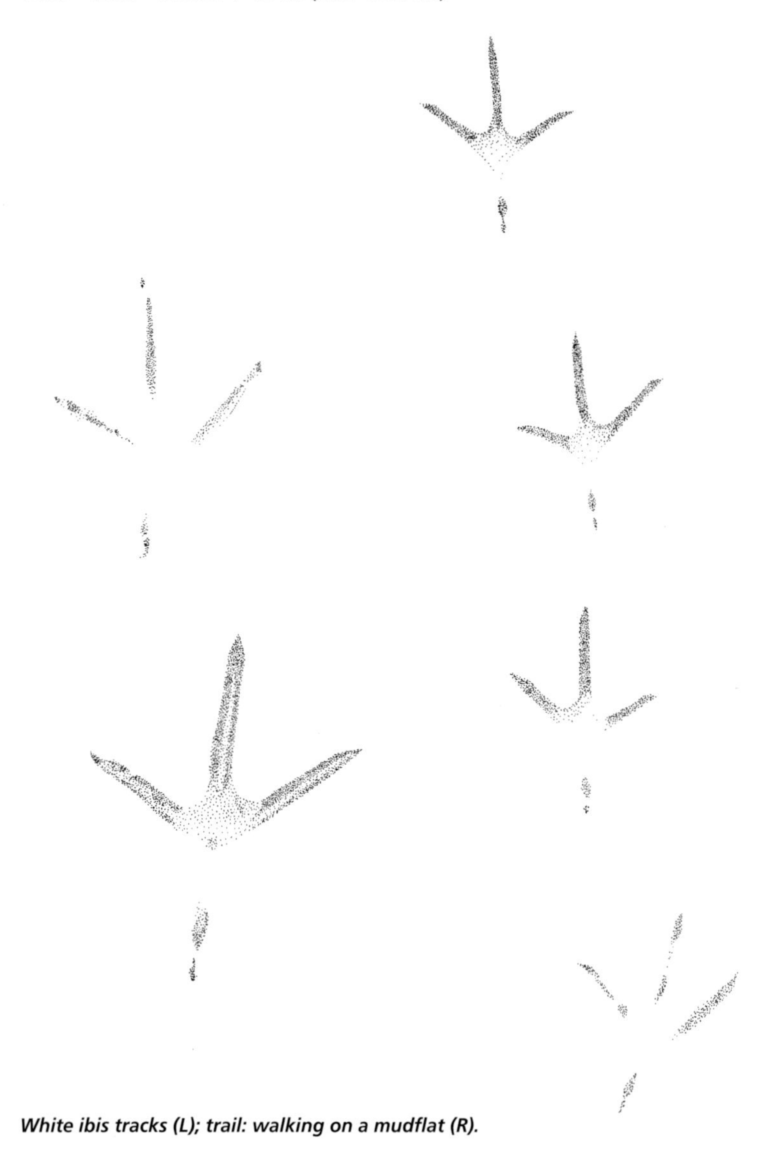

White ibis tracks (L); trail: walking on a mudflat (R).

Bald eagle tracks (L); trail: walking in sand (R).

Bald Eagle *(Haliaeetus leucocephalus)*

Track: 6–8¼ in. (16.2–21 cm) L x 3¼–5¾ in. (8.2–14.6 cm) W

Classic bird track. Very large. Anisodactyl. Metatarsal weakly registers or is absent altogether. Toe pads very bulbous and rough.

Similar species: Heron tracks have slender toes, lacking the bulbous pads used for gripping prey. Other raptors, except the golden eagle, have smaller tracks.

Trail: Walk Strides 4–11 in. (10.2–28 cm)

Notes: Predatory birds don't often stroll along on the ground. If an eagle is on the ground, there is food nearby. Look for signs of feeding, such as fish or carrion remains. *(See photo on next page.)*

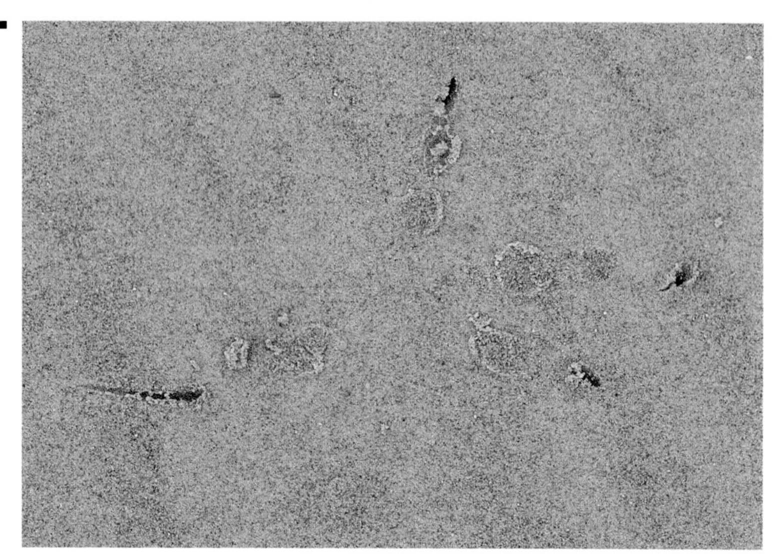

Bald eagle track. (OR)

A great blue heron patrols the dunes, where it takes shelter from the winds and hunts rodents. (TX)

Great blue heron tracks (L); trail: walking in mud (R).

Great Blue Heron *(Ardea herodias)*

Track: 6½–8½ in. (16.5–21.6 cm) L x 4–6 in. (10.2–15.2 cm) W

Classic bird track. Very large. Anisodactyl. Webbing is present between toes 3 and 4. Metatarsal weakly registers or is absent altogether. Toes slim and of an even width for their entire length.

Similar species: Raptor tracks, such as those of eagles, have bulbous pads used for gripping prey. Great egret tracks tend to register slimmer toes, but always consider the effects of substrate.

Trail: Walk Strides 10–18½ in. (25.4–47 cm)

Game Bird Tracks

Piping Plover *(Charadrius melodus)*

Track: 3/4–15/16 in. (1.9–2.4 cm) L x 5/8–7/8 in. (1.6–2.2 cm) W

Game bird track. Very small. Anisodactyl. Incumbent foot structure. Webbing is present between toes 3 and 4. Toe 1 is absent altogether, and the metatarsal pad tends to register lightly or not at all.

Similar species: Sandpipers and dunlins register a toe 1 and have more symmetrical tracks. Semipalmated plover tracks are smaller, and killdeer tracks are larger.

Trail: Walk/Run Strides 1 1/2–4 3/4 in. (3.8–12.1 cm)

Notes: The piping plover is endangered in parts of the country and threatened in others. Plovers are extremely pigeon-toed.

Piping plover track (L); trail: running in moist sand (R).

High Stepping

Certain bird behaviors are easily read in the trail patterns left behind. High stepping, a mating behavior observed in plovers, is easily read in the tracks and trails you find in the field. A courting male approaches a female, taking short, stiff steps, with the feet raised higher than normal. After proceeding in this manner for a short distance, the male slams both feet down, flying straight up into the air. All of these behaviors are accompanied by vocalizations.

Short strides are easy enough to recognize in a track pattern, and found in isolation, they do not always mean high stepping. Also look for an increased trail width, and for the tracks to register less pigeon-toed. The final two tracks, from the takeoff, should be deeper and clearer than the rest.

Diane Boretos has studied this behavior intensively in the endangered piping plover. Taking many measurements, Diane created stride parameters that indicated that the plovers were engaged in this behavior, which a field researcher could then use to know that the plovers would be nesting in the area. She found that strides of piping plovers between 1¾ and 2 inches (4.4 and 5.1 cm) were high stepping. Strides of 2¼ to 3½ inches (5.7 to 8.9 cm) were general walking and feeding. And strides of 4 to 4¾ inches (10.2 to 12.1 cm) were running strides, which would include the defending of territories.

With some time in the field, such parameters could be made for all the plover species. This jump from track pattern identification to behavior study is especially rewarding.

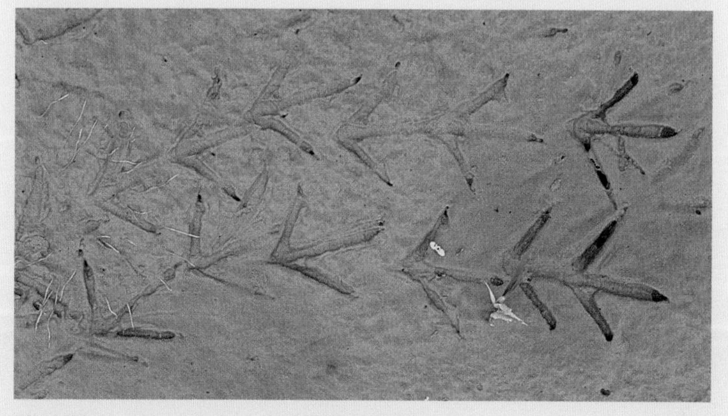

The high-stepping display of a killdeer. (NH)

Least Sandpiper *(Calidris minutilla)*

Track: 3/4–7/8 in. (1.9–2.2 cm) L x 7/8–1 1/8 in. (2.2–2.8 cm) W

Game bird track. Very small. Anisodactyl. Incumbent foot structure. Toe 1 tends to register well in tracks. The metatarsal pad also tends to register well.

Similar species: Difficult to distinguish from other peeps. Sanderlings and plovers do not have a toe 1. Dunlins and most other sandpipers have larger tracks.

Trail: Walk Strides 1 1/4–2 3/4 in. (3.2–7 cm)

Least sandpiper track (L); trail: walking in mud (R).

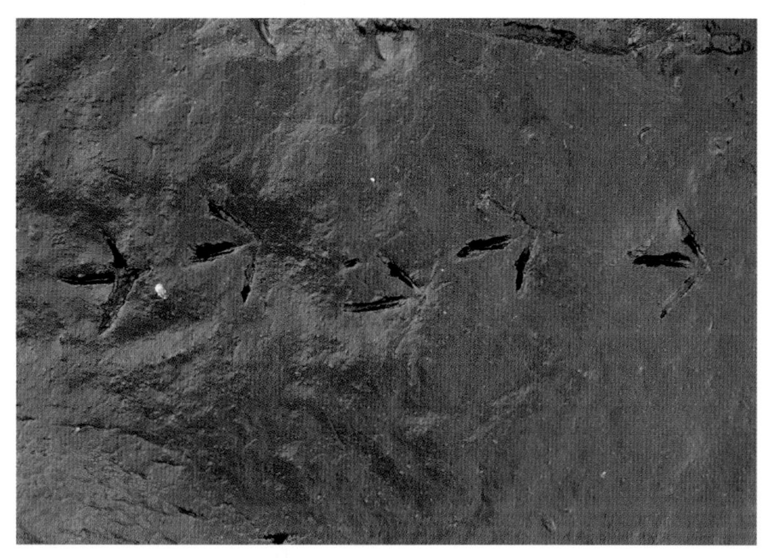

The walking trail of a least sandpiper. (MA)

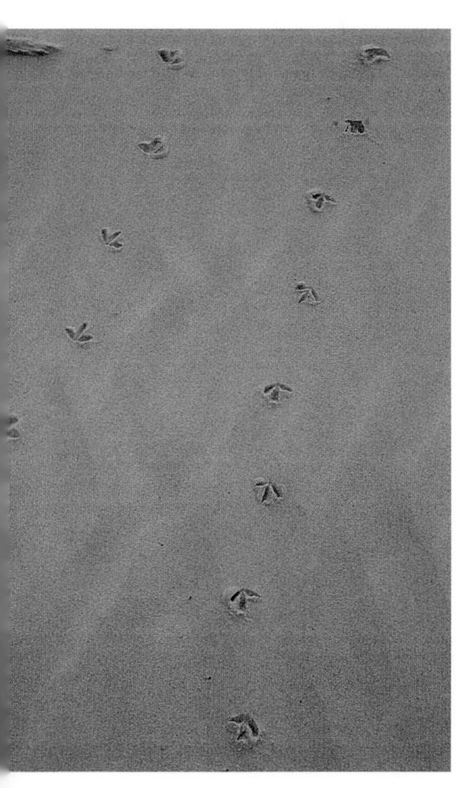

Sanderling *(Calidris alba)*

Track: 3/4–7/8 in. (1.9–2.2 cm) L x 1–1 1/8 in. (2.5–2.8 cm) W

Game bird track. Very small. Anisodactyl. Incumbent foot structure. There is no toe 1, and the metatarsal tends to register lightly or not at all.

Similar species: Sandpipers register toe 1. Plovers have extremely asymmetrical tracks, very pigeon-toed trails.

Trail: Run — Strides 4–5 5/8 in. (10.2–14.3 cm)

Walk — Strides 1 1/2–2 3/4 in. (3.8–7 cm)

(See illustration on next page.)

Running sanderling trails. (TX)

Sanderling track (L); trail: running in moist sand (R).

A spotted sandpiper trail. (MA)

Spotted sandpiper tracks (L); trails: walking in deep mud (M), walking in mud (R).

Spotted Sandpiper *(Actitis macularia)*

Track: 7/8–1 in. (2.2–2.5 cm) L x 1–1 3/16 in. (2.5–3 cm) W

Game bird track. Very small. Anisodactyl. Incumbent foot structure. Webbing is present between toes 3 and 4. Although toe 1 registers quite reliably, the metatarsal pad does not always show in tracks.

Similar species: Plovers, including killdeer, do not register a toe 1 and are very pigeon-toed.

Trail: Walk Strides 1 1/4–4 3/4 in. (3.2–12.1 cm)

Notes: A common track along inland freshwater bodies. Often found with killdeer tracks. In deep substrate, the hallux may appear more passerine, albeit still very short.

Semipalmated Plover
(Charadrius semipalmatus)

Track: 7/8–1 1/16 in. (2.2–2.7 cm) L x 1–1 1/4 in. (2.5–3.2 cm) W

Game bird track. Very small. Anisodactyl. Incumbent foot structure. Webbing is present between toes 3 and 4. Toe 1 is absent altogether, and the metatarsal pad tends to register lightly or not at all.

Similar species: Sandpipers and dunlins register a toe 1 and have more symmetrical tracks. Piping plover tracks are smaller, and killdeer tracks are larger.

Trail: Walk Strides 1 1/4–5 3/4 in. (3.2–14.6 cm)

Semipalmated plover track (L); trail: running in moist sand (R).

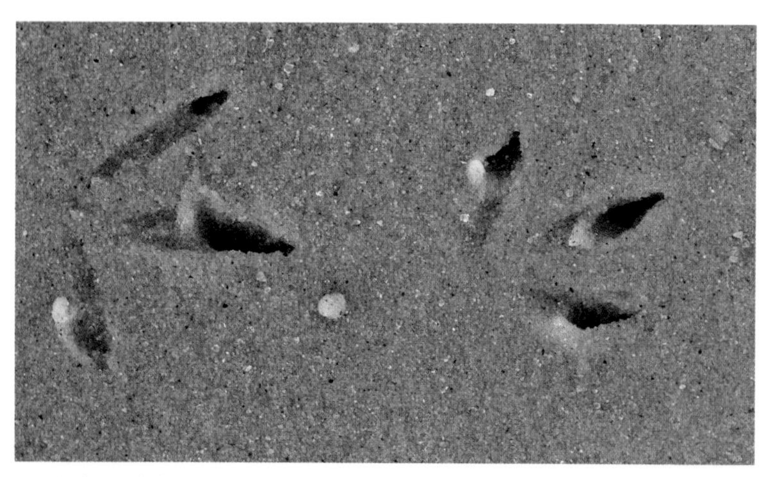

Semipalmated plover tracks. (MA)

Dunlin track (L); trail: walking and feeding on a mudflat (R).

Dunlin *(Calidris alpina)*

Track: 7/8–1 3/16 in. (2.2–3 cm) L x 1 1/8–1 1/2 in. (2.8–3.8 cm) W

Game bird track. Very small. Anisodactyl. Incumbent foot structure. Both toe 1 and the metatarsal pad show reliably in tracks.

Similar species: Sandpiper tracks are similar. Plovers and sanderlings do not show a toe 1 in their tracks.

Trail: Walk Strides 2–5 in. (5.1–12.7 cm)

Killdeer tracks (L); trail: walking in mud (R).

Killdeer *(Charadrius morinellus)*

Track: 1–1 3/16 in. (2.5–3 cm) L x 1 1/8–1 7/16 in. (2.8–3.6 cm) W

Game bird track. Very small. Anisodactyl. Incumbent foot structure. Webbing is present between toes 3 and 4. Toe 1 is absent altogether, and the metatarsal pad tends to register lightly or not at all.

Similar species: Sandpipers and dunlins register a toe 1 and have more symmetrical tracks. Piping and semipalmated plovers have smaller tracks.

Trail: Walk/Run Strides 2–7 in. (5.1–17.8 cm)

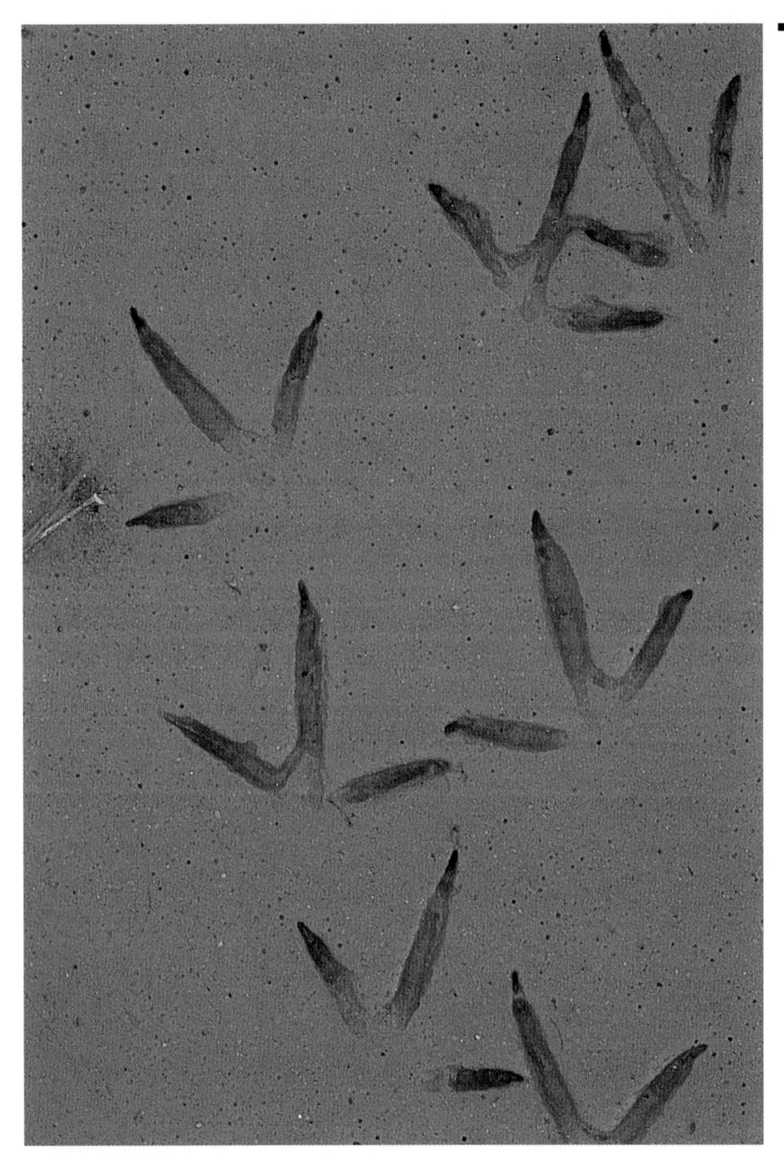

Killdeer tracks. (NH)

California Quail *(Callipepla californica)*

Track: 1 1/4–1 1/2 in. (3.2–3.8 cm) L x 1 1/4–1 3/4 in.(3.2–4.4 cm) W

Game bird track. Small. Anisodactyl. Incumbent foot structure. Toe 1 and the metatarsal pad tend to be visible in tracks, unless the bird is moving at speed.

Similar species: Mountain quail and other quail species. Grouse tracks are larger and significantly wider.

Trail: Walk Strides 1 1/2–5 1/2 in. (3.8–14 cm)

Where Shorebirds Feed

Although high-tide situations force bird species together, which produces a chaotic carpet of tracks, they begin to separate out into preferred feeding habitats as the tide recedes. Understanding this tendency, and learning the effects of water on the track, will make shorebird tracking easier. Were the tracks made in shallow water, before the water receded? Were they made in really wet conditions, near the current waterline? Were they made high on the mudflat, when the water was far below?

Certain birds tend to feed at different margins of the tide and prefer different levels of sociability. Avocets and greater yellowlegs are often in deep enough water to obscure tracks altogether. Short-billed dowitchers wade very near to shore. Dunlins and red knots forage in small flocks on the open wet mud next to the waterline. Least sandpipers and long-billed curlews wander the drier algae-covered mud above.

While many shorebirds are flock feeders, plovers generally are not. Their short bills are ideal for catching exposed prey rather than probing for well-hidden invertebrates. In this way plovers hunt and run at prey alone; a group might alert the prey before it is caught.

Territorial displays are also written in the sand for those who care to look closely. In a 1988 essay, Ehrlich et al. contend that territoriality is most pronounced in times of medium food abundance. If there is too little food, the energy wasted on territorial disputes is not worth it. In times of abundance, energy wasted on territoriality would be poorly spent. Here is field ecology at work, from reading the bird trail to reading about food abundance in the surrounding environment.

A few trips to the beach will help these lessons sink in. I remember one perfect day in Texas. In the morning, I visited a massive mudflat, exposed at low tide. As I stood alone where water had been hours before, tiny peeps and massive long-billed curlews moved about me, probing the earth and wading in small pools left behind. Farther along, a congregation of white pelicans relaxed on the exposed mud, leaving scat, feathers, and tracks to tell the story for those who would take notice later on. Closer to the water, several species of plovers ran about solitarily, snatching at things on the surface that I could not see from where I stood. Godwits worked beyond them, in very shallow water, and farther out, avocets bobbed up and down, their legs fully submerged.

As the tide moved up the mudflat toward me, I retired to the beach to the south, where I spent the afternoon watching and interpreting the territorial trails and displays of sanderlings. They would rush at each other, then stop, turn sideways, and squat. One might rush the other or chase the intruder away. Another might flee, still another threaten a retaliation. Hours and hours of reading pleasure are laid out in the sands.

Gambel's Quail *(Callipepla gambelii)*

Track: 1 3/8–1 5/8 in. (3.5–4.1 cm) L x 1 1/2–1 3/4 in. (3.8–4.4 cm) W

Game bird track. Small. Anisodactyl. Incumbent foot structure. Toe 1 and the metatarsal pad tend to be visible in tracks, unless the bird is moving at speed.

Similar species: Other quail species. Grouse tracks are larger and significantly wider.

Gambel's quail track. (AZ)

Ruddy Turnstone *(Arenaria interpres)*

Track: 1 1/8–1 3/8 in. (2.8–3.5 cm) L x 1 1/4–1 1/2 in. (3.2–3.8 cm) W

Game bird track. Small. Anisodactyl. Incumbent foot structure. Toe 1 registers quite reliably, and the metatarsal pad is visible in most tracks, as well.

Similar species: Difficult to distinguish from tracks of black turnstones and surfbirds. Dunlins have smaller tracks with slimmer toes. Sandpipers also have slimmer toes.

Trail: Walk Strides 1 5/8–3 1/2 in. (4.1–8.9 cm)

Run Strides up to 7 in. (17.8 cm)

Notes: Drags toe 3 when running.

Ruddy turnstone track (L); trail: running in moist sand (R).

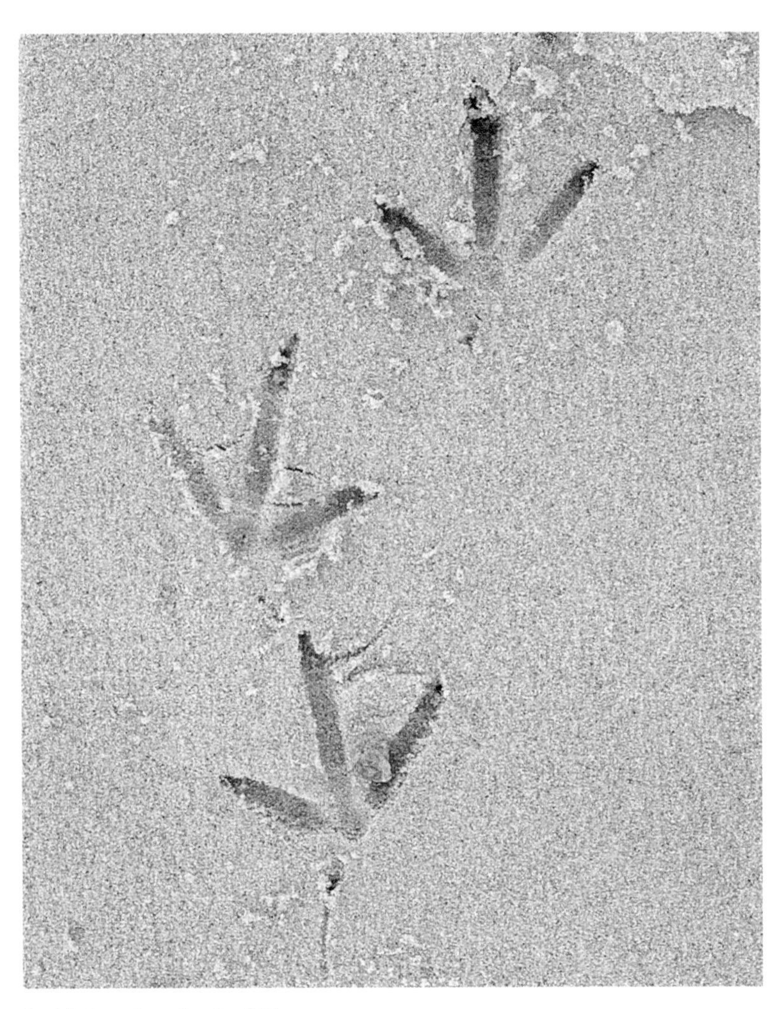

Ruddy turnstone tracks. (TX)

American Golden-plover *(Pluvialis dominica)*

Track: $1^3/_8$–$1^5/_8$ in. (3.5–4.1 cm) L x $1^3/_4$–$1^7/_8$ in. (4.4–4.8 cm) W

Game bird track. Small. Anisodactyl. Incumbent foot structure. Webbing is present between toes 3 and 4. Toe 1 is absent altogether, and the metatarsal pad tends to register lightly, if at all, in tracks.

Similar species: Difficult to distinguish from black-bellied plover tracks.

Trail: Walk Strides $3^1/_2$–5 in. (8.9–12.7 cm)

(Parameters created from small data pool.)

Short-billed Dowitcher *(Limnodromus griseus)*

Track: 1¼–1½ in. (3.2–3.8 cm) L x 1⅝–2 in. (4.1–5.1 cm) W

Game bird track. Small. Anisodactyl. Incumbent foot structure. Webbing is present between toes 3 and 4. Although toe 1 is very reliable in tracks, the metatarsal pad does not show unless the substrate is deep.

Similar species: Red knot tracks are similar in size.

Trail: Walk Strides 2–4½ in. (5.1–11.5 cm)
Run Strides 5–7½ in. (12.7–19.1 cm)

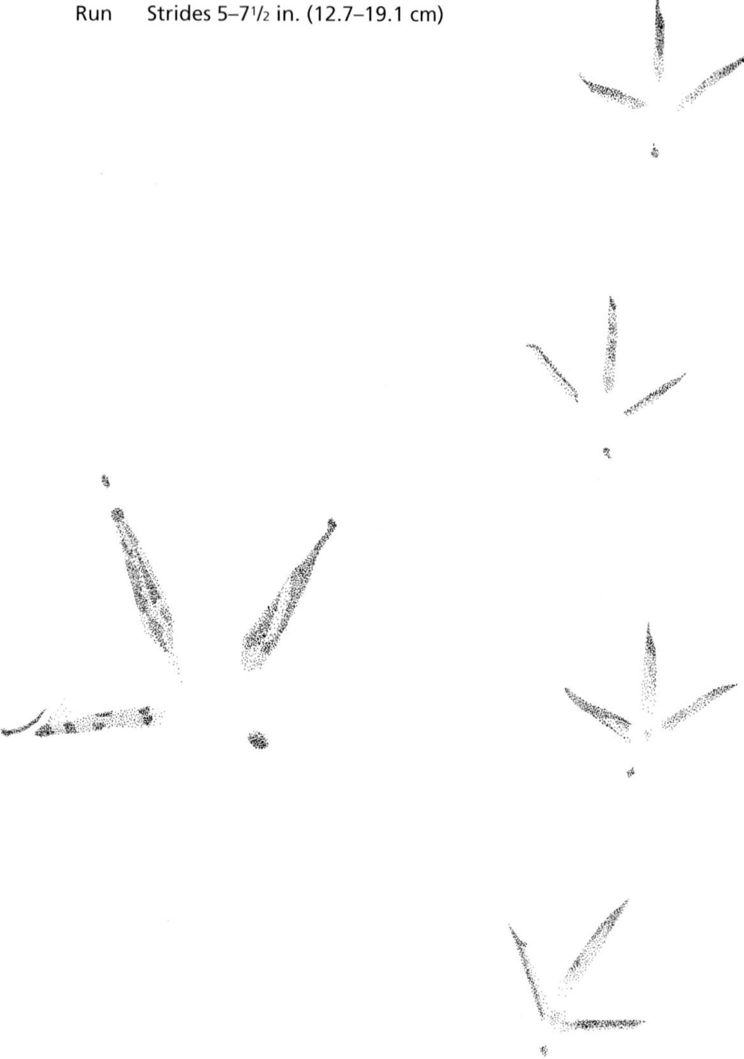

Short-billed dowitcher track (L); trail: walking across a mudflat (R).

Black-bellied plover track (L); trail: running in moist sand (R).

Black-bellied Plover *(Pluvialis squatarola)*

Track: 1 3/8–1 5/8 in. (3.5–4.1 cm) L x 1 3/4–1 7/8 in. (4.4–4.8 cm) W

Game bird track. Small. Anisodactyl. Incumbent foot structure. Webbing is present between toes 3 and 4. Toe 1 is absent altogether, and the metatarsal pad tends to register lightly, if at all, in tracks.

Similar species: Difficult to distinguish from golden-plover tracks.

Trail: Walk Strides 3 1/2–5 in. (8.9–12.7 cm)

Northern Bobwhite *(Colinus virginianus)*

Track: $1\frac{3}{8}$–$1\frac{5}{8}$ in. (3.5–4.1 cm) L x $1\frac{7}{16}$–$1\frac{11}{16}$ in. (3.6–4.3 cm) W

Game bird track. Small. Anisodactyl. Incumbent foot structure. Webbing is present between toes 3 and 4. Toe 1 and the metatarsal pad tend to be visible in tracks, unless the bird is moving at speed.

Similar species: Other quail species. Grouse tracks are larger and significantly wider.

Trail:	Walk	Strides $1\frac{3}{4}$–5 in. (4.4–12.7 cm)
	Run	Strides 8–12 in. (20.3–30.5 cm)

Notes: Tracks in loose sand appear larger, and this bird frequents dune environments in parts of the country.

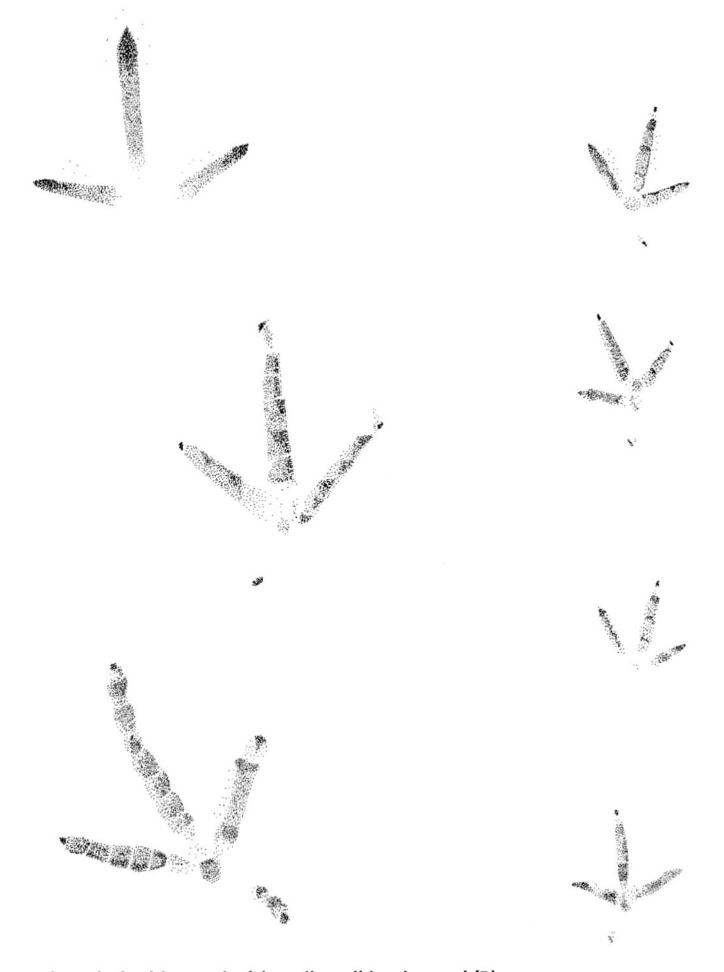

Northern bobwhite tracks (L); trail: walking in sand (R).

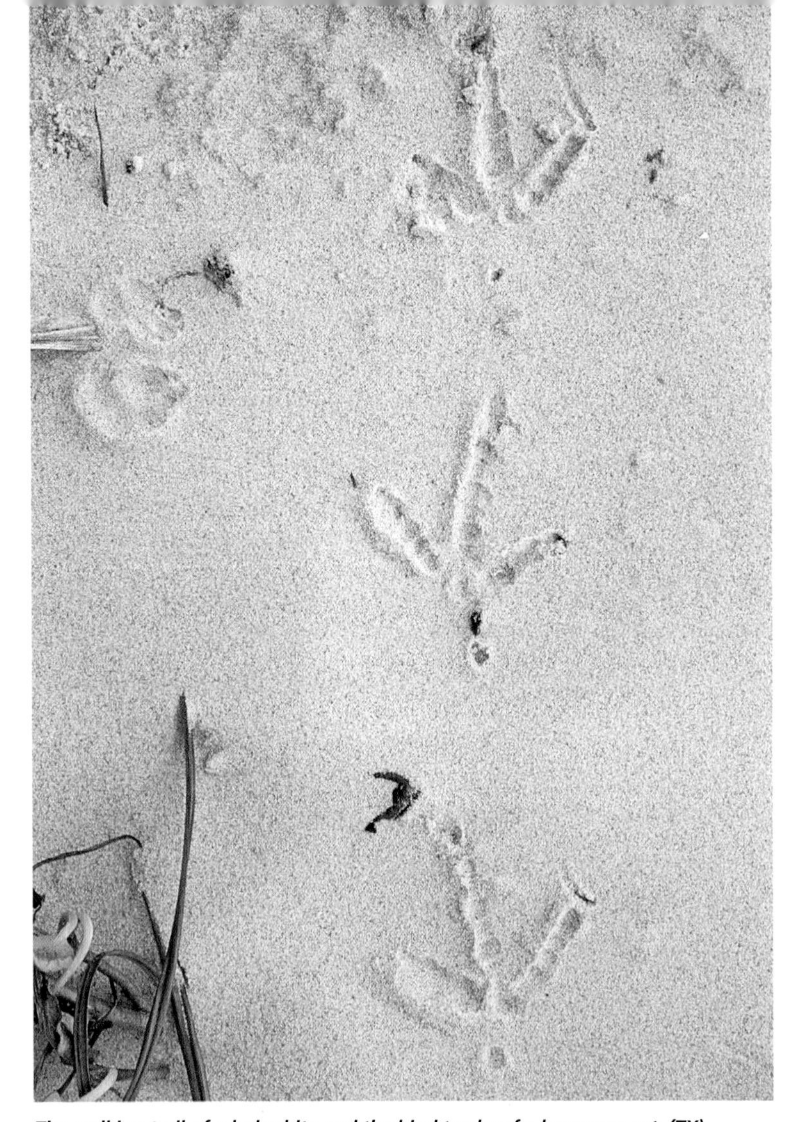

The walking trail of a bobwhite and the hind tracks of a kangaroo rat. (TX)

Scaled Quail *(Callipepla squamata)*

Track: $1\frac{3}{8}$–$1\frac{11}{16}$ in. (3.5–4.3 cm) L x $1\frac{5}{8}$–$1\frac{7}{8}$ in. (4.1–4.8 cm) W

Game bird track. Small. Anisodactyl. Incumbent foot structure. Toe 1 and the metatarsal pad tend to be visible in tracks, unless the bird is moving at speed.

Similar species: Other quail species. Grouse tracks are larger and significantly wider.

Trail: Walk Strides $2\frac{3}{4}$–4 in. (7–10.2 cm)
Run Strides 8–11 in. (20.3–28 cm)

Notes: Tracks in loose sand appear larger.

White-tailed Ptarmigan *(Lagopus mutus)*

Track: 1 3/8–1 3/4 in. (3.5–4.4 cm) L x 1 1/2–1 3/4 in. (3.8–4.4 cm) W

Game bird track. Small. Anisodactyl. Incumbent foot structure. The appearance of toe 1 in the track is dependent upon the depth of substrate—toe 1 shows most clearly in deep tracks. However, the metatarsal pad regularly shows in the track.

Similar species: Other ptarmigan species and grouse have larger, significantly wider tracks.

Trail: Walk Strides 2 1/4–4 in. (5.7–10.2 cm)

Notes: Consider snow conditions, which influence the appearance of tracks and trails significantly. In winter, the birds are found roosting and feeding in groups.

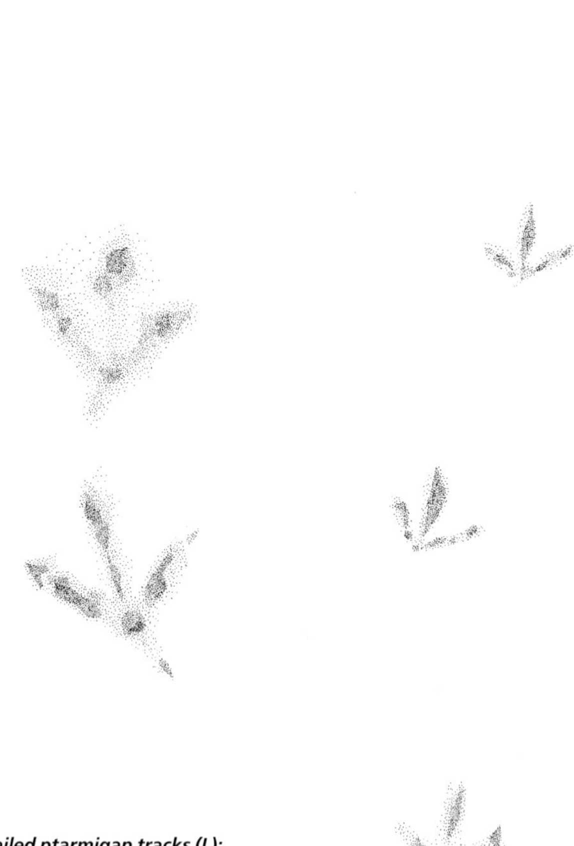

White-tailed ptarmigan tracks (L); trail: walking in snow (R).

A white-tailed ptarmigan trail in powdery snow looks similar to a mammal trail. (CO)

Marbled Godwit *(Limosa fedoa)*

Track: 1 3/4–1 15/16 in. (4.4–4.9 cm) L x 2 1/4–2 5/8 in. (5.7–6.7 cm) W

Game bird track. Small to medium. Anisodactyl. Incumbent foot structure. Webbing is present between toes 3 and 4. Although toe 1 is very common in tracks, the metatarsal pad does not show unless the substrate is deep. Toe 1 registers very close to the metatarsal.

Similar species: Greater yellowlegs have slimmer toes, and willets have bulbous toes. Whimbrel tracks are larger.

Trail: Run   Strides 7–12 in. (17.8–30.5 cm)

Notes: The natural gait for godwits is the walk.

Common Snipe *(Gallinago gallinago)*

Track 1 3/8–1 3/4 in. (3.5–4.4 cm) L x 1 5/8–2 in. (4.1–5.1 cm) W

Game bird track. Small to medium. Anisodactyl. Incumbent foot structure. Both toe 1 and the metatarsal pad tend to show well in tracks.

Similar species: Difficult to distinguish from American woodcock tracks; use habitat as an aid. Short-billed dowitcher is smaller. Willets have bulbous toes and tend not to register their metatarsal pads.

Trail: Walk Strides 2–4 in. (5.1–10.2 cm)

Common snipe track (L); trail: walking in mud (R).

A snipe trail in soft mud. (MA)

Virginia Rail *(Rallus limicola)*

Track: 1¾–2 in. (4.4–5.1 cm) L x 1½–2¼ in. (3.8–5.7 cm) W

Game bird track. Small to medium. Anisodactyl. Although the hallux is lower on the foot than in other game birds, rails still have an incumbent foot structure. Webbing is present between toes 3 and 4. Toe 1 tends to register in tracks, although the metatarsal may be absent altogether at speed.

Similar species: Use habitat and behavior to help distinguish among different types of rail tracks.

Trail: Run Strides 7–12 in. (17.8–30.5 cm)

Notes: Rail tracks are an occasional lucky find in substrate dividing two wetlands. Rails tend to run when exposed, but would be expected to walk under cover.

(Parameters created from a small data pool.)

This Virginia rail crossed a dusty road between two wetlands. (CA)

American woodcock tracks (L); trail: walking in an inch of snow (R).

American Woodcock *(Scolopax minor)*

Track: 1½–2 in. (3.8–5.1 cm) L x 1⅝–2 in. (4.1–5.1 cm) W

Game bird track. Small to medium. Anisodactyl. Incumbent foot structure. Both toe 1 and the metatarsal pad tend to show well in tracks.

Similar species: Difficult to distinguish from common snipe tracks; use habitat as an aid.

Trail: Walk Strides 2–4 in. (5.1–10.2 cm)

Notes: A common track in the forests and fields of the Northeast. This bird lives almost entirely on earthworms; check habitat to help confirm identification.

Greater Yellowlegs *(Tringa melanoleuca)*

Track: 1⅝–1⅞ in. (4.1–4.7 cm) L x 2¼–2⅝ in. (5.7–6.7 cm) W

Game bird track. Small to medium. Anisodactyl. Incumbent foot structure. Webbing is present between toes 3 and 4. Although toe 1 is reliable in tracks, the metatarsal pad does not show unless the substrate is deep. Toes appear very slim in track registrations.

Similar species: Willet has bulbous toes.

Trail: Walk Strides 3⅝–9 in. (9.2–22.9 cm)

Greater yellowlegs tracks (L); trail: fast walk in moist sand (R).

Willet *(Catoptrophorus semipalatus)*

Track: 1 5/8–1 15/16 in. (4.1–4.9 cm) L x 2 1/4–2 1/2 in. (5.7–6.3 cm) W

Game bird track. Small to medium. Anisodactyl. Incumbent foot structure. Webbing is present between toes 3 and 4 and toes 2 and 3. Although toe 1 is very common in tracks, the metatarsal pad does not show unless the substrate is deep. Willets have distinctly bulbous, cigar-shaped toes.

Similar species: Greater yellowlegs have slimmer toes. Godwits and whimbrels have larger tracks.

Trail: Walk/Run Strides 4 1/4–12 in. (10.8–30.5 cm)

Willet tracks (L); trail: walking in mud (R).

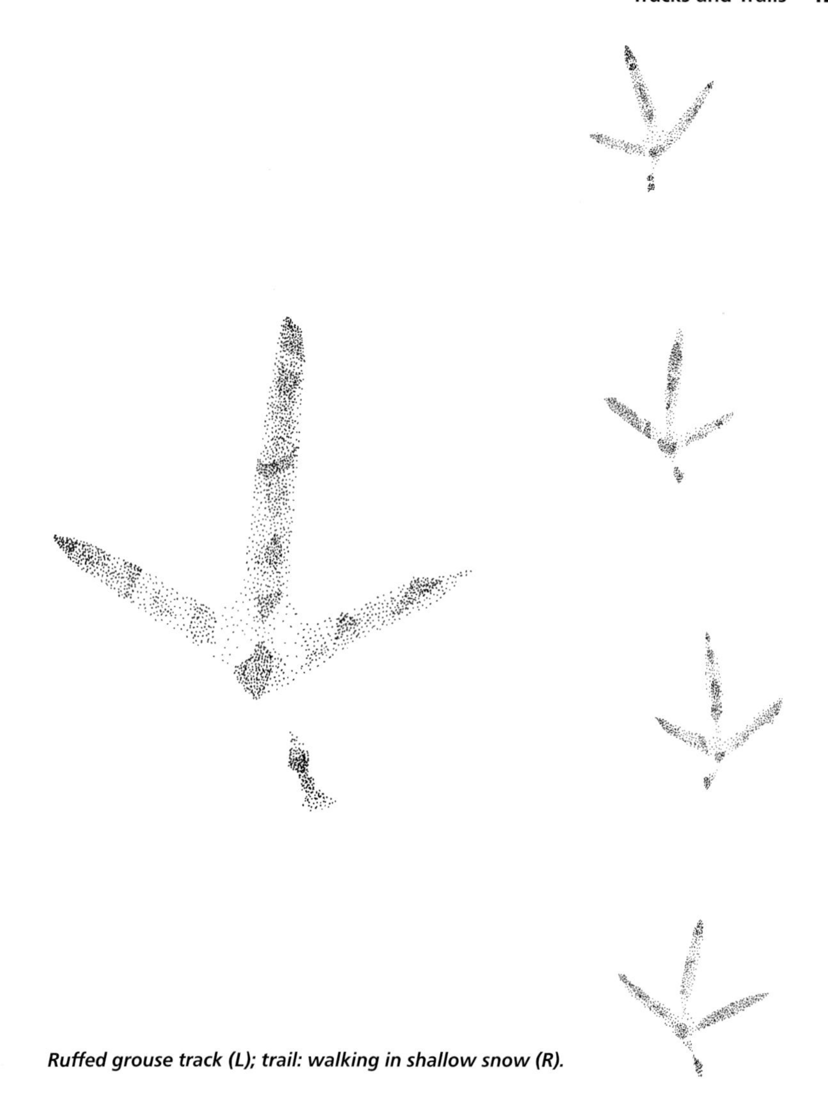

Ruffed grouse track (L); trail: walking in shallow snow (R).

Ruffed Grouse *(Bonasa umbellus)*

Track: 1 5/8–2 1/4 in. (4.1–5.7 cm) L x 1 3/4–2 3/8 in. (4.4–6 cm) W

Game bird track. Medium. Anisodactyl. Incumbent foot structure. Webbing is present between toes 3 and 4 and toes 2 and 3. Both toe 1 and the metatarsal pad tend to show well in tracks.

Similar species: Other grouse species. Pheasant tracks are larger. Quail tracks are smaller, especially in width.

Trail: Walk/Run Strides 3 1/4–10 1/2 in. (8.3–26.7 cm)

Notes: The grouse is primarily a solitary creature.

Black-necked Stilt *(Himantopus mexicanus)*

Track: 1 3/4–2 1/16 in. (4.4–5.2 cm) L x 2 1/8–2 5/8 in. (5.4–6.7 cm) W

Game bird track. Medium. Anisodactyl. Incumbent foot structure. Webbing is present between toes 3 and 4. Toe 1 is absent altogether. The metatarsal pad shows quite regularly, although it is small in size.

Similar species: Avocet tracks are webbed and larger. Willets and yellowlegs register a toe 1.

Trail: Walk Strides 4 3/4–9 in. (12.1–22.9 cm)

Notes: The tracks of black-necked stilts have been misrepresented in other tracking literature. Webbing is only significant between toes 3 and 4.

Black-necked stilt track (L); trail: walking in mud (R).

Spruce Grouse
(Falcipennis canadensis)

Track: $1^{1}/_{2}$–$2^{1}/_{4}$ in. (3.8–5.7 cm) L x 2–$2^{3}/_{8}$ in. (5.1–6 cm) W

Game bird track. Medium. Anisodactyl. Incumbent foot structure. Webbing is present between toes 3 and 4 and toes 2 and 3. Both toe 1 and the metatarsal pad tend to show well in tracks.

Similar species: Other grouse species. Pheasants are larger.

Trail: Walk/Run Strides $3^{1}/_{4}$–$10^{1}/_{2}$ in. (8.3–26.7 cm)

A whimbrel trail. (NC)

Whimbrel
(Numenius phaeopus)

Track: $1^{13}/_{16}$–$2^{1}/_{16}$ in. (4.6–5.3 cm) L x $2^{3}/_{8}$–$2^{5}/_{8}$ in. (6–6.6 cm) W

Game bird track. Medium. Anisodactyl. Incumbent foot structure. Webbing is present between toes 3 and 4 and toes 2 and 3. Although toe 1 is very common in tracks, the metatarsal pad does not show unless the substrate is deep.

Similar species: Long-billed curlew tracks are significantly wider. Willets and yellowlegs have smaller tracks.

Trail: Walk Strides 4–8 in. (10.2–20.3 cm)

The Grouse's Snowshoes

Each fall, ruffed grouse grow small, fingerlike projections from their toes. These small, fringelike or comblike additions, which measure around $^{1}/_{16}$ in. (2 mm) each, are then worn until the next spring, when they will be shed irregularly over a given amount of time. It is believed that these "fringes" aid in winter travel, acting like snowshoes and distributing the weight of the bird over a greater surface area. This idea is supported by the fact that the fringes grow longest on birds in the northern part of the species range. These fringes are often visible in tracks in the spring, especially if they are made in shallow mud puddles, which produce perfect tracks.

Long-billed curlew track (L); trail: walking on a mudflat (R).

Long-billed Curlew *(Numenius americanus)*

Track: 1⅞–2⅛ in. (4.7–5.4 cm) L x 2⅝–2⅞ in. (6.7–7.3 cm) W

Game bird track. Medium. Anisodactyl. Incumbent foot structure. Webbing is present between toes 3 and 4 and toes 2 and 3. Although toe 1 is very common in tracks, the metatarsal pad does not show unless the substrate is deep.

Similar species: Whimbrel tracks are significantly narrower.

Trail: Walk Strides 3–8½ in. (7.6–21.6 cm)

Notes: Curlews tend to feed high on the mudflat.

American Oystercatcher *(Haematopus palliatus)*

Track: 2–2 1/2 in. (3.8–6.3 cm) L x 2–2 5/8 in. (5.1–6.7 cm) W

Game bird track. Medium. Anisodactyl. Incumbent foot structure. Webbing is present between toes 3 and 4. Toe 1 is absent altogether, and the other toes are very robust.

Similar species: Long-billed curlews have narrow toes and register a hallux.

Trail: Walk Strides 3–5 1/2 in. (7.6–14 cm)

Run Strides 5 3/4–7 1/2 in. (14.6–19 cm)

American oystercatcher tracks (L); trail: walking in sand (R).

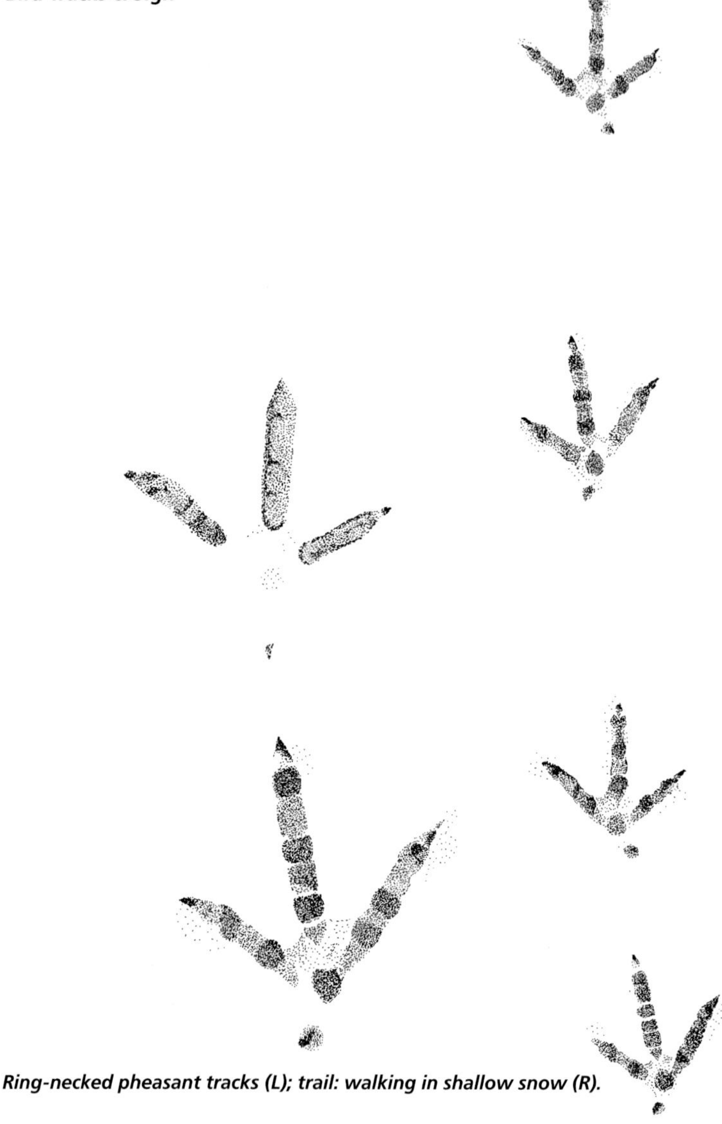

Ring-necked pheasant tracks (L); trail: walking in shallow snow (R).

Ring-necked Pheasant *(Phasianus colchicus)*

Track: 2–2 15/16 in. (3.8–7.5 cm) L x 2–2 3/4 in. (5.1–7 cm) W

Game bird track. Medium. Anisodactyl. Incumbent foot structure. Both toe 1 and the metatarsal pad tend to show well in tracks.

Similar species: Grouse tracks are smaller, and turkey tracks are larger.

Trail: Walk/Run Strides 2 1/2–9 in. (6.3–22.9 cm)

Can run up to 24 in. (61.4 cm)

Notes: Ring-necked pheasants are comfortable in the open, whereas many grouse species stick to cover.

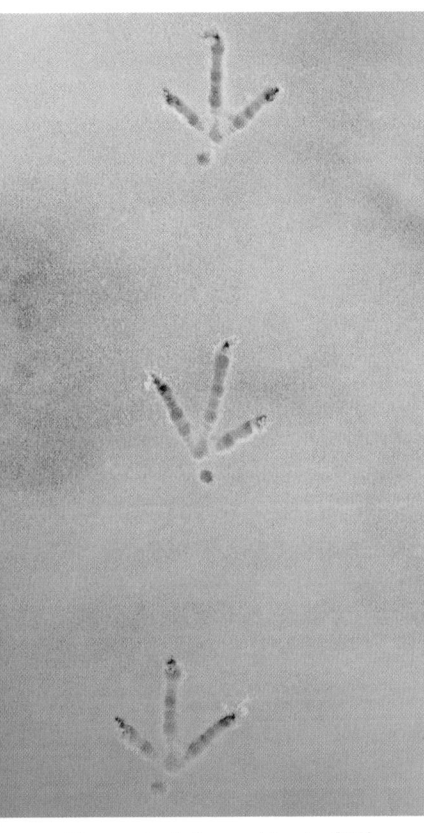

Ring-necked pheasant tracks. (NH)

American Coot
(Fulica americana)

Track: 3 1/8–3 5/8 in. (7.9–9.2 cm) L x 3–4 1/2 in. (7.6–11.4 cm) W

Game bird track. Large. Anisodactyl. Lobate. Incumbent foot structure. Although toe 1 is very common in tracks, the metatarsal pad may be absent in shallow substrates. At a run, toes 2 and 4 register closer to toe 3, and toe 1 often does not show at all, or registers very lightly.

Similar species: Hard to confuse with tracks of other species, due to the lobed toes. Moorhen tracks are similar in dimensions but have very thin toes.

Trail:	Walk	Strides 4 1/2–8 in. (11.4–20.3 cm)
	Run	Strides 8–15 in. (20.3–38.1 cm)

(See illustration on next page.)

The running trail of an American coot. (CA)

American coot tracks (L); trail: moving from a walk to a run in mud (R).

Sandhill crane tracks (L); trail: walking in mud (R).

Sandhill Crane *(Grus canadensis)*

Track: 3¾–4¾ in. (9.5–12.1 cm) L x 4¼–5¾ in. (10.7–14.6 cm) W

Game bird track. Large to very large. Anisodactyl. Incumbent foot structure. Webbing is present between toes 3 and 4. The presence of both toe 1 and the metatarsal pad is dependent upon depth of substrate. In shallow substrates, the hallux tends to be absent altogether, and the metatarsal will show lightly, if at all.

Similar species: The toe pads of turkeys are much more bulbous and rough, and turkey tracks tend to register toe 1. Turkeys also walk with shorter strides.

Trail: Walk Strides 12–21 in. (30.7–53.4 cm)
Run Strides up to 31 in. (78.7 cm)

Notes: In the summer months, sandhill cranes are often found in pairs, and they are eventually followed by smaller and younger cranes. *(See photo on next page.)*

The walking trails of sandhill cranes. A Canada goose crosses in the background. (WY)

Wild Turkey
(Meleagris gallopavo)

Track: 3¾–5 in. (9.5–12.7 cm) L x 4–5¼ in. (10.2–13.3 cm) W

Game bird track. Large to very large. Anisodactyl. Incumbent foot structure. Webbing is found between toes 2 and 3 and toes 3 and 4. Both toe 1 and the metatarsal pad tend to register clearly in tracks.

Similar species: Sandhill cranes tend not to register their toe 1 or the metatarsal pad. They also have less bulbous toe pads and take longer walking strides.

Trail:	Walk	Strides 5–13 in. (12.7–38.1 cm)
	Run	Strides up to 33 in. (83.8 cm)

Notes: The turkey is a flock species, except at the start of nesting, when laying hens will be on their own.

Wild turkey tracks. (VT)

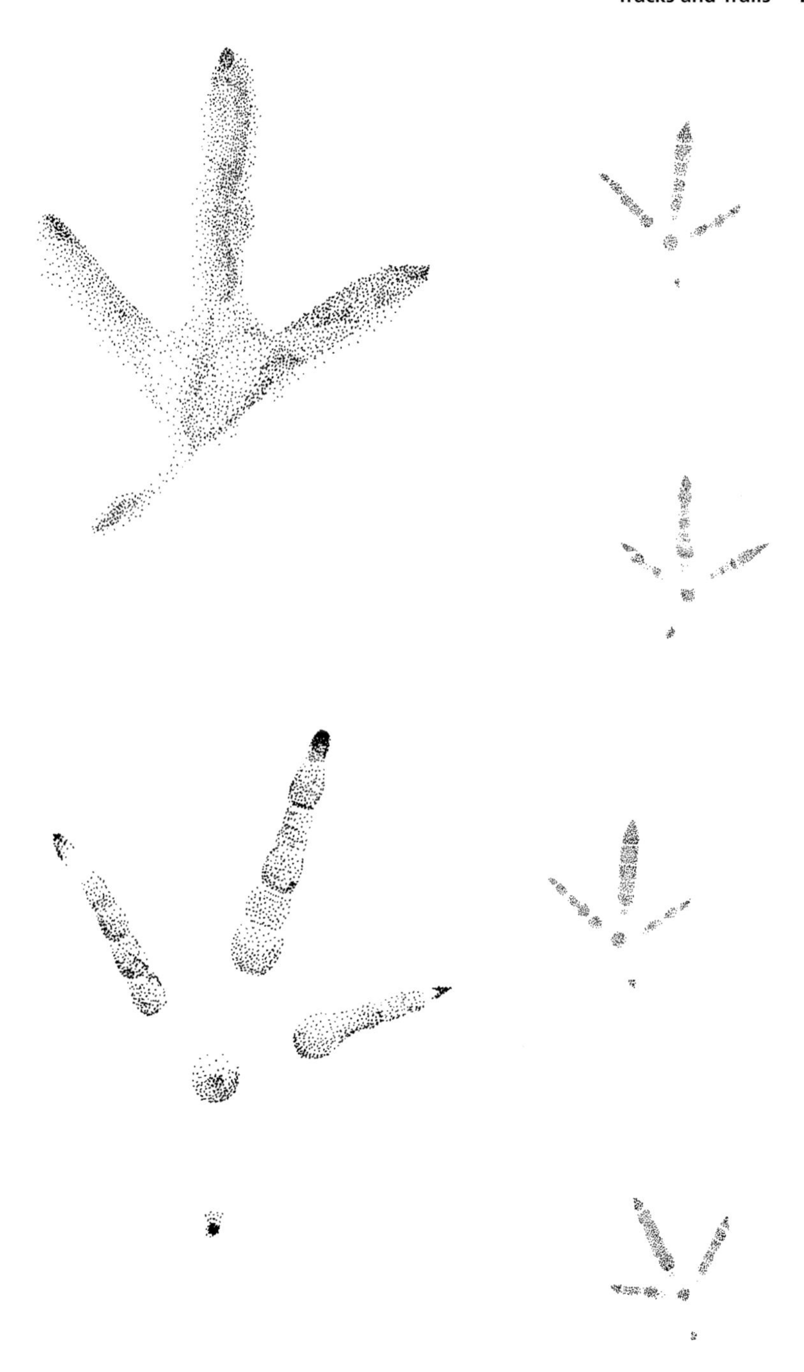

Wild turkey tracks (L); trail: walking in moist sand (R).

Webbed, or Palmate, Tracks

Least Tern *(Sterna antillarum)*

Track: 7/8–1 1/16 in. (2.2–2.7 cm) L x 7/8–1 in. (2.2–2.5 cm) W

Webbed track. Small. Anisodactyl. Palmate. Incumbent foot structure. Both toe 1 and the metatarsal show reliably in tracks. Toe 3 is noticeably longer than the others, the end of which protrudes beyond the webbing, which gives the impression of a long claw on toe 3 in the track.

Similar species: Gull tracks are much larger and don't often show toe 1. Other terns usually have larger tracks. In those terns that do overlap in size, such as the common tern, consider track width.

Trail: Walk Strides 1 1/2–2 5/8 in. (3.8–6.7 cm)

Notes: The least tern is endangered in its range.

Least tern tracks (L); trail: walking on a mudflat (R).

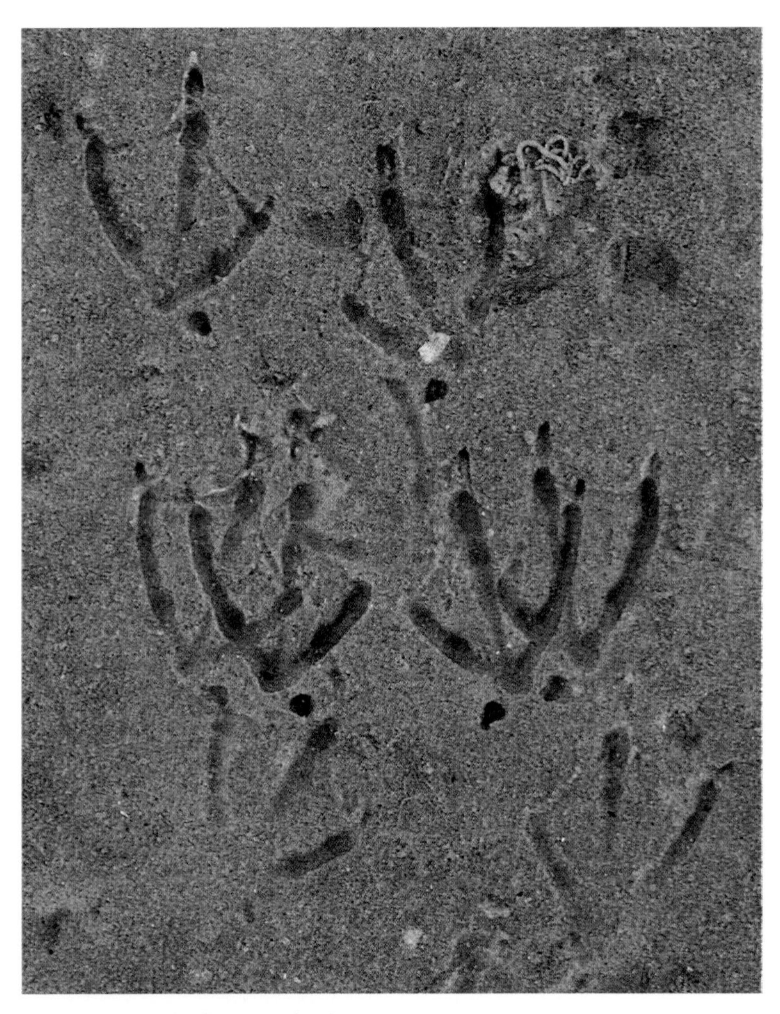

A least tern at a loafing area. (MA)

Common Tern *(Sterna hirundo)*

Track: 1–1 3/8 in. (2.5–3.5 cm) L x 7/8–1 3/8 in. (2.2–3.5 cm) W

Webbed track. Small. Anisodactyl. Palmate. Incumbent foot structure. Both toe 1 and the metatarsal show reliably in tracks. Toe 3 is noticeably longer than the others, the end of which protrudes beyond the webbing, which gives the impression of a long claw on toe 3 in the track.

Similar species: Gull tracks are larger and don't often show toe 1. Least terns have smaller tracks. Other tern tracks are similar.

Trail: Walk Strides 1 1/2–3 in. (3.8–7.6 cm)

Forster's Tern *(Sterna forsteri)*

Track: $1\frac{1}{8}$–$1\frac{1}{2}$ in. (2.8–3.8 cm) L x $1\frac{1}{8}$–$1\frac{1}{2}$ in. (2.8–3.8 cm) W

Webbed track. Small. Anisodactyl. Palmate. Incumbent foot structure. Both toe 1 and the metatarsal show reliably in tracks. Toe 3 is noticeably longer than the others, the end of which protrudes beyond the webbing, which gives the impression of a long claw on toe 3 in the track.

Similar species: Gull tracks are larger and don't often show toe 1. Least terns have smaller tracks. Other tern tracks are similar.

Trail: Walk Strides $2\frac{1}{2}$–4 in. (6.3–10.2 cm)

Royal Tern *(Sterna maxima)*

Track: $1\frac{1}{4}$–$1\frac{3}{4}$ in. (3.2–4.4 cm) L x $1\frac{1}{4}$–$1\frac{7}{8}$ in. (3.2–4.7 cm) W

Webbed track. Small. Anisodactyl. Palmate. Incumbent foot structure. Both toe 1 and the metatarsal show reliably in tracks. Toe 3 is noticeably longer than the others, the end of which protrudes beyond the webbing, which gives the impression of a long claw on toe 3 in the track.

Similar species: Other tern tracks are similar. Least, common, and Forster's terns have smaller tracks. Gull tracks are larger and do not often show toe 1.

Trail: Walk Strides $2\frac{1}{4}$–5 in. (5.7–12.7 cm)

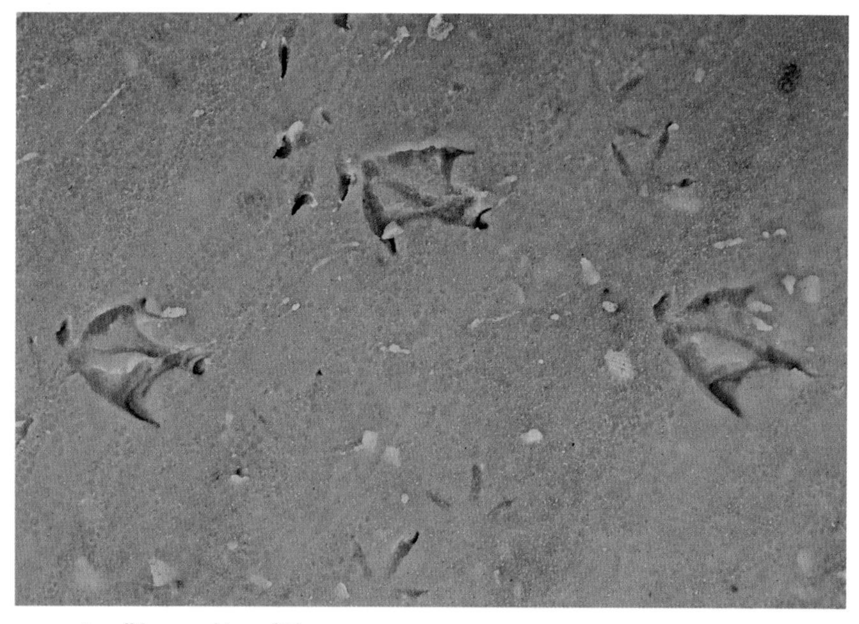

A walking royal tern. (FL)

Royal tern tracks (L); trail: walking in moist sand (R).

Black Skimmer *(Rynchops niger)*

Track: 1 1/2–1 3/4 in. (3.8–4.4 cm) L x 1 1/4–1 9/16 in. (3.2–4 cm) W

Webbed track. Small. Anisodactyl. Palmate. Incumbent foot structure. Both toe 1 and the metatarsal show reliably in tracks. Toe 3 is noticeably longer than the others, the end of which protrudes beyond the webbing, which gives the impression of a long claw on toe 3 in the track.

Similar species: Tern tracks are similar, especially those of royal terns. Least tern tracks are smaller. Gulls have larger tracks and do not often register toe 1.

Trail: Walk Strides 3–4 1/2 in. (7.6–11.4 cm)

Green-winged Teal *(Anas crecca)*

Track: $1\frac{7}{16}$–$1\frac{3}{4}$ in. (3.6–4.4 cm) L x $1\frac{1}{2}$–$1\frac{7}{8}$ in. (3.8–4.7 cm) W

Webbed track. Small. Anisodactyl. Incumbent foot structure. Both toe 1 and the metatarsal pad show unpredictably in teal tracks.

Similar species: Other small ducks, including blue-winged teal, ruddy ducks, and goldeneyes, have similar tracks.

Trail: Walk Strides 3–$4\frac{1}{2}$ in. (7.6–11.4 cm)

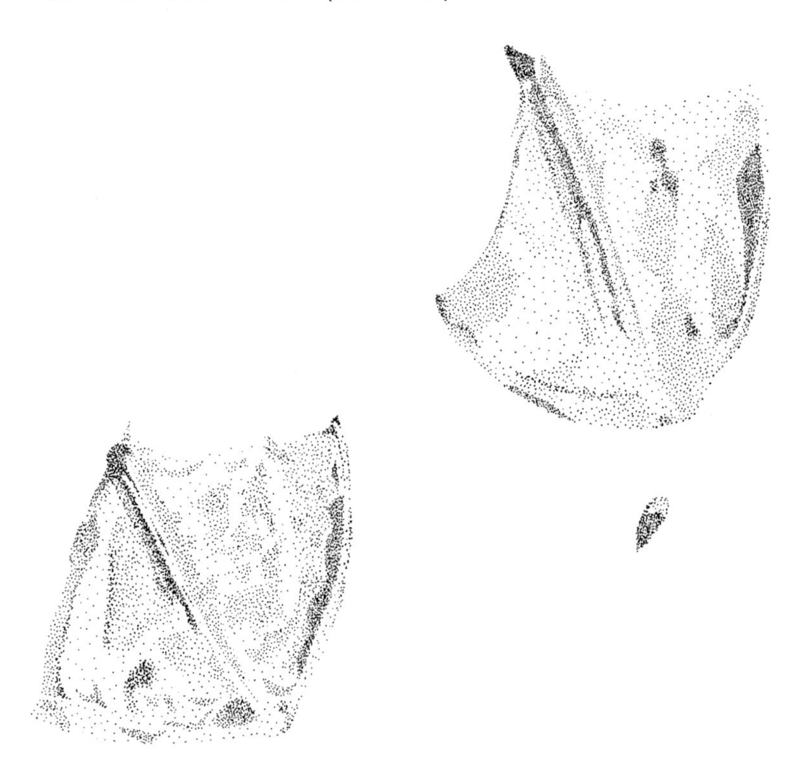

Green-winged teal tracks.

California Gull *(Larus californicus)*

Track: $1\frac{11}{16}$–2 in. (4.3–5.1 cm) L x $1\frac{7}{8}$–$2\frac{1}{4}$ in. (4.7–5.7 cm) W

Webbed track. Medium. Anisodactyl. Palmate. Incumbent foot structure. The hallux is greatly reduced and tucked in behind the metatarsal in gulls; therefore it tends not to register, unless the substrate is deep. The metatarsal was present in about half the tracks analyzed.

Similar species: Difficult to distinguish from ring-billed gull tracks, and tracks of other small gulls also overlap in size.

Trail: Walk Strides $2\frac{3}{4}$–$5\frac{1}{2}$ in. (7–14 cm)

Laughing Gull *(Larus altricilla)*

Track: $1\frac{5}{8}$–2 in. (4.1–5.1 cm) L x $1\frac{3}{4}$–$2\frac{1}{4}$ in. (4.4–5.7 cm) W

Webbed track. Medium. Anisodactyl. Palmate. Incumbent foot structure. Toe 1 tends not to register, unless the substrate is deep. The metatarsal was present in about half the tracks analyzed.

Similar species: Other small gulls have similar tracks. Small ducks tend to register toe 1 more often.

Trail: Walk Strides $2\frac{3}{4}$–$7\frac{1}{4}$ in. (7–18.4 cm)

Notes: This gull's trail is slightly pigeon-toed.

Northern Pintail *(Anas acuta)*

Track: $2\frac{1}{8}$–$2\frac{9}{16}$ in. (5.4–6.4 cm) L x 2–$2\frac{3}{4}$ in. (5.1–7 cm) W

Webbed track. Medium. Anisodactyl. Palmate. Incumbent foot structure. Both toe 1 and the metatarsal are reliably found in most duck tracks. Males are larger than females.

Similar species: Other medium-sized ducks have similar tracks. Gull tracks tend not to show toe 1.

Trail: Walk Strides $2\frac{1}{2}$–7 in. (6.3–17.8 cm)

Laughing gull trail: walking in mud.

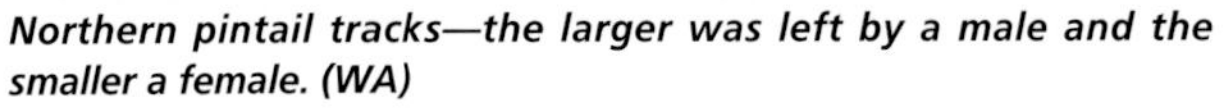

Northern pintail tracks—the larger was left by a male and the smaller a female. (WA)

Ring-billed Gull *(Larus delawarensis)*

Track: 1 7/8–2 1/4 in. (4.7–5.7 cm) L x 1 15/16–2 3/8 in. (4.9–6 cm) W

Webbed track. Medium. Anisodactyl. Palmate. Incumbent foot structure. The hallux is greatly reduced and tucked in behind the metatarsal in gulls; therefore it tends not to register, unless the substrate is deep. The metatarsal was present in about half the tracks we analyzed.

Similar species: Difficult to distinguish from California gull tracks, and tracks of other small gulls also overlap in size.

Trail: Walk Strides 2 3/4–7 in. (7–17.8 cm)

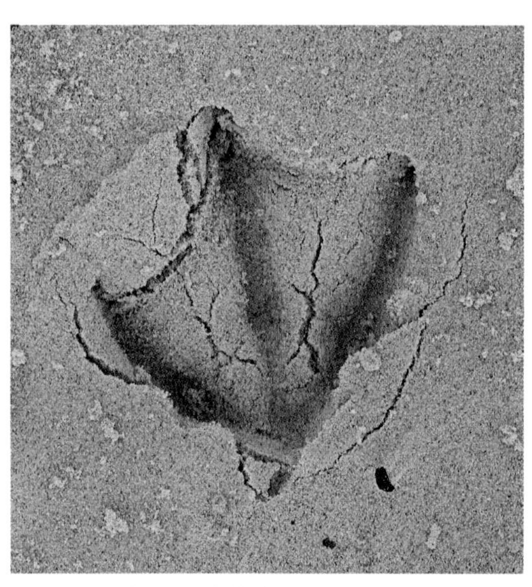

A perfect ring-billed gull track. Note how small and tucked in the hallux is. (TX)

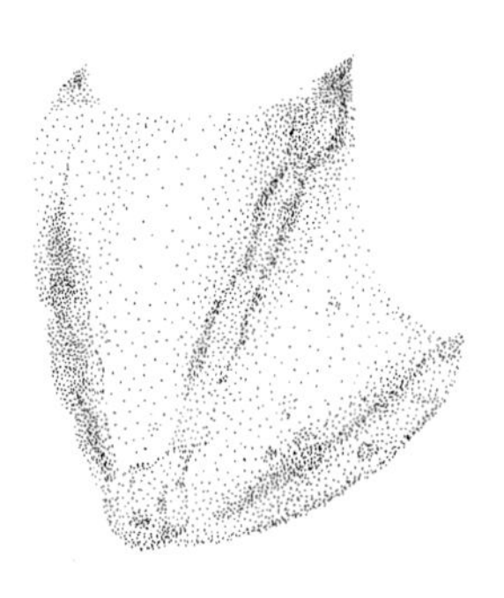

Ring-billed gull track (L); trail: walking in loose sand (R).

The walking trail of an American black duck. (MA)

American Black Duck *(Anas rubripes)*

Track: 2½–2⅞ in. (6.3–7.3 cm) L x 2⅜–2¾ in. (6–7 cm) W

Webbed track. Medium. Anisodactyl. Palmate. Incumbent foot structure. Both toe 1 and the metatarsal are reliably found in most duck tracks.

Similar species: Other medium-sized ducks have similar tracks. Gull tracks tend not to show toe 1.

Trail: Walk Strides 3½–7 in. (8.9–17.8 cm)

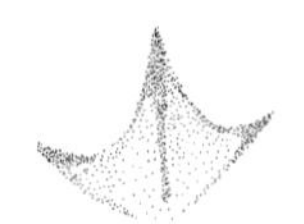

American avocet tracks (L); trail: walking on a mudflat (R).

American Avocet *(Recurvirostra americana)*

Track: 2–2 1/4 in. (5.1–5.7 cm) L x 2 1/4–2 3/4 in. (5.7–7 cm) W

Webbed track. Medium. Anisodactyl. Palmate. Incumbent foot structure. Toe 1 does not register in tracks, as it is found very high on the leg and is greatly reduced. The metatarsal registers unpredictably in tracks.

Similar species: Godwit and curlew tracks reliably show toe 1, and their toes 2 and 4 are straight, rather than the curved toes of species with webbed tracks. Gulls and ducks are pigeon-toed and take smaller strides.

Trail: Walk Strides 4 1/2–8 1/2 in. (12.4–21.6 cm)

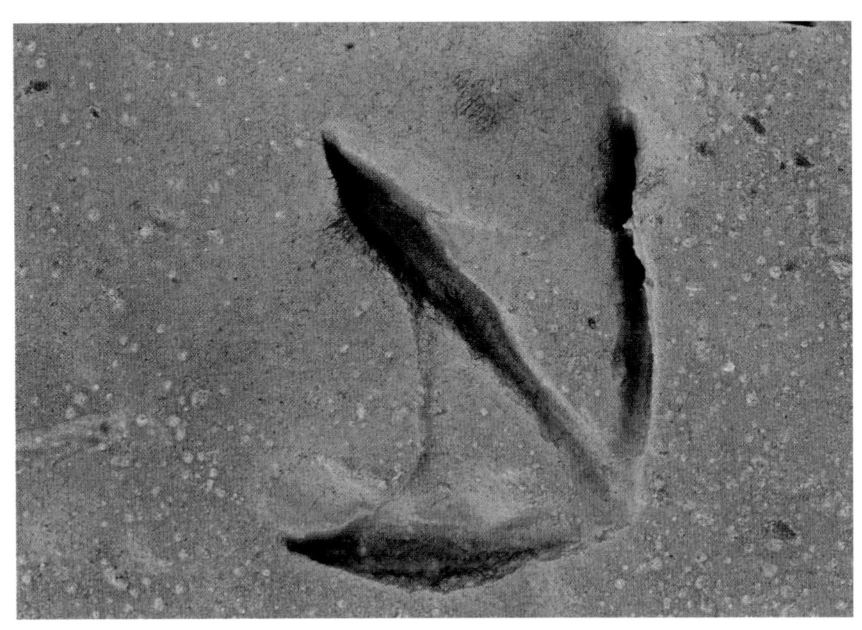

American avocet track. (CA)

Brant *(Branta bernicla)*

Track: 2⁵/₈–3¹/₈ in. (6.7–7.9 cm) L x 2³/₄–3³/₁₆ in. (7–8.1 cm) W

Webbed track. Medium. Anisodactyl. Palmate. Incumbent foot structure. Toe 1 registers unpredictably, but the metatarsal tends to register clearly.

Similar species: Width is a more important comparison than length when differentiating from large duck tracks. Other geese have larger tracks.

Trail: Walk Strides 5–8 in. (12.7–20.3 cm)

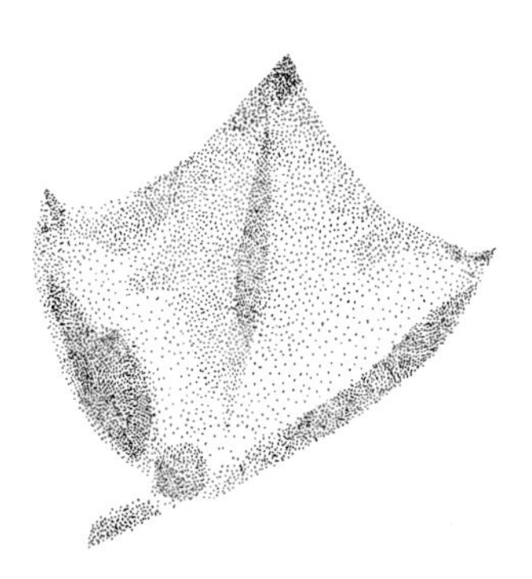

Brant track.

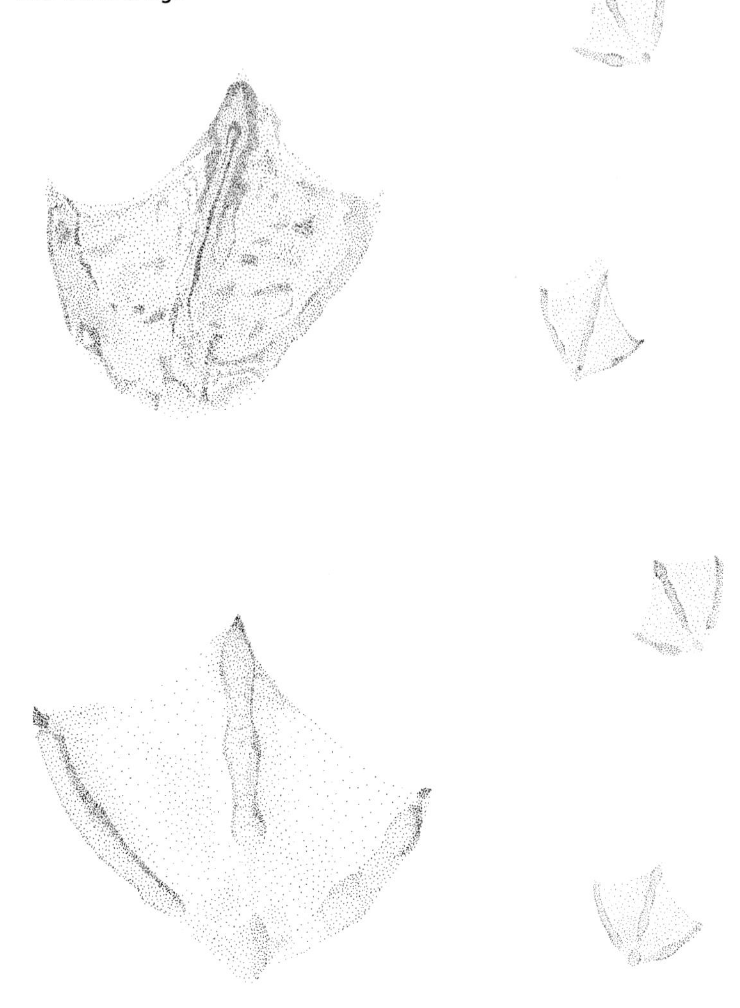

Herring gull tracks (L); trail: walking on a mudflat (R).

Herring Gull *(Larus argentatus)*

Track: 2 1/2–2 7/8 in. (6.3–7.3 cm) L x 2 3/4–3 1/4 in. (7–8 1/4 cm) W

Webbed track. Medium. Anisodactyl. Palmate. Incumbent foot structure. The hallux is greatly reduced and tucked in behind the metatarsal in gulls; therefore it rarely registers. The metatarsal tends to lightly register in tracks.

Similar species: Medium-sized ducks tend to register toe 1 and to be more pigeon-toed.

Trail: Walk Strides 3 1/2–10 1/2 in. (8.9–26.7 cm)

Mallard *(Anas platyrhynchos)*

Track: 2 3/8–2 15/16 in. (6–7.5 cm) L x 2 1/4–3 1/16 in. (5.7–7.8 cm) W

Webbed track. Medium. Anisodactyl. Palmate. Incumbent foot structure. Both toe 1 and the metatarsal are reliably found in most duck tracks.

Similar species: Other medium-sized ducks have similar tracks. Gull tracks tend not to show toe 1.

Trail: Walk Strides 3 3/4–7 1/2 in. (9.5–19.1 cm)

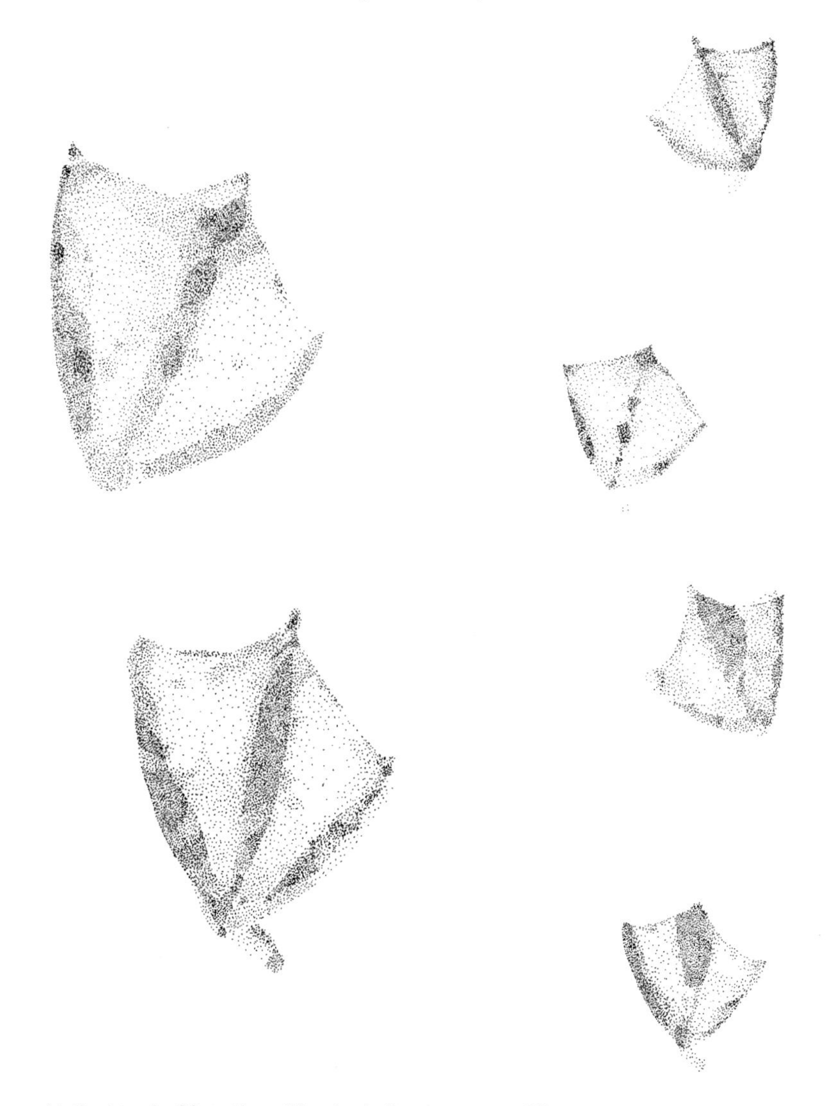

Mallard tracks (L); trail: walking in shallow wet snow (R).

Takeoff Patterns

There is a typical pattern associated with the takeoffs of many bird species, and it is a pattern worth knowing. Certain birds require this running takeoff to become airborne, but others do not, and therefore the trail patterns may aid in species identification. Ravens must run to take off, whereas crows can lift straight into the air from any given spot. Doves are well known for their lightning-fast takeoffs, which do not need a running start. This is also true of many game birds. Now bird tracking has a new twist, as you add your observations of whether a bird you are watching on the ground needs to run to lift off, or whether it can shoot straight into the air in alarm.

The semipalmated plover trail at right illustrates the classic takeoff pattern.

A Canada goose track, flanked by a portion of spotted sandpiper trail. (WY)

Canada Goose *(Branta canadensis)*

Track: 3 7/8–4 3/4 in. (9.8–12.1 cm) L x 3 3/4–5 in. (9.5–12.7 cm) W

Webbed track. Medium to large. Anisodactyl. Palmate. Incumbent foot structure. Toe 1 tends not to register, and the metatarsal pad tends to register lightly.

Similar species: Other gulls and dabbling ducks have smaller tracks. Swan tracks are larger.

Trail: Walk Strides 3 1/2–11 in. (8.9–28 cm)

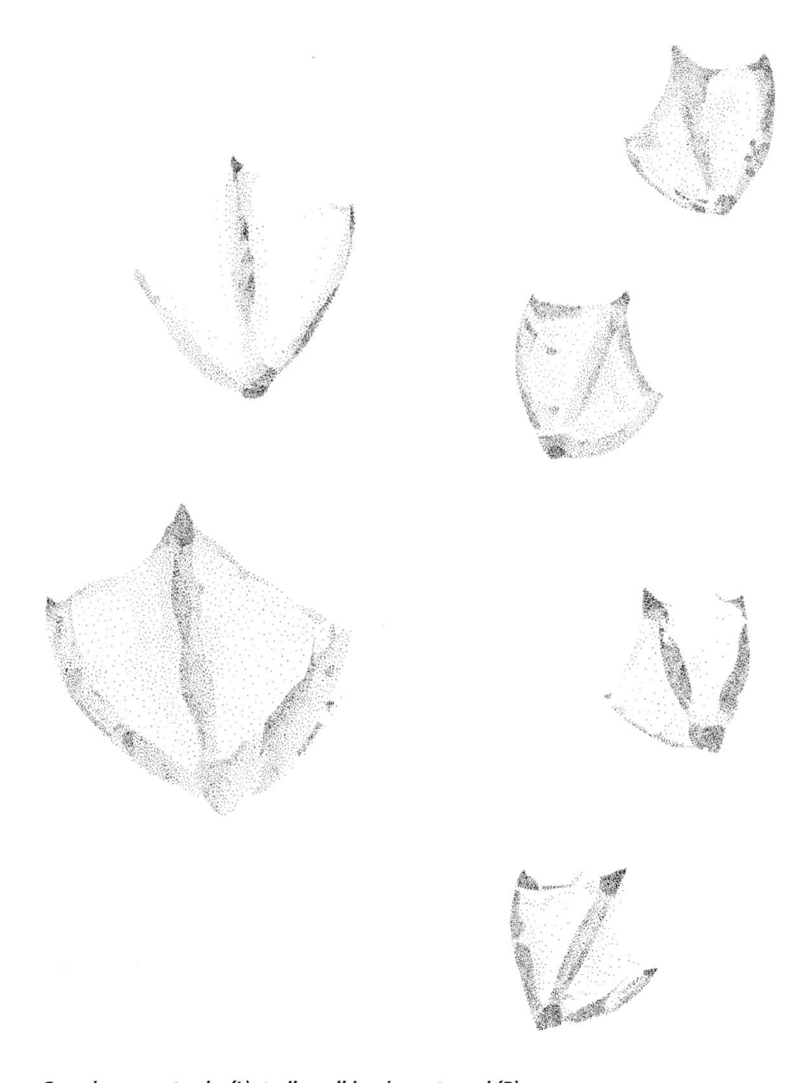

Canada goose tracks (L); trail: walking in wet sand (R).

Great Black-backed Gull *(Larus marinus)*

Track: 3–3¾ in. (7.6–9.5 cm) L x 3½–4¼ in. (8.9–10.8 cm) W

Webbed track. Medium to large. Anisodactyl. Palmate. Incumbent foot structure. The hallux is greatly reduced and tucked in behind the metatarsal in gulls; therefore it rarely registers at all, and when it does, it is small and tucked in close to the metatarsal. The metatarsal pad tends to register lightly, if at all.

Similar species: Other gulls and dabbling ducks have smaller tracks. Canada geese have larger tracks.

Trail:	Walk	Strides 5–11 in. (12.7–28 cm)
	Run	Strides 11–14 in. (28–35.6 cm)

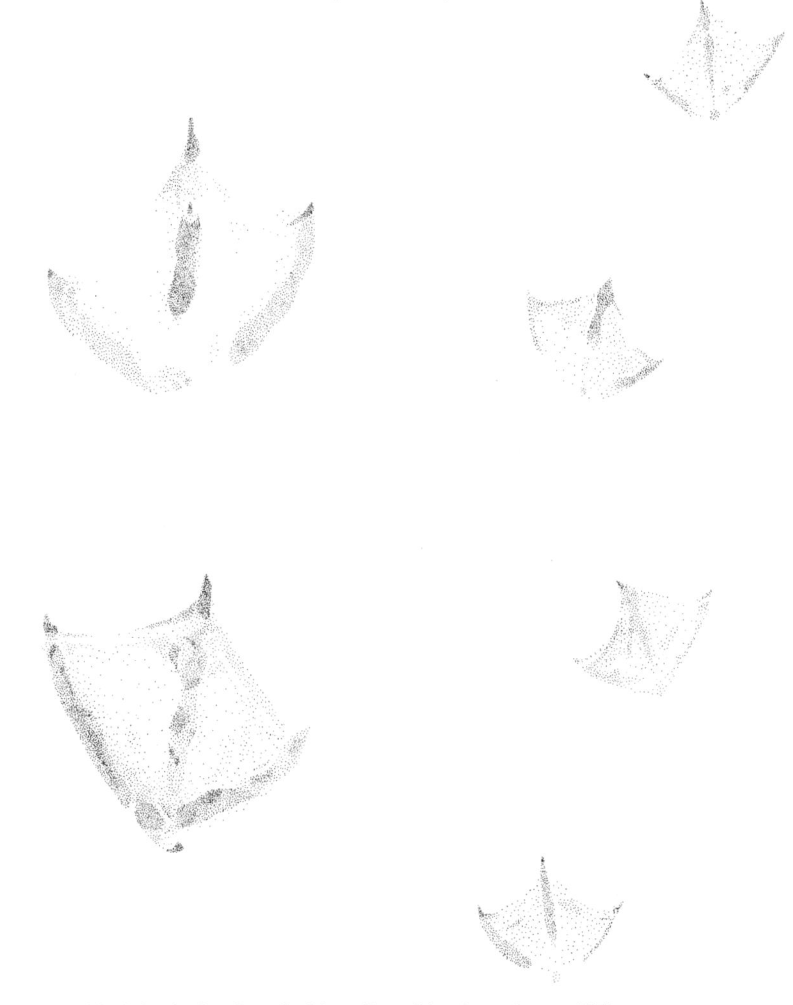

Great black-backed gull tracks (L); trail: walking in moist sand (R).

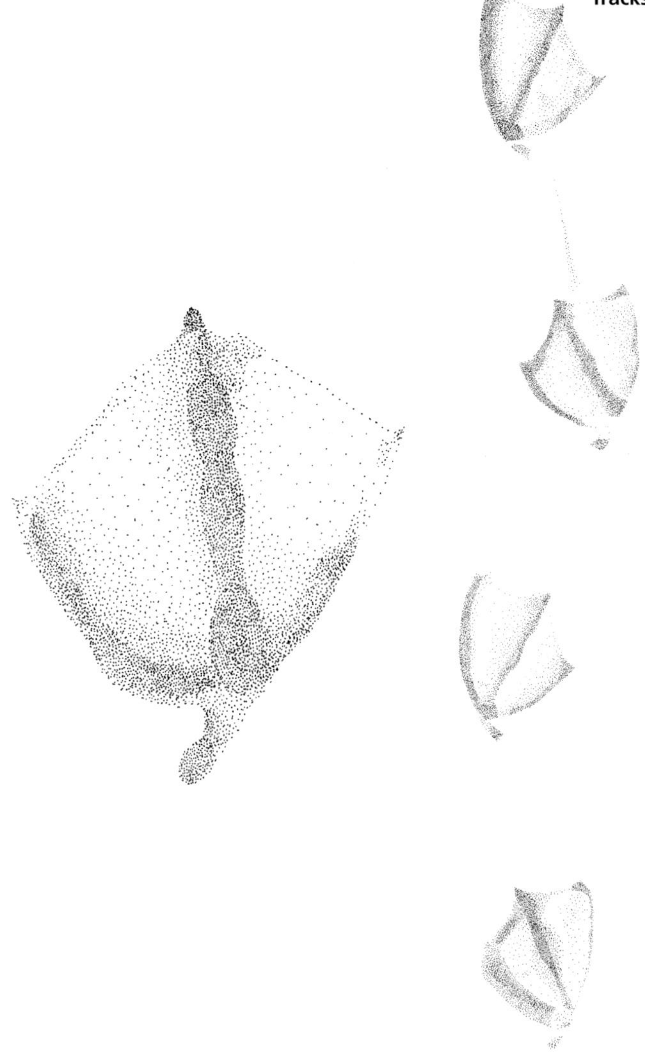

Trumpeter swan track (L); trail: walking in snow (R).

Trumpeter Swan *(Cygnus buccinator)*

Track: 5¾–7 in. (14.6–17.7 cm) L x 5¾–7 in. (14.6–17.7 cm) W

Webbed track. Medium to large. Anisodactyl. Palmate. Incumbent foot structure. Toe 1 registers sporadically, depending upon the depth of substrate.

Similar species: Other swan tracks are similar. Canada geese have smaller tracks.

Trail: Walk Strides 7–15 in. (17.8–38.1 cm)

Totipalmate Tracks

Double-crested Cormorant
(Phalacrocorax penicillatus)

Track: 4½–5⅝ in. (11.4–14.3 cm) L x 2½–3¼ in. (6.3–8.3 cm) W

Totipalmate. Large to very large. Metatarsal tends to show. Webbing registers inconsistently in tracks and may not be apparent at all.

Similar species: Other cormorants and boobies have similar tracks. Pelican tracks are larger. Loon tracks are larger and, although similar, are palmate.

Trail: Walk Strides 5–11 in. (12.7–28 cm)
Hop Strides 15–45 in. (38.1–114.3 cm)

Notes: Loafing areas, where terns, gulls, cormorants, and pelicans may congregate at various times of the day, are easily detected by the presence of tracks, scats, and feathers. The hop trail is also used to take off. Cormorants leave stream lines of scat, whereas gulls and pelicans tend to leave plops.

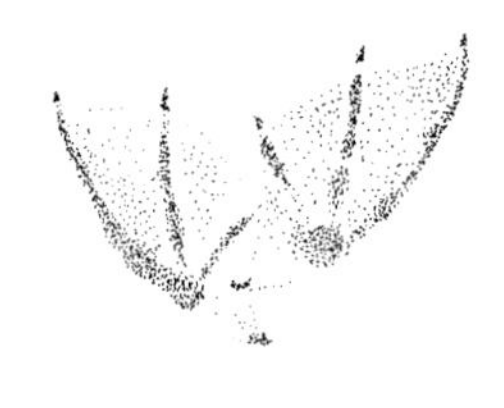

Double-crested cormorant trails: hopping on a mudflat (L), walking on a mudflat (R).

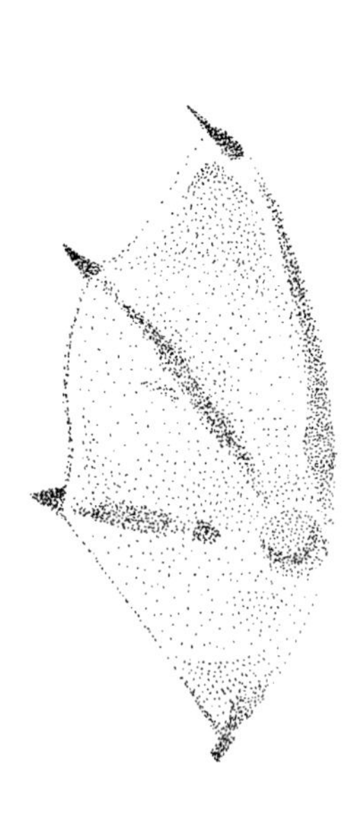

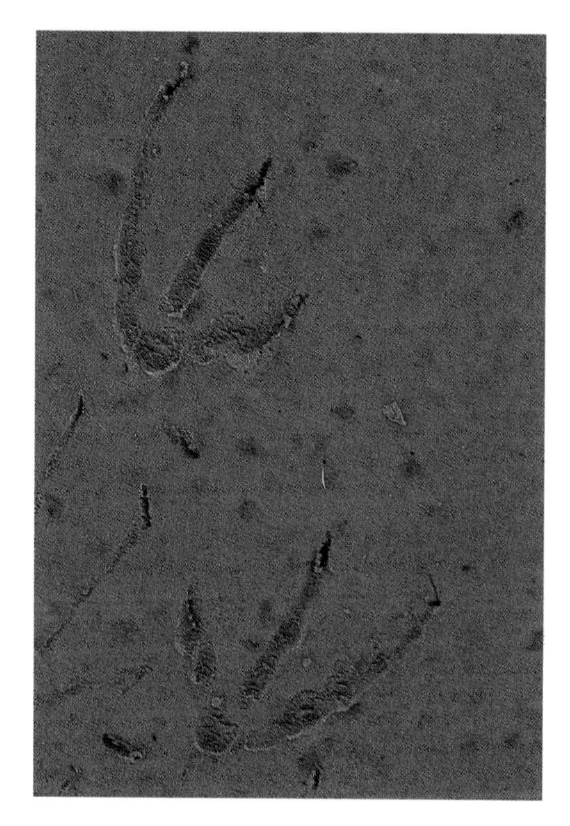

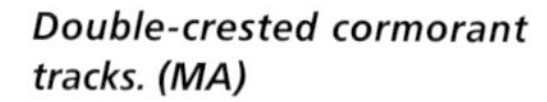

Double-crested cormorant tracks. (MA)

Brown Pelican *(Pelecanus occidentalis)*

Track: $6^{1}/_{4}$–7 in. (15.9–17.8 cm) L x $3^{1}/_{2}$–$4^{1}/_{2}$ in. (8.9–11.4 cm) W

Totipalmate. Very large. Metatarsal tends to show. Webbing registers inconsistently in tracks and may not be apparent at all.

Similar species: Cormorants tracks are smaller, and white pelican tracks are significantly wider. Loons have similar tracks, but they are palmate. Loons are also very uncomfortable on land.

Trail: Walk Strides 7–13 in. (17.8–38.1 cm)

Notes: An endangered species. Loafing areas, where terns, gulls, cormorants, and pelicans may congregate at various times of the day, are easily detected by the presence of tracks, scats, and feathers. The hop trail is also used to take off. Cormorants leave stream lines of scat, whereas gulls and pelicans tend to leave plops.

Brown pelican tracks.

White Pelican *(Pelecanus erythrorhynchos)*

Track: 6½–7½ in. (16.5–19 cm) L x 4¼–5⅜ in. (10.8–13.7 cm) W

Totipalmate. Very large. Metatarsal tends to show. Webbing registers inconsistently in tracks, and may not be apparent at all.

Similar species: Cormorant tracks are smaller, and brown pelican tracks are significantly narrower. Also consider differences in strides between pelican species if the bird was moving steadily along.

Trail: Walk Strides 9–16½ in. (22.9–42 cm)

Notes: Loafing areas, where terns, gulls, cormorants, and pelicans may congregate at various times of the day, are easily detected by the presence of tracks, scats, and feathers. The hop trail is also used to take off. Cormorants leave stream lines of scat, whereas gulls and pelicans tend to leave plops.

White pelican track (L); trail: walking on a mudflat (R).

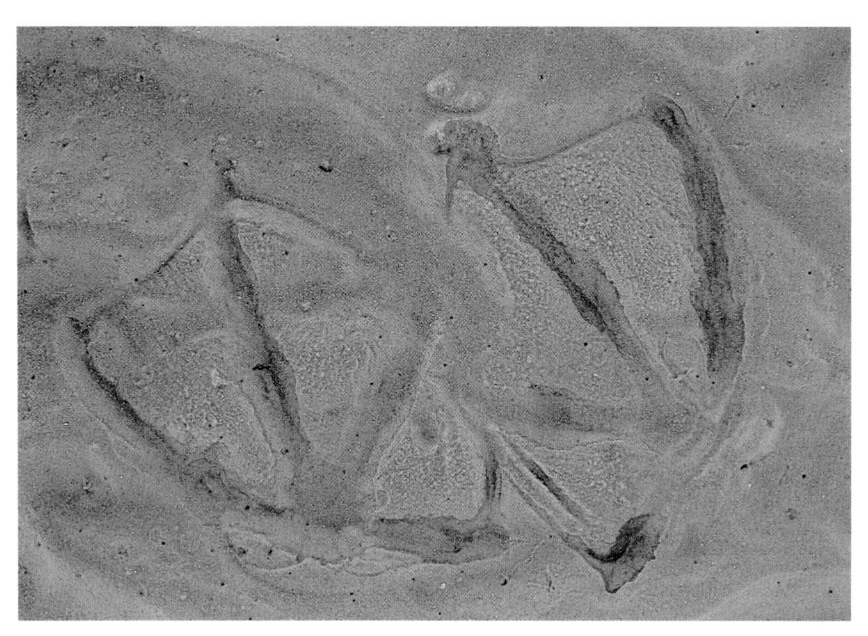

White pelican tracks. (TX)

Zygodactyl Tracks

Eastern Screech-owl (*Otus asio*)

Track: 1 1/2–2 in. (3.8–5.1 cm) L x 15/16–1 1/2 in. (2.4–3.8 cm) W

Zygodactyl. Small to medium. Metatarsal does not always show.

Similar species: Tracks of small owls overlap in size—consider ecological information.

(Parameters created from small data pool.)

Eastern screech-owl tracks.

Northern Flicker *(Colaptes auratus)*

Track: 1¾–2⅜ in. (4.4–6 cm) L x ⅜–⅝ in. (1–1.6 cm) W

Zygodactyl. Medium. Metatarsal does not always show. Toes 1 and 2 are much shorter than toes 3 and 4, which is true for all woodpeckers.

Similar species: Owls have more robust toes, and toe proportions are radically different.

Trail: Hop TW 1½–2¾ in. (3.8–7 cm)
Strides 3½–18 in. (8.9–45.8 cm)

Notes: Tracks are common in areas of feeding, especially in areas of high ant density.

Northern flicker tracks (L); trail: hopping in sandy soil (R).

Burrowing Owl *(Athene cunicularia)*

Track: 1¾–2 3/16 in. (4.4–5.5 cm) L x 1¼–1½ in. (3.2–3.8 cm)

Zygodactyl. Medium. Metatarsal does not always show.

Similar species: Tracks around burrows should not be confused with those of other owl species. Roadrunners have larger tracks.

Trail: Walk Strides 1–7 in. (2.5–17.8 cm)

Notes: Tracks around burrows are usually accompanied by scats and pellets, as well as other signs of feeding.

Pileated woodpecker tracks.

Pileated Woodpecker *(Dryocopus pileatus)*

Track: 2¼–2⅝ in. (5.7–6.6 cm) L x 1–1¾ in. (2.5–4.4 cm) W

Zygodactyl. Medium. Metatarsal does not always show. Toes 1 and 2 are much shorter than toes 3 and 4, which is true for all woodpeckers.

Similar species: Owls have more robust toes, and toe proportions are radically different.

Trail: Hop TW 3¾–6½ in. (9.5–16.5 cm)
Strides 3½–18 in. (8.9–45.8 cm)

(Parameters created from small data pool.)

Greater Roadrunner
(Geococcyx californianus)

Track: $2\frac{5}{8}$–$3\frac{1}{2}$ in. (6.7–8.9 cm) L x $1\frac{1}{8}$–$1\frac{5}{8}$ in. (2.8–4.1 cm) W

Zygodactyl. Medium. Metatarsal does not always show. Track registration varies considerably. Toes 1 and 4 often register lightly or are absent from the track.

Similar species: Woodpeckers often have smaller tracks, spend far less time on the ground, and have slender toes. They also don't run. Small owls have a different toe configuration.

Trail:	Run	Strides $5\frac{1}{2}$–$14\frac{1}{2}$ in. (14–36.9 cm)
	Skip	Strides 13–19 in. (33–48.2 cm)

Greater roadrunner tracks (L); trail: running in dust (R).

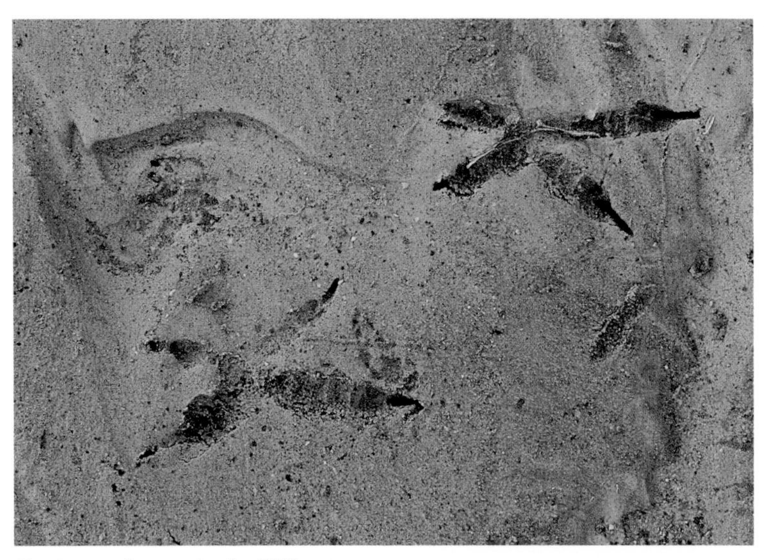

Greater roadrunner tracks. (CA)

Great Horned Owl
(Bubo virginianus)

Track: 3 1/4–4 5/8 in. (8.2–11.7 cm) L x 2 3/8–3 in. (6–7.6 cm) W

Zygodactyl. Large. Metatarsal does not always show.

Similar species: Hawks and eagles have a very different foot configuration. Snowy owls have feathered feet, which alters the track. Many other owls are not as comfortable on the ground.

Trail: Walk Strides 3–11 in. (7.6–28 cm)

(See illustration on next page.)

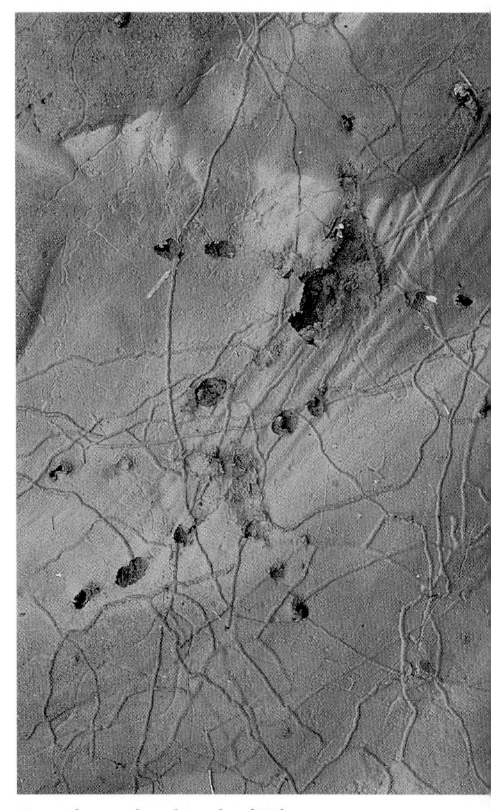

Great horned owl tracks. (CO)

Great horned owl track (L); trail: walking in mud (R).

Comparing Zygodactyl Tracks

Many zygodactyl tracks may be compared to the letter K in that two toes often form a straight line, and the other toes point at angles away from the center of this line. Owl tracks are perfect examples of the K configuration. It is also interesting to note where the straight side of the K is found: on the inside of the track, close to the centerline of the trail, or on the outside of the foot and falling to the outer edges of the trail.

Compare the trails left by roadrunners and owls. Roadrunner tracks have the straight line of the K on the outside of the foot, created by toes 3 and 4, while owls are the opposite, the straight line created by toes 1 and 2 and on the inside of the foot. Note the K shape and how it relates to the overall trail patterns to help differentiate among bird families.

Encountering an owl trail in the field, while always exciting, is not as unusual as one might imagine. Nonetheless, my first discovery led to a certain perplexity, as the left and right feet appeared reversed—the trail would seem to hold more in common with those of other birds if the straight line of the K were on the outside of the track, between toes 3 and 4. Perhaps for this reason, many previous and current tracking books have owl trails drawn in such a reversed way, wrongly extrapolating a left or right foot from a single, unlabeled track found in another book or a small pocket of soft substrate in the field. Carefully study the correctly presented photos here—in owls, the straight line created by toes 1 and 2 falls on the inside of each foot and thus each track.

The walking trail of a great horned owl moving from left to right. (CO)

Snowy owl track (L); trail: hopping in shallow snow (R).

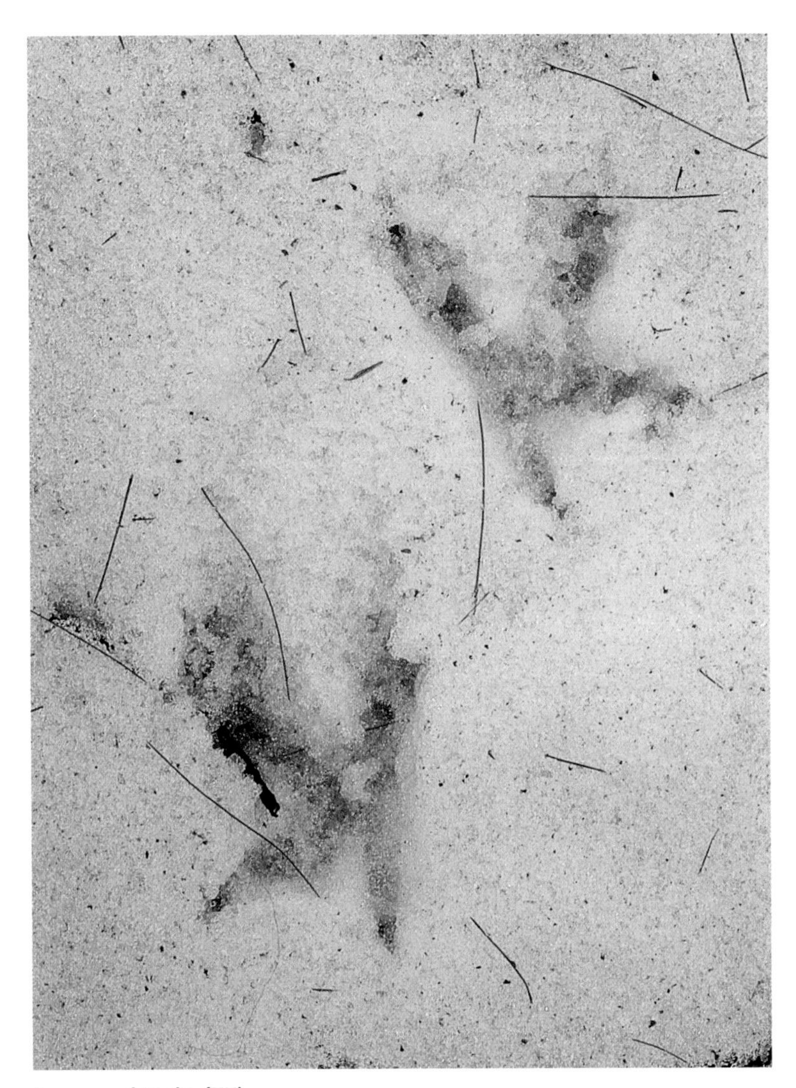

Snowy owl tracks. (NH)

Snowy Owl *(Nyctea scandiaca)*

Track: 3¾–5¼ in. (9.5–13.3 cm) L x 2¼–3 in. (5.4–7.6 cm) W
Zygodactyl. Large. Metatarsal does not always show.
Similar species: Great horned owl overlaps in size. Consider ecological clues.
Trail: Skip Strides 18–45 in. (45.8–114.3 cm)
Notes: Snowy owls commonly roost on the ground out in the open.
(Parameters created from small data pool.)

Pellets

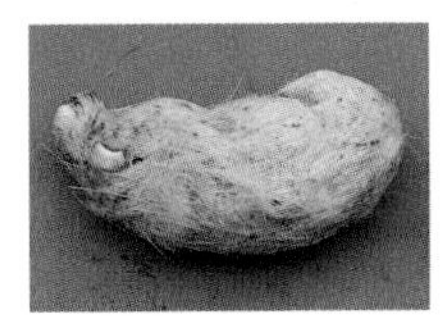

E.M./M.E.

Pellets are a fascinating source of information about birds and their intimate ecological relationship with habitat through feeding. *Pellets* are the compressed masses of indigestible food remains that birds regurgitate; they are composed of hair, bones, feather scraps, undigested sheaths of seeds and nuts, fish scales, small mammal skulls, insect parts, mollusk shells, and other odd bits such as plastic or garbage. Although owl pellets are the most well known, in fact most bird species in North America are capable of producing this sign; according to Leahy (1982), in addition to owls, "species of grebes, cormorants, vultures, hawks and eagles, herons, rails, shorebirds, gulls and terns, swifts, flycatchers, swallows, crows and jays, dippers, thrushes, catbirds, kinglets, wagtails, shrikes, and wood warblers" produce pellets. Those of the larger species last longer, however, and, depending on the contents and the weather, are more likely to be found in the field.

The contents of a pellet and the weather are key variables in determining the pellet's longevity. Fur and bone pellets retain their shape much longer and far outlast their insect, fish, and mollusk counterparts, although garbage pellets beat them all. Weather and associated moisture affect the longevity of a pellet: The wetter the environment, the faster it will break down. You are most likely to find pellets in places where they are somewhat protected from the elements. Mark, for example, found some minute pellets, the size of BB shot, of insectivorous phoebes nesting in the rafters of a huge, open porch by looking in the cracks of floorboards in a sheltered spot directly below the nest.

Pellets beneath a great horned owl roost under a bridge. (CO)

When searching for pellets, be aware that large fur and bone pellets—often associated with nests or roosts—are the type most likely to be found, especially if they are partially protected by an overhanging structure, such as downed limbs, a thicket, or a barn roof. Pellets exposed to tides or rain will rapidly deteriorate, so if you are searching for gull or cormorant pellets, look after high tide or in higher, more protected areas of dunes or marshes where gulls may be loafing; try checking dry land perches, such as a seldom-used dock, where cormorants have roosted, wings outspread, to dry their plumage.

In birds, much as in humans, a simple esophagus—a long, muscular tube—runs from the bird's mouth to its two-part stomach. This esophagus, sometimes called a *gullet,* is flexible and can temporarily expand to hold food so that birds can quickly gorge when supply is abundant, essentially allowing them to cache internally. Some species, including gallinaceous, or chickenlike, birds, parrots, doves and pigeons have a storage pouch called a crop at the midpoint of the esophagus for short-term extra food storage. This enables birds to fill up quickly and then move to the safety of cover to digest or feed young. Apparently birds can tolerate a stuffed gullet much more comfortably than can mammals. Proctor and Lynch (1993) note that common terns "often feed chicks on fish so large that the tails stick out of the chicks' mouths for most of a day, yet the chicks appear to suffer no ill effects from the experience."

Birds don't have teeth. Some swallow food whole, as owls, or tear it into large pieces, as hawks; some seedeaters, as the grosbeaks, have strong enough bills to at least partially crush nuts or seeds. The design of their two-part stomachs evolved as an alternative to incisors and grinding molars, and many species also rely on ingested grit or sand to help break down foods. The authors once watched Clark's nutcrackers avidly cache nuts in a high alpine forest in the Rocky Mountains. These gregarious birds energetically buried single nuts for over an hour, averaging one nut per minute, caching four to eight nuts in individual holes grouped within inches of one another before moving to another group site. They occasionally prodded and pushed a twig until it carefully covered the tiny hole in which they had secreted a nut. This extra landscaping did not appear to be needed as camouflage, because the cache sites had already been covered by the fastidious birds with loose dirt and duff and were imperceptible before the twig decoy was employed. After burying one group site of nuts, one nutcracker stopped to eat three or four bits of gravel, and then returned to rapidly caching. This large nutcracker ate grit at least three times during this foraging session.

The esophagus leads from the bird's mouth to a two-part stomach where the grit is stored. In the first part, the proventriculus, digestive juices are secreted to begin digestion; then the coated whole chunk of food proceeds to the second part of the stomach, the ventriculus, what we commonly call the gizzard. In birds that feed on tough substances like seeds, hard grains, or nuts, the strong, sandpaperlike muscular wall of the gizzard serves as an efficient grinder, aided by bits of sand and gravel stored in its deeply furrowed walls. Items that the gizzard cannot completely break down, such as bones, fur, or fish scales, are separated from the digestive material and compressed into a pellet. The crushed, mostly liquid remainder, called chyme, moves on to the small intestine to be further digested. Muscular contractions move the pellet back up to the first part of the stomach, where it is stored for several hours, until hunger stimulates the bird to hunt or feed again, and the pellet is then expelled. Although the process of expelling the pellet may appear to be painful, because the bird looks superficially as though it is choking, it does no harm. Because the pellet is composed of indigestible bits and is coated with mucus and hair or feather vanes, it seems to safely help scrape the esophagus clean as it exits.

Owls, which typically swallow small prey like mice or voles whole, produce one pellet per small animal eaten; with multiple hunts, this can easily add up to several pellets per day, usually regurgitated about six hours after hunting. Pellets from the same bird may be found both at the daytime roost or nest and at night in the vicinity of the hunt. On a dark, rainy afternoon, I once startled a great horned owl from its day roost on an overhanging, shoulder-height branch of a downed white pine. Underneath this ideal perch were scattered at least nine large pellets, averaging 4 to 4 1/2 inches (10.2–11.4 cm) long, indicating that the owl had used this day roost repeatedly. Although in this case the rain made all the pellets glisten uniformly, ordinarily a freshly expelled specimen will be coated with shiny mucus, whereas an older pellet will have a dry, matte surface.

The size and shape of a pellet reflect the anatomy of the bird that produced it. These characteristics, when considered along with the contents of the pellet and the habitat in which it was found, are all important clues to bird identification, and to a greater understanding of the prevalence of particular species in the environment around you. Both the diameter and much of the shape of the pellet are determined by the size of the esophagus, which in turn corresponds roughly to the overall size of the bird. Thus, great horned owls can be expected to produce long, relatively thick pellets, whereas gull and barn owl pellets are often round, irregular spheres.

The pellets of owls and similar-sized hawks and falcons can often be differentiated by contents. After a session of partial plucking, which can leave irregular "fairy rings" of feathers on the ground below their feeding perches, hawks typically tear bite-size chunks from their prey, whereas owls often swallow their prey whole. (This depends on the size of the owl and the prey, however. Mark has watched tiny northern pygmy owls tear at white-footed mice on several occasions while working in Glacier National Park in Montana.) Owls have weak digestive juices, but hawks have stronger acid levels, so owl pellets contain many more bones than hawk pellets and will also include entire small bones, whereas hawk pellets usually contain only bone fragments.

The photographs in this chapter, in combination with your own field experience, will help you begin distinguishing major patterns in pellet formation by species of birds. Jotting down diameters and

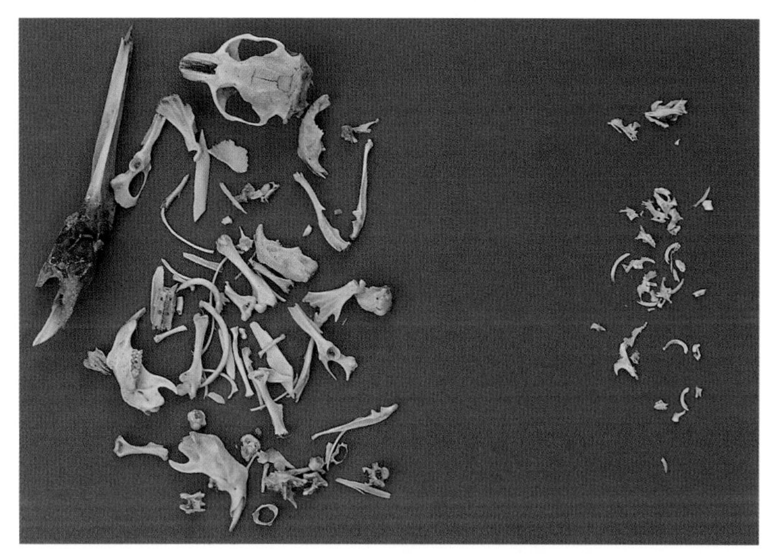

Two equal-size pellets were soaked and dissected to compare bone content. The bones on the left were pulled from the pellet of a great horned owl, and the bones on the right were removed from the pellet of a red-tailed hawk. The weaker stomach acid of owls and their tendency to swallow prey whole become very apparent when their pellets are compared with those of hawks or falcons.

lengths of pellets you collect and quickly sketching their shapes in your field notes will, in time, help you identify the likely species that produced them, especially when you think ecologically, considering the habitat surrounding the location where you found the pellet, the birds typically found in that area, and their preferred feeding and roosting habits.

I once watched a crow working the muddy low-tide edge of a salt marsh that was dominated by immense flats of spartina grass and blue mussel colonies, where the intent bird was gorging on small periwinkles, about 1/2 inch (1.2 cm) long and 3/8 inch (1 cm) in diameter, scattered among the barnacle-laced mussels. Although this large adult crow watched my approach carefully, it continued gulping one periwinkle almost every five seconds, eating at least thirty-eight shells, until its throat bulged almost comically with the bounty. I was not able to locate the roost of this adult in order to search for pellets, though I hope to find this type of feeding sign and still search for it. Someday I may be walking in a forest miles from a coastal salt marsh and come upon a disintegrating pellet made of crushed periwinkle shells, showing glints of iridescence in a shaft of late-afternoon sun. Obviously in this case, we as trackers need to be able to imagine the various habitats that may be part of the wide foraging range of a particular species when trying to make sense of the prey remains in a pellet.

Avian pellets can resemble the scat of some mammals, and the usual accompanying bird sign that the tracker would normally rely on, such as whitewash under roosts, may be subtle or altogether absent because of rain. Therefore, when you search for pellets in the field, contents and shape are important diagnostic characteristics.

Any scats with fruit remains, such as skins or seeds, are easily distinguished from pellets. The twisted and tapered scats of canines and mustelids are unlikely to be confused with pellets, which are generally round or cylindrical and smooth. However, raccoons and cats can produce smooth, cylindrical scats. In such cases, contents as well as odor will often help you differentiate them from pellets. (*Caution:* Raccoon scat should not be handled, as it is may contain the parasitic roundworm *Baylisascaris procyonis.*) Cat scats have a very strong, acrid odor, unlike the odorless pellets of raptors, and they are usually deposited as linked segments, whereas pellets, when found in groups, will clearly be disconnected. Also, aged raccoon scat tends to break across clean and dry when you crack it open, whereas pellets are usually densely packed and woven together and therefore more difficult to break apart. You may be more likely to misidentify the small, round insect-containing pellets of owls and kestrels for rabbit scat, although a close inspection of contents will separate the two. (Rabbit pellets, approximately the size of a chickpea, are odorless and entirely composed of compressed vegetative forage.)

The greatest concentrations of pellets are usually found where the birds roost for a while after feeding or at nest sites. These accumulations are particularly visible on snow and often reveal the nests of raptors that lay eggs in late winter. Never approach an active nest to collect pellets. This would disturb the parents and possibly affect their success at fledging young. After the nest is abandoned, however, feel free to return and collect pellets.

Soaking pellets in warm water and then dissecting them is a fascinating way to begin to inventory some of the prey of pellet producers in your area. Owl pellets are especially useful and diagnostic in this regard because of the inability of their weak stomach acids to digestively dissolve bones. You may even find entire disassembled skeletons of small vertebrates, including the most delicate bones and intact skulls still showing the sign of a bite to the back of the head, which is the owl's preferred method of killing small prey.

Pellet Comparisons

Differentiating the pellets of various species in the field can be a difficult task. Pellets in association with nests, regular roosts, and tracks are the most easily identified.

The size of pellets produced by a single bird varies dramatically depending on diet and the amount of food eaten. Therefore, there is substantial overlap across species, leading to such large measurement parameters that the numbers lose some differentiating value to the tracker and naturalist. Certainly, a bird will have an upper limit, meaning a maximum size above which a pellet would simply be too large for its anatomy to produce; thus a great horned owl pellet is unlikely to be confused with the pellet of a screech-owl. From our research, the width of the pellet, rather than the length, is a more reliable measure for making comparisons across species. Owl species typically produce a pellet that measures larger in length than in width, and the width is more reliably diagnostic because its dimension is mostly predetermined by the internal digestive structure of the bird.

Please note that the following parameters may be skewed slightly for several reasons. We measured only whole pellets for this project. Longer pellets tend to break, and therefore we disqualified them. Because we eliminated them from our sample, the length parameters may not include the absolute upper limits. Pellets that contain only insect parts are much smaller than their fur and bone counterparts, even when produced by a single bird. In some cases, we were not able to find both insect and fur and bone pellets for a species. In general, pellets are organized by width, from smallest to largest.

Pellet Descriptions

Elf owl pellets. (TX)

Elf Owl *(Micrathene whitneyi)*

1/4–5/16 in. (.6–.8 cm) W x 1/4–5/8 in. (.6–1.6 cm) L

Only elf owl pellets of insect remains are included in parameters. Pellets of fur and bones are likely larger.

American kestrel. (VT)

American Kestrel *(Falco sparverius)*

5/16–7/16 in. (.8–1.1 cm) W x 5/16–1 1/16 in. (.8–2.7 cm) L

Loggerhead shrike. Note the size difference between pellets composed of fur and those of insect remains. (NV)

Loggerhead Shrike *(Lanius excubitor)*

5/16–1/2 in. (.8–1.2 cm) W x 3/4–1 9/16 in. (1.9–4 cm) L

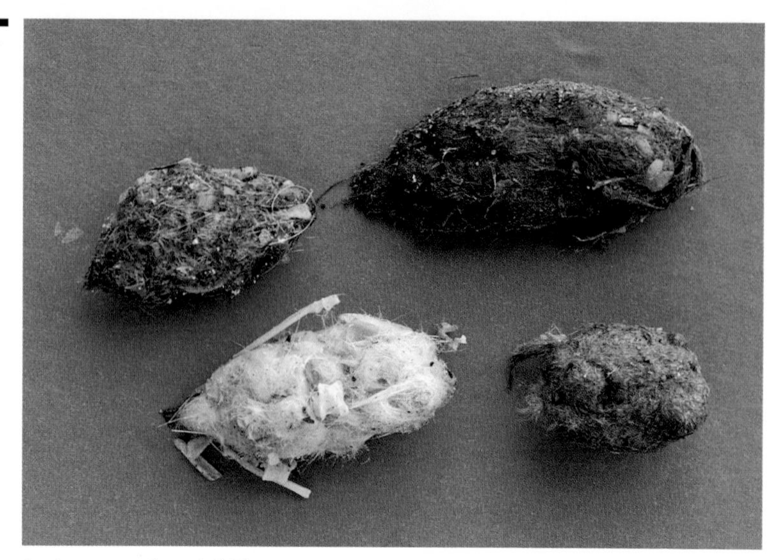

Eastern screech-owl. (VT)

Eastern Screech-owl *(Otus asio)*

3/8–9/16 in. (.9–1.4 cm) W x 5/8–1 1/2 in. (1.6–3.8 cm) L

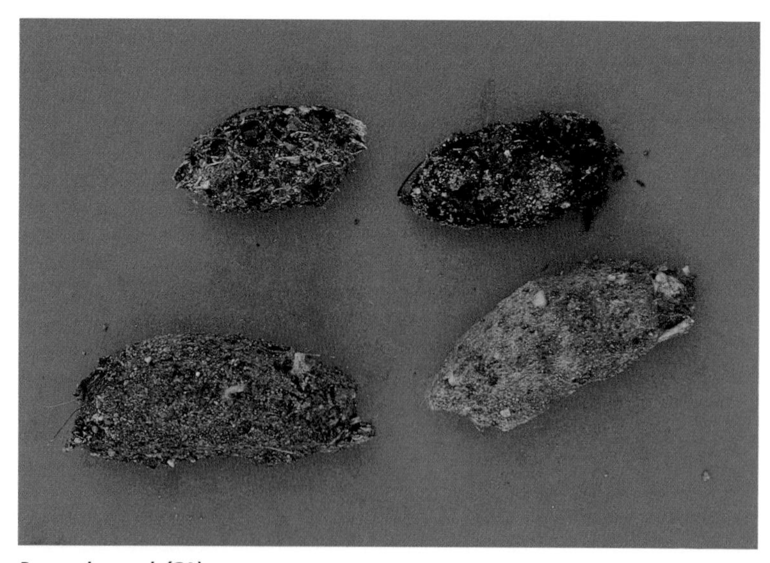

Burrowing owl. (CA)

Burrowing Owl *(Athene cunicularia)*

3/8–9/16 in. (.9–1.4 cm) W x 5/8–1 1/2 in. (1.6–3.8 cm) L

Black-billed magpie. (WY)

Black-billed Magpie
(Pica pica)

3/8–9/16 in. (.9–1.4 cm) W x 11/16–2 in. (1.7–5.1 cm) L

Merlin
(Falco columbarius)

3/8–5/8 in. (.9–1.6 cm) W x 1/2–1 3/8 in. (1.2–3.5 cm) L

Sharp-shinned Hawk
(Accipiter striatus)

3/8–5/8 in. (.9–1.6 cm) W x 1/2–7/8 in. (1.2–2.2 cm) L

Black Oystercatcher
(Haematopus bachmani)

5/8 in. (1.6 cm) W x 3/4 in. (1.9 cm) L

(A single pellet was measured.)

Black oystercatcher. (WA)

Clark's nutcracker. (WY)

Clark's Nutcracker *(Nucifraga columbiana)*

5/16–3/4 in. (.8–1.9 cm) W x 11/16–1 1/8 in. (1.7–2.8 cm) L

Northern Saw-whet Owl *(Aegolius acadicus)*

7/16–11/16 in. (1.1–1.7 cm) W x 5/8–1 1/8 in. (1.6–2.8 cm) L

Cooper's Hawk *(Accipiter cooperii)*

1/2–13/16 in. (1.2–2 cm) W x 3/4–1 1/8 in. (1.9–2.8 cm) L

Northern saw-whet owl. (VT)

Northern goshawk. (VT)

Northern Goshawk *(Accipiter gentilis)*

1/2–13/16 in. (1.2–2 cm) W x 3/4–2 3/4 in. (1.9–7 cm) L

Peregrine Falcon *(Falco peregrinus)*

1/2–7/8 in. (1.2–2.2 cm) W x 1/2–1 7/8 in. (1.2–4.7 cm) L

Short-eared owl. (MA)

Short-eared Owl *(Asio flammeus)*

1/2–7/8 in. (1.2–2.2 cm) W x 7/8–1 7/8 in. (2.2–4.7 cm) L

Long-eared Owl
(Asio otus)

7/16–1 in. (1.1–2.5 cm) W x
3/4–1 1/4 in. (1.9–3.2 cm) L

Broad-winged Hawk
(Buteo platypterus)

9/16–3/4 in. (1.4–1.9 cm) W x
3/4–1 1/2 in. (1.9–3.8 cm) L

Barred Owl
(Strix varia)

5/8–3/4 in. (1.6–1.9 cm) W x
1 1/8–2 in. (2.7–5.1 cm) L

Broad-winged hawk. (VT)

Barred owl. (VT)

Rough-legged Hawk *(Buteo lagopus)*

9/16–3/4 in. (1.4–1.9 cm) W x 11/16–1 3/4 in. (1.7–4.4 cm) L

Red-tailed hawk. (VT)

Red-tailed Hawk *(Buteo jamaicensis)*

9/16–15/16 in. (1.4–2.4 cm) W x 7/8 –2 1/4 in. (2.2–5.7 cm) L

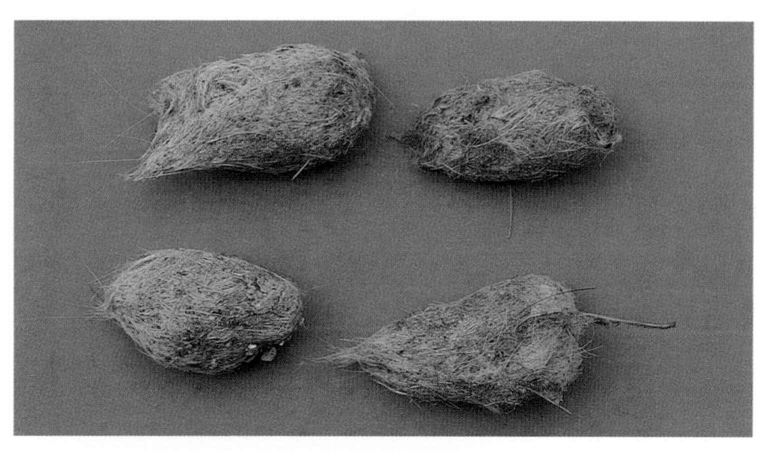

Turkey vulture. Note the leaf litter, caused by feeding on the ground. (VT)

Turkey Vulture *(Cathartes aura)*

13/16–1 in. (2–2.5 cm) W x 1 1/2–2 7/8 in. (3.8–7.3 cm) L

Common raven. (NV, WY)

Common Raven
(Corvus corax)

$\frac{9}{16}$–$\frac{15}{16}$ in. (1.4–2.4 cm) W x
$1\frac{3}{8}$–$2\frac{1}{8}$ in. (3.5–5.4 cm) L

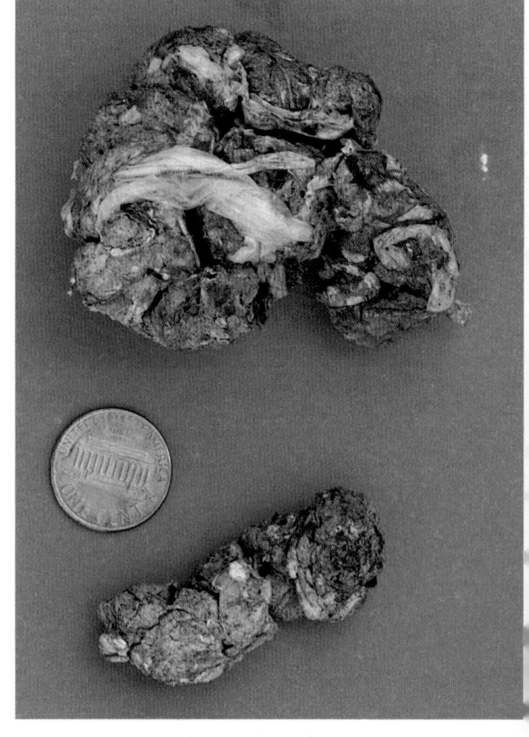

Common raven garbage pellets. (NY)

Barn owl. (MA)

Barn Owl *(Tyto alba)*

$^{11}/_{16}$–1$^{1}/_{8}$ in. (1.7–2.8 cm) W x
$^{7}/_{8}$–2 in. (2.2–5.1 cm) L

Cattle egret pellets filled with insect remains. (Photo by Chris and Tilde Stuart.)

Cattle Egret *(Bubulcus ibis)*

$^{3}/_{4}$–1$^{1}/_{8}$ in. (1.9–2.8 cm) W x 1$^{11}/_{16}$–2$^{1}/_{2}$ in. (4.3–6.3 cm) L

Snowy owl. (VT)

Snowy Owl *(Nyctea scandiaca)*

$\frac{11}{16}$–$1\frac{3}{16}$ in. (1.7–3 cm) W x $1\frac{3}{8}$–$3\frac{1}{2}$ in. (3.5–8.9 cm) L

Great gray owl. (VT)

Great Gray Owl *(Strix nebulosa)*

$\frac{3}{4}$–$1\frac{1}{4}$ in. (1.9–3.2 cm) W x $\frac{7}{8}$–$2\frac{3}{4}$ in. (2.2–7 cm) L

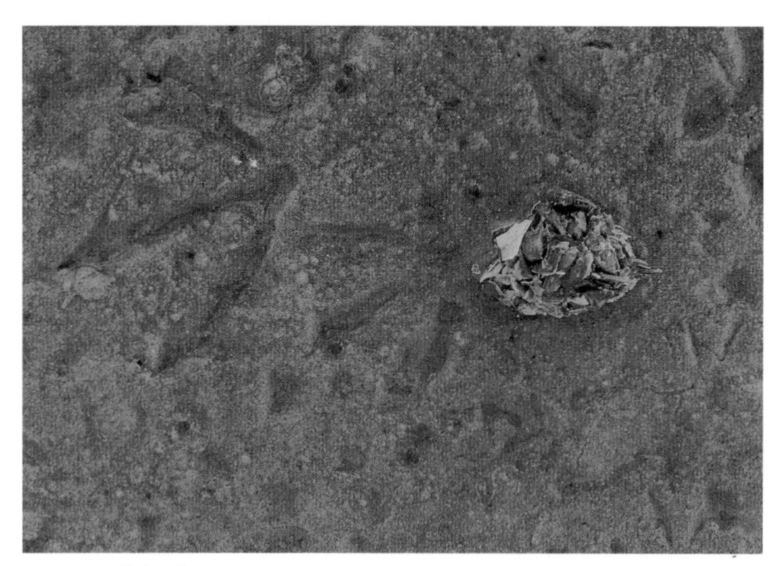

Herring gull. (MA)

A cormorant pellet for comparison with that of a gull. Note that the cormorant's is made entirely of feathery fish bones. (Used with permission of Gads Forlag. Photo by Erik Thomsen.)

Herring Gull *(Larus argentatus)*

3/4–1 3/16 in. (1.9–3 cm) W x 1 3/16–2 in. (3–5 cm) L

Great horned owl. (CO)

Great Horned Owl *(Bubo virginianus)*

$\frac{9}{16}$–$1\frac{1}{2}$ in. (1.4–3.8 cm) W x $\frac{5}{8}$–$4\frac{1}{2}$ in. (1.6–11.4) L

Golden eagle. (VT)

Golden Eagle *(Aquila chrysaetos)*

$\frac{11}{16}$–$1\frac{1}{2}$ in. (1.7–3.8 cm) W x $1\frac{3}{8}$–$3\frac{3}{4}$ in. (3.5–9.5 cm) L

Droppings

3

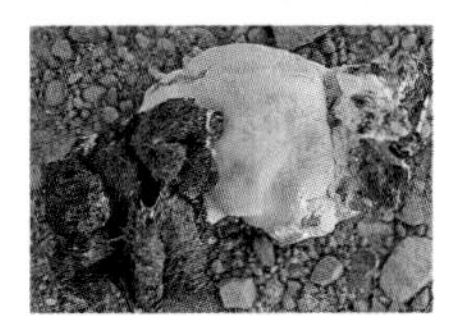

M.E./E.M.

S*cat* is often the word used by trackers and field researchers to describe animal feces of solid material expelled from the intestinal gut, separate and in addition to the liquid urine. Ornithologists prefer the distinctive term *droppings* when referring to bird excrement because, unlike mammals, birds mix their undigested gut material with urine sent from the kidneys and expel them together through the anal vent or cloaca. Nevertheless, scat and droppings are often used interchangeably. Many other colorful, and not so colorful, words are used to describe this sign, including feces, pellets, and turds. But despite the rich language, there are still cultural taboos which influence people to draw the line when it comes to getting close to scat. On any number of occasions, just before I begin teaching a class or workshop, I've been approached by a person who tentatively asked, "We're not going to be looking at a lot of turds, are we? We aren't going to pull any apart, right?" And I always smile—"As much as we can find." Then I attempt to sell them on the educational wonders of scatology.

Scats offer a tremendous amount of information to the ecological tracker. It is possible to identify many species from droppings, and thus they are just as valuable as any other sign for identification purposes. Bird scat is more difficult to differentiate in regard to particular species than is mammal scat. It also has not received the attention that mammal scat has in terms of field research as a useful sign for trackers. But as I've studied this new component of tracking, I've found that there are many more easily recognizable droppings than I

Droppings litter the ground at the lek sites (where mating dances occur) of lesser prairie chickens and thus are a good indicator of such activity. (CO)

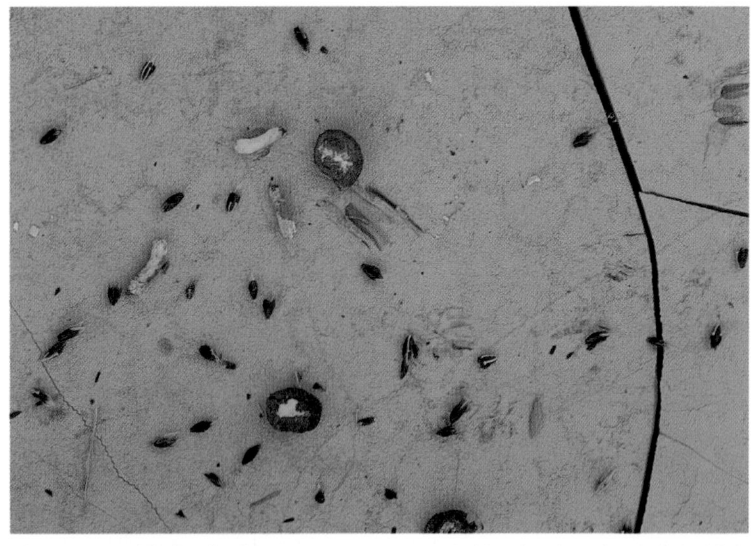

The open seed hulls of tall grasses are accompanied by the droppings of the birds that eat them: white-crowned sparrows and mourning doves. Look carefully for the old mouse tracks baked in place. (TX)

had anticipated. For example, pileated woodpeckers leave tubes of carpenter ants, dipped in white at one end, and often these are found at the base of trees beneath great excavations, where wood chips on the ground confirm the identification. Even as the scats crumble, their contents and location make identification easy.

Beyond mere observation of the droppings, you can investigate the contents, which may tell you where a bird was before passing through that spot and what it has been eating.

I was sitting near a beaver pond not too long ago, watching flycatchers do what they do best. A robin landed nearby and began moving debris and feeding on insects. The bird noticed me in short order and perched atop an old stone wall for a better look. While studying me further, the bird left some droppings behind. They were full of black cherry pits, and in my mind I moved in ever-expanding concentric rings from where I sat, thinking of every fruiting black cherry within a 3-mile radius. This was the limit of the area I knew well, an area I'd been roaming, exploring, and tracking for over a year. Black cherries were not common in the immediate area. In fact, the two old trees that were the most likely source stood surrounded by horse fields of wild rose and hawthorn, in my own backyard. That robin may have been one I had watched earlier that morning, gorging itself on cherries in those trees.

Droppings are left within an information-packed ecological environment. Although there are limitations to bird scat analysis, there is still much that can be learned and then used systematically for identification purposes, especially for families of birds, and sometimes even down to the individual species. This aspect of bird sign deserves greater attention and further study in the future.

We will group bird scat into three categories, similar to those found in Brown et al. (1987), based upon diet and the subsequent visual characteristics: carnivores and scavengers, whose droppings take the form of splats and sprays; insectivores, granivores, and frugivores, which produce loose tubes of semisolid materials; and herbivores, whose scats are very tubular, or cylindrical. There are many variations and gradations within each category, as well as exceptions. Scat deposits are highly variable in size and visual characteristics, depending upon a particular species' preferred diet and seasonal availability of food. This is also true of mammal scat, but it seems to be even more the case with birds.

Carnivores and Scavengers

Larger birds, such as gulls, birds of prey, and large corvids (birds of the crow family), produce much of their solid waste material as regurgitated pellets; their droppings often have no hard materials and appear on the landscape as splats or sprays of white fluid. On occasion, these white droppings are stained with brown tracings or contain a bit of hard, undigested material, which occurs in gulls, especially when they have been feeding on garbage.

Single instances of droppings of scavengers and carnivores appear as splats or sprays of white liquid. Such a site may be found along a streambank where a great blue heron hunted or near the remains of an owl kill. Collections of such splats and sprays along the shore indicate loafing areas, places where gulls, terns, or cormorants congregate to rest. Many such sites are used only at low tide and, as the waters rise, are washed clean of the white sign before the next low tide. The presence of cormorants is easily detected at these sites, as they spray uric acid, leaving 2- to 4-foot streaks of white, whereas gulls and pelicans leave splats of varying sizes.

Splats and sprays at kill sites can help you determine the presence of an owl versus a hawk or falcon. Owls splat or puddle, drop-

Least tern droppings. (MA)

Typical woodcock droppings. Area of concentrated feeding can be identified by the numerous liquid droppings 1¼ to 2½ in. (3.1 to 6.3 cm) in diameter. (MA)

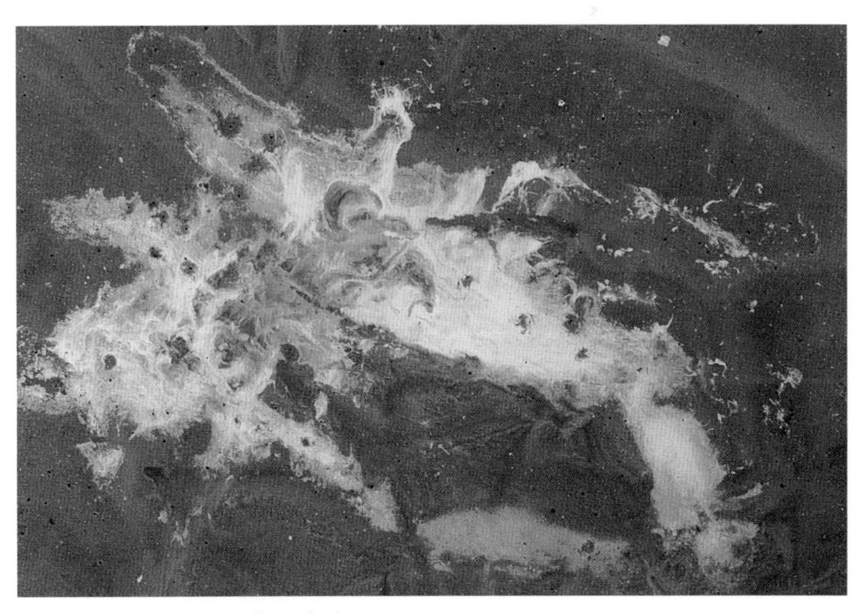

The splat of a white pelican. (TX)

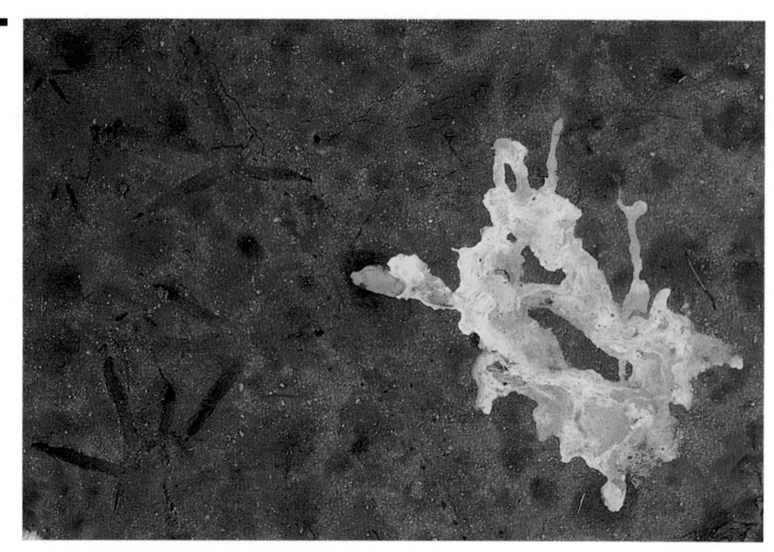

The tracks and splat of a great blue heron. (MA)

ping scat straight down, whereas hawks and falcons raise the tail to eject a spray of uric acid. I measured such a spray from a red-tailed hawk, which was perched high in the naked limbs of an old oak one fine February day. The stream separated into two lines as it descended, and then landed in the leaf litter. From tip to tip, it measured 13 feet, including a foot and a half of space between the two lines. On other occasions, I've measured red-tailed hawk sprays that were shot from the cloaca as the bird perched on the ground, and they measured 3 to 3¾ feet.

Impressive accumulations of sprays and splats are common and indicate daily or nocturnal roosts

Owl scat drops straight down, as seen on this roadside perch, where the owl likely watched this desert wetland in the background. (CO)

The uric spray, ecological information, and details of the kill all make it likely that this squirrel was eaten by a red-tailed hawk. (MA) (Reprinted by permission of HarperCollins Publishers, Inc. Photo by Paul Rezendes.)

or nests. These accumulations, termed whitewash, are distinctive, looking like enormous white canvases hung across cliffs, splattered down tree trunks, and dripped down the sides of buildings and inside barns. Other, equally impressive displays of whitewash are created by birds that nest or roost in colonies, their collective pigment painting bold designs across cliff faces (swifts and seabirds), underneath bridges on girders (swallows) and concrete abutments (dippers), and even upon vegetation and rocks below regular hunting roosts

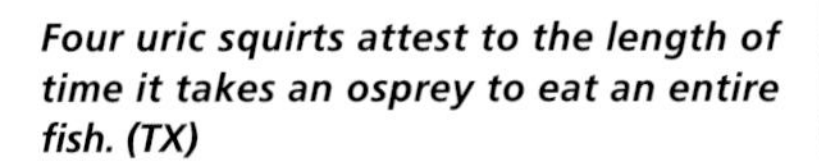

Four uric squirts attest to the length of time it takes an osprey to eat an entire fish. (TX)

A great horned owl roost near its nest under a lone bridge in a vast, rolling, dry plains. (CO)

Active white-throated swift nests are easy to locate thanks to whitewash. (CO)

Here the vegetation below a regular hunting roost of a green kingfisher is covered by numerous uric sprays. (TX)

(kingfishers). In some cases, such as below heron rookeries or cormorant nesting areas in trees, the acidic excrement cakes so thickly on the foliage below that it kills not only adjacent vegetation, but also eventually the nesting tree itself.

Splotches of whitewash can, however, be extremely hard to detect when they are minimal or camouflaged by dense tree limbs and leaves or needles. A friend and I once went in search of jaguarundi sign in the canyons of southwest Texas. The beauty of

This spot is called Gull Rock by the locals who spend time on Fourth Lake in the Adirondacks. It is the creation of both roosting herring and ring-billed gulls. (NY)

Elf owl whitewash. (TX)

this setting was indescribable, as we slid down pink and yellow slickrock, stopping frequently to peer at mud patches at water holes where gray foxes and scaled quail came in to drink. Ringtail and fox scats appeared at regular intervals, and we always stopped to investigate. At a turn in the canyon, set back a bit, well off any marked trail, we found an owl roost. I'm not sure what attracted me to the particular mescal bean shrub—maybe its dense foliage, hidden interior, or beautiful peapod shapes—but as I approached, I caught just a glimpse of white, which stood out against the overall dull interior of shade and woven branches. As we drew closer, the whitewash became clear, dripped and draped over the branches, hardened in place. Looking for pellets, we soon picked them out from the leaf litter below the shrub—tiny pellets of insect remains that resembled cottontail scat. We had spooked an elf owl from its day roost, and we quickly moved on to allow it to return to its secret place in the corner of the canyon.

Insectivores, Granivores, and Frugivores

The droppings of insectivores (insect eaters), granivores (seedeaters), and frugivores (fruit eaters), unlike those of scavengers and carnivores, offer ample sign for the tracker to study, including the undigested chitin of insect prey, carapaces of beetles, and seeds from berries, weeds, and backyard feed which they have been eating.

Droppings composed of insect remains are often not perfectly cylindrical, but have bulbous ends and protruding sections. Fruit

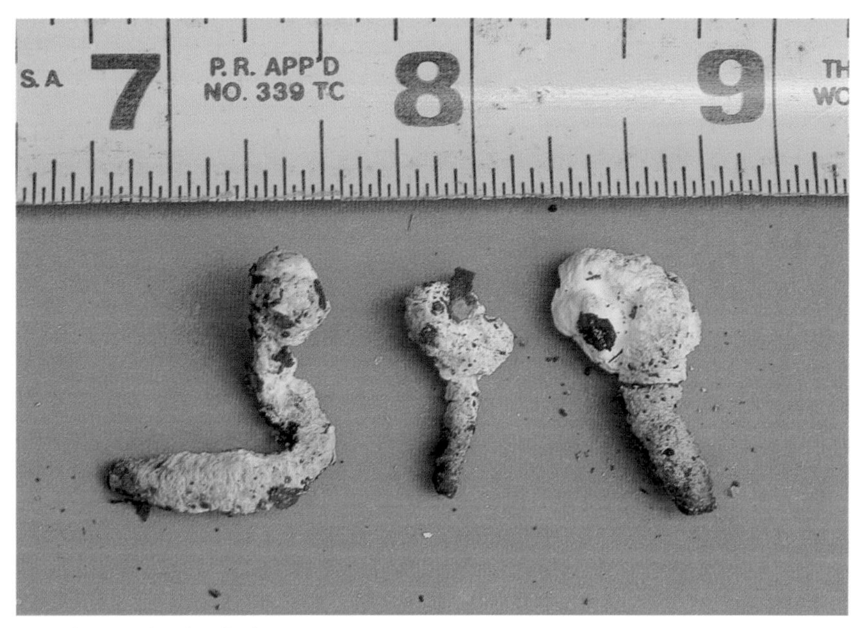

Hairy woodpecker. (NY)

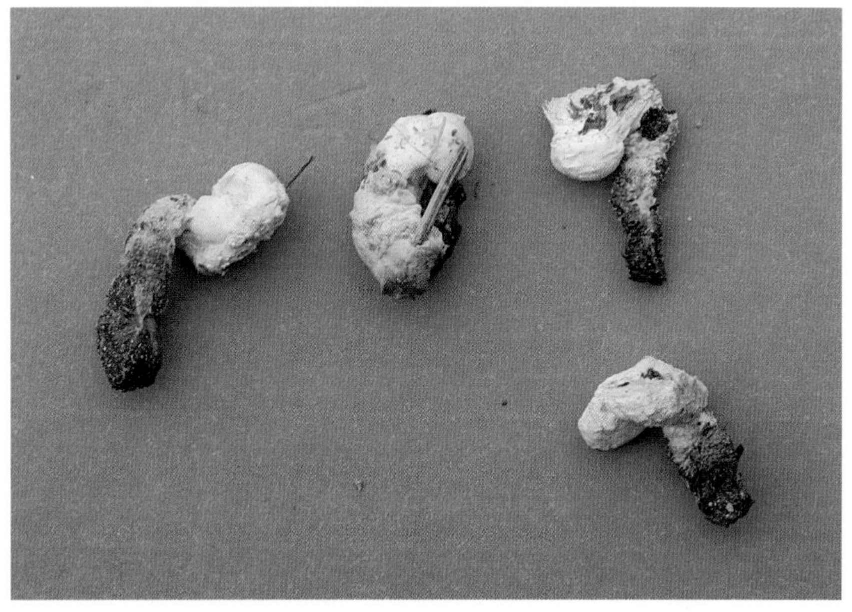

***Eastern phoebe:** 5/16 to 3/4 in. (.7 to 1.9 cm) L x 1/8 to 5/16 in. (.3 to .7 cm) diameter. (NH)*

Pileated woodpecker. (NH)

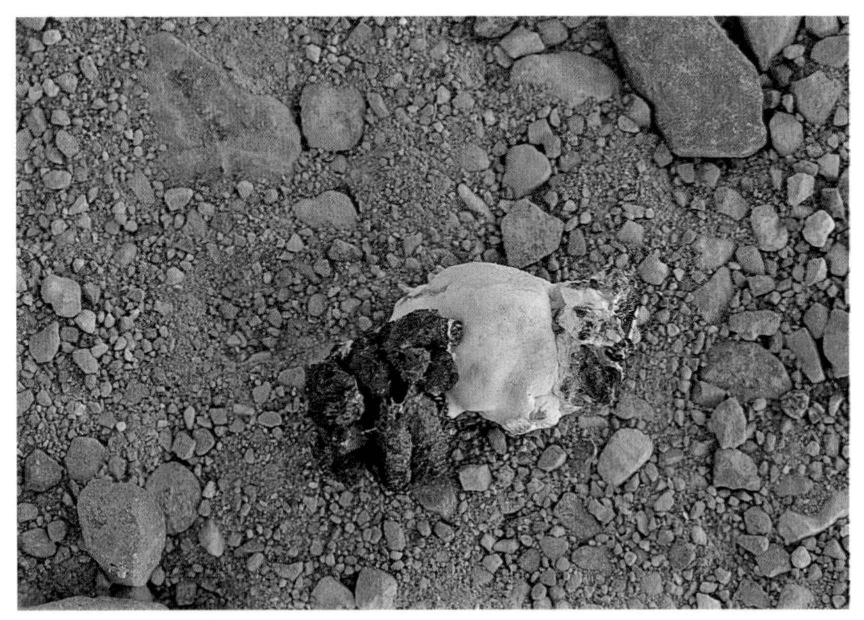

Greater roadrunner scats are highly variable but usually have the very round white cap at one end: $1\frac{1}{8}$ to $1\frac{5}{8}$ in. (2.8 to 4.1 cm) L. (TX)

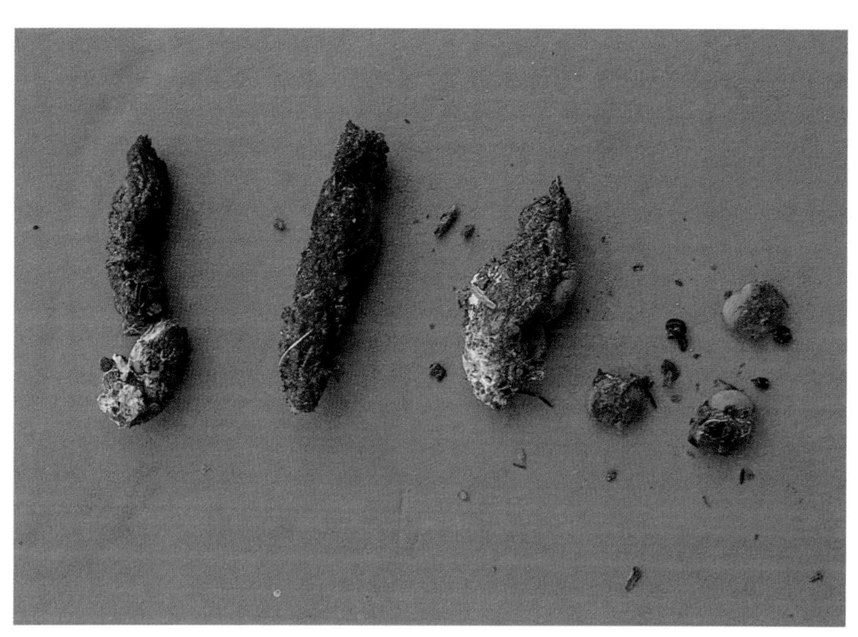

Northern flicker dropping filled with skunkbush seeds: 3/4 to 1 1/4 in. (1.9 to 3.2 cm) L x 1/4 to 3/8 in. (.6 to 1 cm) diameter. (CO)

Northern flicker. Note the amount of sand mixed with the remains of ants. (NY)

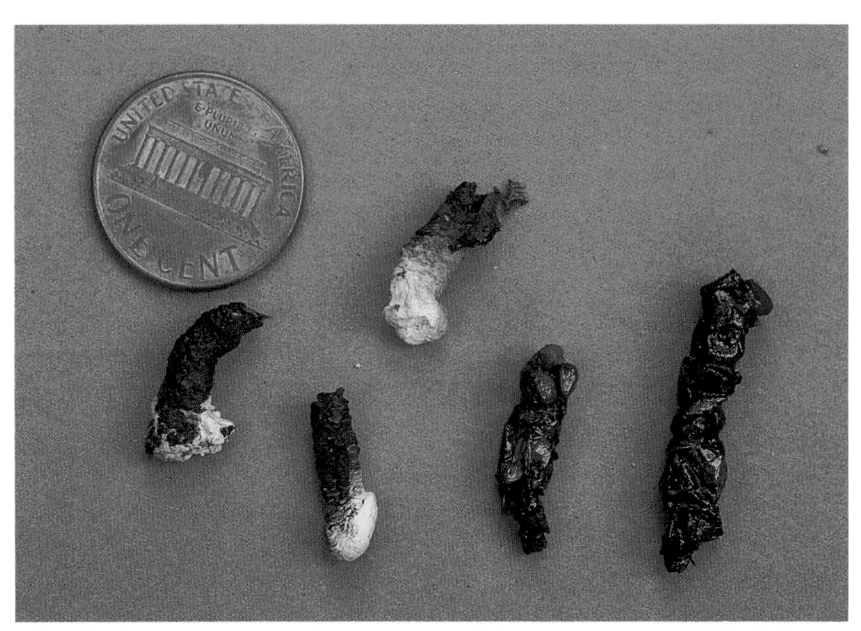

American robin: 1/2 to 1 in. (1.2 to 2.5 cm) L x 1/8 to 3/16 in. (.3 to .5 cm) diameter. (MA)

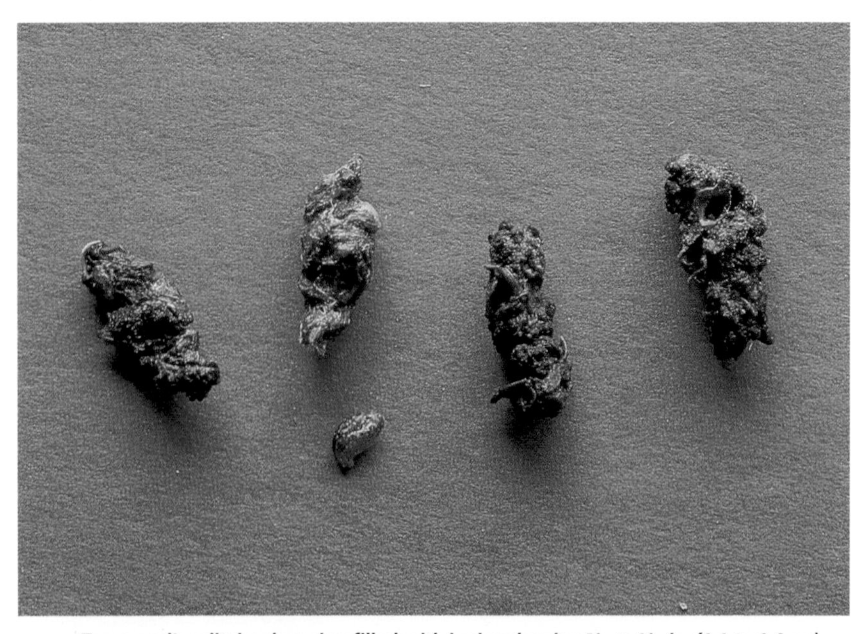

Townsend's solitaire dropping filled with juniper berries: 7/16 to 9/16 in. (1.1 to 1.4 cm) L x 1/8 in. (.3 cm) diameter. (WY)

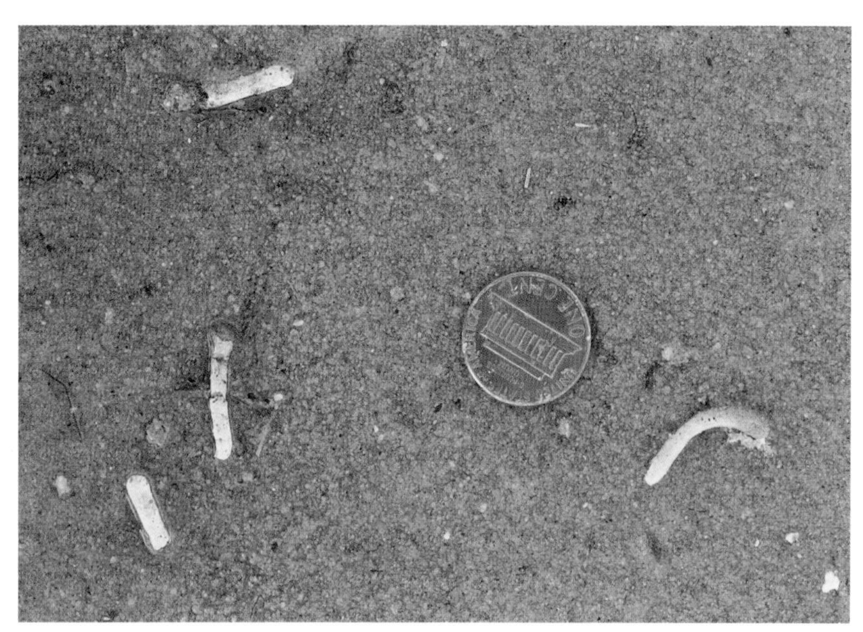

Snow bunting. The penny is 3/4 in. (1.9 cm) in diameter. (MA)

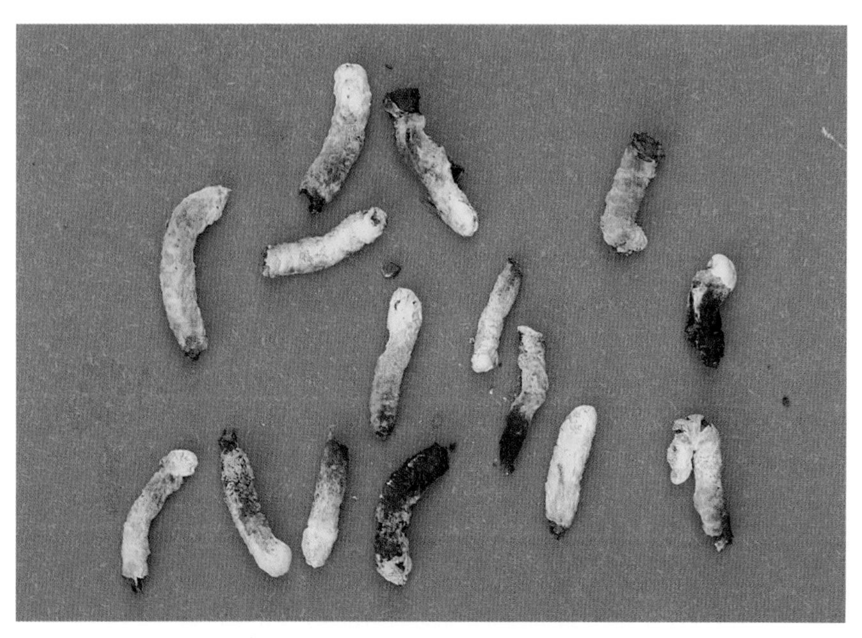

Sparrows: 3/8 to 9/16 in. (.9 to 1.4 cm) L x 1/8 in. (.3 cm) diameter. (MA)

American crow scat varies tremendously with diet. Here the scat was composed solidly of digested cucumbers, upon which the crows were feasting. The penny is 3/4 in. (1.9 cm) in diameter. (MA)

The easy-to-recognize droppings of mourning doves: 1/2 to 5/8 in. (1.2 to 1.6 cm) in diameter. (NH)

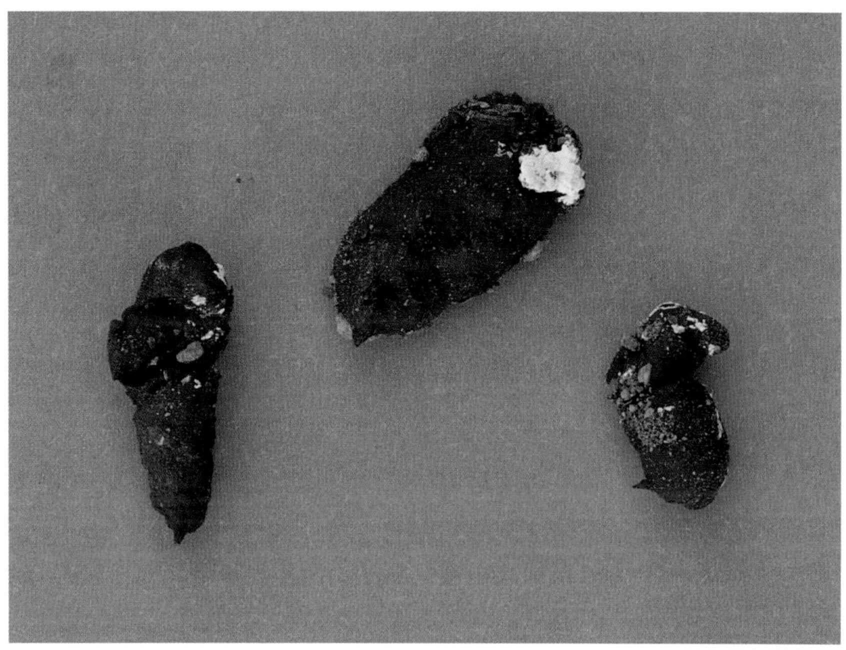

California quail: $^{7}/_{16}$ to $^{3}/_{4}$ in. (1.1 to 1.9 cm) L x $^{1}/_{4}$ to $^{3}/_{8}$ in. (.6 to 1 cm) diameter. (CA)

and seed scats tend to be more uniform in diameter for their entire length and may or may not be accompanied by a small splat of uric acid. The bright colors of berries will dye the droppings, providing an additional clue to the birds' feeding habitats.

Adult songbirds produce semifirm droppings, tubular or oblong in shape. The droppings of their young are enveloped in tiny mucous containers called fecal sacs, which are shiny and relatively firm—firm enough that the parent, after feeding the chick, can pluck the sac from the chick's cloaca and remove it from the nest, carrying it away in its beak; some species ingest it to maximize the calorie benefits of their foraging. This housecleaning behavior protects the chicks' plumage from becoming matted and thus losing its insulating capability; it probably also helps reduce the incidence of parasites. Naturalist Edwin Way Teale underscored the wariness of parent birds that secretively remove all signs of their fledglings by noting that tree swallows dropped fecal sacs in his pond, while song sparrows carried them to trees some distance from the nest and camouflaged them by attaching them to twigs.

Herbivores

Herbivores produce droppings that are cylindrical, firm, dry, and slightly curved. All herbivorous birds that feed primarily on vegetation, including swans, geese, ducks, turkeys, pheasants, and quail, create relatively firm cylinders of chopped-up grazed or browsed plant material, which give obvious clues to the particular food source, such as grasses, sedges, or ferns. *Grazing* describes feeding on herbaceous materials, whereas *browsing* refers to feeding upon woody materials. Whether a bird grazes or browses may be due to preferred dietary habits, as in geese, or to seasonal abundance, as in grouse.

Turkey scat, like that of all birds, varies with diet, and it does not always conform to our categories. But research has shown that different shaped scats are also caused by differences in the intestinal systems of male and female turkeys. Male scat tends to be straighter or J-shaped, while female scat tends toward tight, twisted clumps. When turkeys are eating a diet that results in classic turkey scats, this is a wonderful clue to the sex of the bird.

***Mallard:** 1¾ to 2¼ in. (4.4 to 5.7 cm) L x ⅜ to ⅝ in. (1 to 1.6 cm) diameter. Mallard scat can also appear as small, loose patties. (NH)*

Trumpeter swan: $2^{3}/_{4}$ to 6 in. (7 to 15.2 cm) L x $^{9}/_{16}$ to 1 in. (1.4 to 2.5 cm) diameter. (WA)

Canada goose: $2^{1}/_{8}$ to $3^{7}/_{8}$ in. (5.3 to 9.8 cm) L x $^{7}/_{16}$ to $^{5}/_{8}$ in. (1.1 to 1.6 cm) diameter. (MA)

Canada geese create these "cairns" of droppings in the immediate vicinity of their nests. (WY)

Snow goose: $2^1/_8$ to $3^1/_8$ in. (5.3 to 8 cm) L x $^5/_{16}$ to $^7/_{16}$ in. (.7 to 1.1 cm) diameter. (DE)

White-tailed ptarmigan: 7/8 to 1 1/4 in. (2.2 to 3.1 cm) L x 3/16 to 1/4 in. (.5 to .6 cm) diameter. (CO)

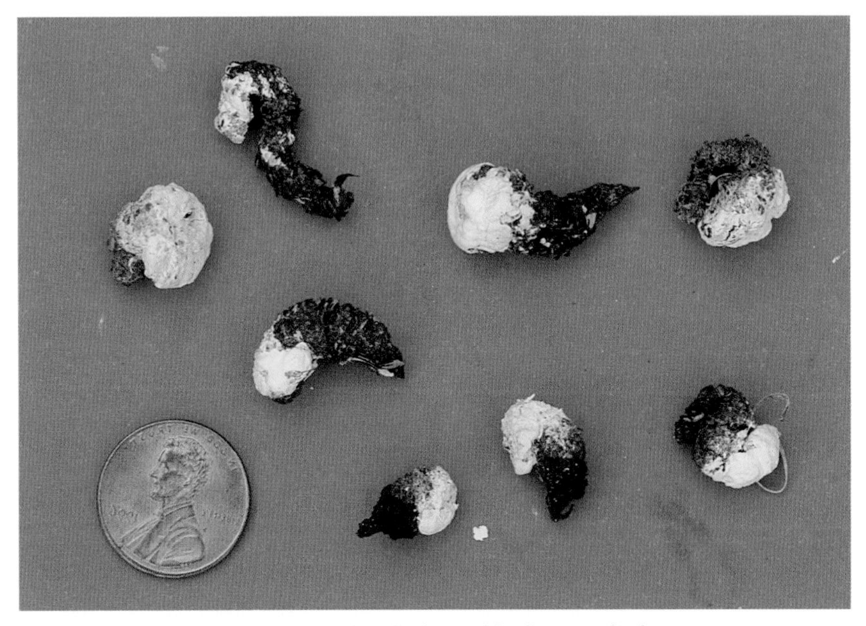

Lesser prairie chicken. The penny is 3/4 in. (1.9 cm) in diameter. (CO)

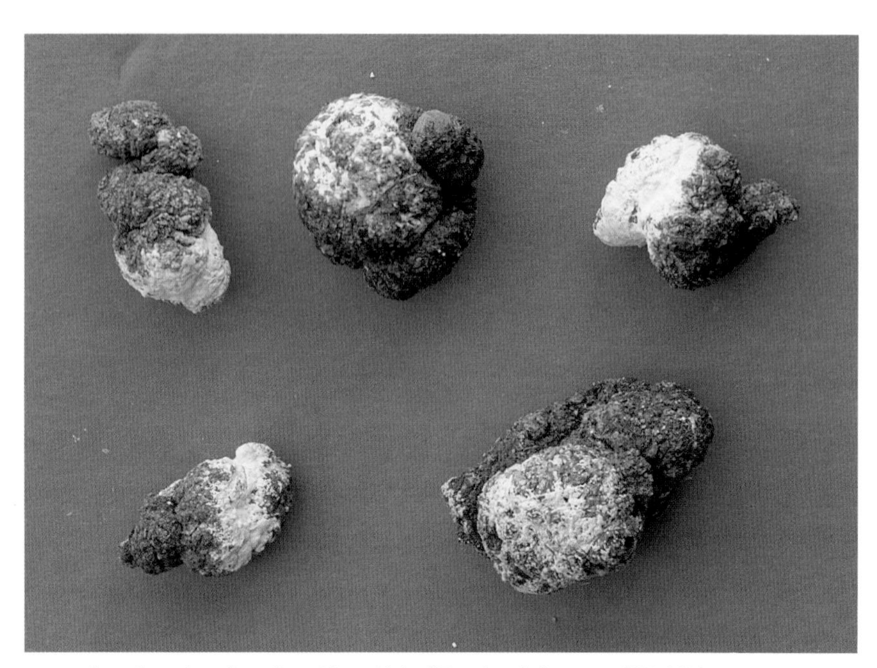

Female turkey droppings: 3/16 to 3/8 in. (.5 to 1 cm) diameter. (NH, MA)

Male turkey dropping: up to 3 in. (7.6 cm) L x 3/8 to 5/8 in. (1 to 1.6 cm) diameter. (MA) (Reprinted by permission of HarperCollins Publishers, Inc. Photo by Paul Rezendes.)

Visiting a Rookery

A group of us was canoeing a swollen river in eastern Kansas. The farm matrix of land that surrounded us was broken by the river we canoed and its narrow buffer of trees on either side. No one could forget such a field day easily. We saw close to forty owls, all barred and great horned, roosting in the trees next to the river, and I wonder how many we missed. We saw two bobcats dozing on tree limbs, as well as beavers and otters. And we also passed under a heron rookery of about twenty birds. As the two canoes passed beneath them, they all lifted into the air and circled low above us like vultures. In a short time, the bombing began, their multiple droppings plummeting straight down and looking like great faucet spews of uric acid, up to 15 feet in length. These lines sliced the water all around us, marking the canoes as if shrapnel had scraped them. The noise of the droppings hitting the water was reminiscent of machine-gun fire, and in the end, one of us was hit twice—first cut in half across the waist, and then across the wrist. There was no doubt in our minds that the birds had been aiming for us. As we rounded the next turn of the river, we put the herons safely behind us. I'm sure our peals of laughter could be heard miles away.

Ruffed grouse roost. Individual droppings are 3/4 to 1 1/4 in. (1.9 to 3.2 cm) L. (VT)

This grouse roost shows the typical fibrous scat, followed by the softer cecal droppings. (MA)

Interestingly, after producing these lower-gut-generated solid evacuations, some game birds, such as grouse, often then evacuate a semiliquid brownish mass from the upper gut, or cecum, with the two types of droppings coming out sequentially; the more liquid, almost liver-colored scat comes out second and is spread on top of the solid matter. In ruffed grouse, it is common to find the hard, fibrous scats at one roost and the soft, brown cecal droppings at another.

4 Signs of Feeding and Other Behaviors

M.E.

As I left the dry desert behind and drove the curves up into the pines and oaks of the Laguna Mountains of Southern California, bird sign began to flank me on either side of the road. So many trees were riddled with holes from base to crown that one might wonder whether there's some great insect infestation in this part of the world. But birds were responsible. When I pulled down a dirt road to find a good spot to park, the acorns were just reaching maturity, and the forests rang with the calls of acorn woodpeckers collecting their crops. Most of the trees I passed were full of empty holes, and until you see an active larder, you might wonder what they were all for.

Try to visualize a pine tree whose girth requires three people with arms outstretched to circle the base, with holes drilled from base to crown, which is 100 feet above your head. The limbs, which begin high up, are all riddled with holes as well, and in every hole, from root to highest limb, there is a single acorn. It is breathtaking in its beauty, and the amount of work that created the storage spaces and then filled them is incredible. A family of acorn woodpeckers creates and maintains such a larder of acorns to last them until the next crop, a year later.

Not all signs of feeding are as obvious as this, but each is beautiful and informative in its own way. From the signs of birds eating grass seeds to the kill sites of great horned owls, a recognition and understanding of each helps us better understand our neighbors the birds, our world, and our role within it. The relationships between birds and their environments become clearer when you study their feeding and behavioral signs.

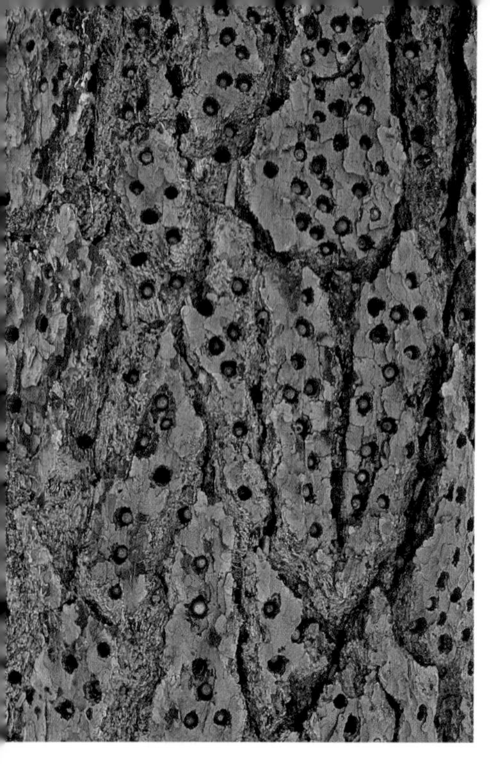

An acorn woodpecker larder in a Jeffrey pine. (CA)

The feeding and behavioral signs of birds are as diverse as the habitats and the birds themselves. The greater your knowledge of the birds in the area and their individual niches, the easier it is to interpret sign in the field, often right down to species.

Representative examples are included in this chapter, along with enough information for an observer to interpret in an array of situations. For example, one type of feeding sign left by birds is the skin, pulp, and flesh of a fruit from which they have removed

This kill site is much easier to interpret when taken within an ecological context. The widest of the three perspectives includes this rocky island off the coast of Maine. You also would be aware that it was the first week of October, and you should know that you were at the right place at the right time to watch peregrines and merlins pass southward in great numbers. As you move closer, a study of the carcass will offer further information. (ME) (Photo by Eleanor Marks)

the seeds. Ravens do this with date palms—quite the contrary of our feeding behaviors. Although we are unable to show examples of this behavior in every fruit species, or explain how every bird species uses a single sort of fruit, the concept is presented clearly enough to allow the observer to look for this behavior on all fruits. The behavior and type of sign are far more important than the specific example used to illustrate the concept. If you want to interpret sign that does not appear to be represented within these pages, we suggest first looking for the most similar sign or behavior in the text, then using the Bibliography to find literature that discusses which birds eat what, and so on.

Bird-Plant Relationships

The seeds of grasses, weeds, and other herbaceous plants, as well as the seeds of shrubs and trees, are vital to birds. Birds and seeds have coevolved together, allowing for greater diversity in each. Myriad beak shapes are directly related to how they are used to obtain and open the seeds of specific plants. Obviously, birds benefit from feeding upon seeds, but in many cases, the plants also benefit; the birds aid in the dispersion and planting of seeds. Certain cases have been well studied over the years, such as the relationships between whitebark pines and Clark's nutcrackers, and between pinyon pines and nutcrackers and pinyon jays.

Nutcrackers have long, strong bills ideal for prying out pine nuts before the cone has opened and a pouch under the tongue in which large quantities of seeds can be carried away to be cached. In the case of whitebark pines, nutcrackers are the only species that cache seeds at just the right depth of forest debris for germination. So dependent is the tree upon nutcrackers for seed dispersion and planting that the seeds have evolved with a very small wing and in some cases have no wing at all. Without adequate wings, the seeds are not able to depend upon wind for dispersion and thus must rely completely upon wild creatures for their distribution.

Pinyon pines have adapted a similar strategy, evolving a larger seed that is irresistible to wildlife rather than a wing for wind dispersion. Nutcrackers and pinyon jays both cache the seeds of pinyon pines far and wide; the pouches evolved by both bird species to exploit the seed crop allow seed burdens to be carried great distances. Both whitebark and pinyon pines, as well as many other tree species, rely upon the fact that not every cache will be used over the course of a winter. It's remarkable how many caches are remembered—which proves that these birds have an impressive memory—but each year, some are forgotten, and the seeds remain neatly planted by the birds.

Grass and Weed Seeds

Grass and weed seeds are of utmost importance to birds in fall migration and while wintering over in a specific locale. Bird sign on grass seed is common, though subtle, and varies depending on the size of the seed. Tiny seeds are often eaten in their entirety, but larger seeds, which can be manipulated with the tongue and bill, are separated from the seed coat, or "chaff." In this case, you may find the chaff accumulating under the grass stems in large piles, blown about by wind and weather. Sparrows, juncos, and doves feed heavily upon grass seeds, especially in wintering areas. Look for their numerous droppings, along with the accumulating chaff under the grasses where they are feeding.

You may also find signs of feeding on the standing seed heads, although this requires greater attention to detail. If the seed head

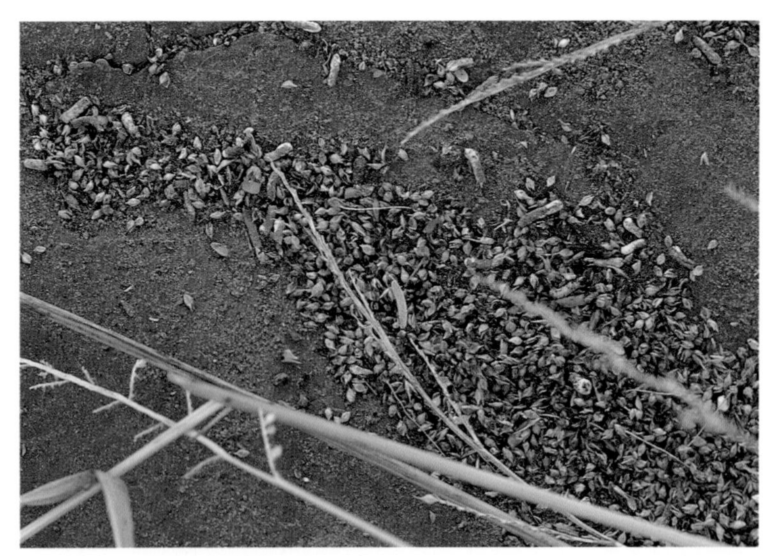

Sure signs of sparrow feeding on grass seed are accumulations of empty seed hulls and droppings carpeting the ground. (AZ)

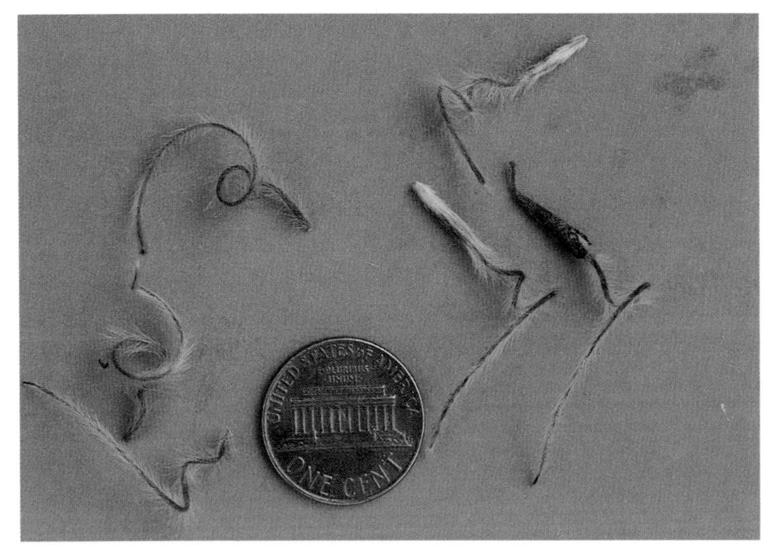

The mountain-mahogany seeds on the left were plucked by a Cassin's finch; on the right are intact seeds and tails. (CO)

and stem are still green, the seeds should still be attached. At this stage, juncos and sparrows hop up, grabbing the stem and several seeds in their beaks, and yank the stem to the ground, removing several seeds at a time. The resulting pattern is incredibly random. As the seeds and stems dry and brown, however, they naturally fall to the ground in an equally haphazard fashion, leaving an almost identical random pattern.

Mountain-Mahogany

Species of mountain-mahogany are common in the foothills and mountain regions of the West. Their seeds are attached to long tails that stick forth from cups that hold seeds. When seeds reach maturity, they tend to fall to the ground, gathering in niches among the debris and soil, where wind or water carries them. It is in these small, natural caches, which are often right below the shrub itself, that you tend to find feeding sign. Cassin's finches, spotted towhees, and blue grouse are among the birds that eat these seeds. Look for natural caches that have been disturbed, noting whether only tails remain or whether seeds are still attached and present. Finches and towhees will work a cache until it is exhausted before moving on to find another such seed source.

Thistles

Thistles, which are common along roadsides and in fields, are excellent places to look for signs of American goldfinches. The slightly long, thin bills of goldfinches allow them to reach between the spines of thistles, as well as teasles and burdocks, to extract the seeds.

When bright purple or white thistle blooms, it attracts a host of insect species from far and wide. The colorful petals protrude from a vase-shaped collection of specialized leaves, called phyllaries. The vase remains green while the flower petals begin to brown, but eventually the phyllaries also begin to dry, and the seeds mature within. At this stage, goldfinches will begin to feed upon the seeds, even though the vase has not completely dried or opened to allow the seeds to disperse by wind. The birds pull out the down, a fluffy, white material, and drop it next to the plant, then remove the seeds. If down and seeds are missing from a closed thistle head, and the plant fibers that help organize the seeds inside the vase-shaped phyllaries are still green, you can be sure a goldfinch has fed upon the seeds. The color is important to note, as a mature seed head with brown plant fibers may open up when the conditions are right for optimal wind dispersion and allow some seeds to disperse, then close the next day to wait until conditions are favorable again for the remaining seeds.

The seed chaffs caught in the open and surrounding flower heads are obvious signs of goldfinch feeding. (NY)

This sunflower has been harvested by both goldfinches and chickadees. The chickadees carried the seeds to a nearby perch, while the goldfinches fed at the source, dropping seed hulls at the base of the plant. (NY)

Feeding sign of goldfinches can be much more obvious than this, though. The finch family has adapted to feed at the food source, rather than carry food some distance before feeding. A goldfinch will surf a thistle flower head in the wind for ages, feeding all the while. In a patch of thistle where goldfinches have fed for some time, the down buildup around the plants can be thick, carpeting the surrounding herbaceous layer. After first assessing the area for down buildup, check for seeds and chaff within the open flower heads, visibly strewn about the internal plant fibers from feeding. The chaff and seeds are also likely to be seen stuck in the Velcrolike phyllaries of the flower heads surrounding the feeding area. If a bird has fed long enough, there will also be scat sticking to the furry leaves and stems. The combination of all these signs makes it very clear that goldfinches have been around. Thistles are also great places to look for goldfinch nests.

Sunflowers

Sunflower seeds are well liked by many birds, which feed on them in different ways. Members of the finch family feed at the source, whereas chickadees and titmice carry their seeds to a perch to feed.

Jays tend to carry their seeds away to cache them. This behavior is easily observed at feeders. Watch chickadees swoop in, grab a seed, and fly to a perch, where they open it. Watch nuthatches and woodpeckers swoop in, grab a seed, and fly to the exact same place over and over again to open seeds. Watch grosbeaks and goldfinches come and settle in, taking over the feeders until they decide to leave or are chased off.

So rather than focusing upon the flower head of the sunflower to find evidence of certain species feeding, it is better to look at the ground below the flower head. Are there opened seed cases, showing evidence of feeding at the source, or have the seeds been carried away? Also check the back of the flower head, which droops with the weight of the maturing seeds. On those flowers that have drooped enough, morning dew and rain collect in the depression around the flower stalk. Many bird species come in to drink at these spots, and they can be differentiated by the ring of scat that surrounds these short-lived puddles.

Dandelions

Dandelions mature in much the same way as sunflowers and are often fed upon in the Rocky Mountain area by flocks of siskins, close relatives of goldfinches, which feed on dandelions in the East. Driving through the Rockies in early summer, I would scare up great flocks of siskins, which were feeding upon dandelions at the roadside.

When the seeds have matured but are still tight within the closed sepals, or flower head, they are ripe for the harvest. Siskins often land in large flocks among the dandelions, pulling off a few sepals to gain entrance to the seed cache within. Their sign thus is quite conspicuous, even after the sepals open and allow the remaining seeds to be dispersed by wind: Sepals will be missing on one side of the plant.

Flowers and Herbaceous Plants

Besides eating the ripened seeds of flowers, birds also eat flowers at every stage of their development. Birds feed upon flower heads as they are developing as well as while they are in bloom. House finches feed on black raspberry flower heads before they have unfurled. Blooming flower petals in the garden, such as crocuses, are occasionally eaten by house sparrows and finches. Cedar waxwings feed on apple blossoms in the spring. Ducks and geese will also consume

Compare the three intact dandelion heads along the top of the picture with the three heads along the bottom of the photo, which show various stages of sepal removal. Behind these flower heads the white fluff has been dropped where siskins cut it free from the seeds. (CO)

flowers if they are within their grazing path.

Once the flowers fade, the fertilized seeds develop and are fed on by many species. Rose hips are a favored food of many birds and animals and are a perfect example of how mature seeds of a flower are utilized. Sparrows, mockingbirds, mice, and martens are among the varied species that partake of the annual feast.

The stems and leaves of herbaceous plants also may be eaten. Ducks and geese are heavy grazers on grasses and other greenery, and if they are feeding in any numbers, they may leave behind what appears to be a manicured lawn.

The developing blackberry flower heads in the back have been opened and partially eaten by house finches. (WA)

Song sparrows feed heavily on Rosa rugosa *rose hips along the coast. (MA)*

In these cases, you will surely find scat, as ducks and geese move grass and other forbs through their digestive systems with incredible speed. They feed continuously, food entering at the bill while food ingested moments before exits at the other end. Researchers believe that quick digestion and excretion is an adapted strategy to aid in the escape from predators. These birds are able to take off more easily and with more speed without the excess weight of undigested grasses and forbs. But poor use of ingested grasses means they must eat large quantities of food to gain adequate nutrition.

The bark of dead plants and trees is often collected for nesting material. Orioles strip the outer layer of the previous season's milkweed plants, and many bird species pull at the shaggy bark of juniper trees.

Apples and Other Fruits

Apples are popular fare among many bird and animal species. Larger mammals, from fishers to bears, tend to eat the fruit in its entirety. Smaller mammals, such as meadow voles and chipmunks, leave obvious incisor marks in the flesh of freshly fallen windfalls. Birds leave clear beak marks and feed upon apples in particular ways.

Blue jays feed on apples in distinct fashions. While the fruit is ripe and still hanging on the tree, jays feed on the flesh, leaving obvious gouges in the sides of the apples. Jays also peck at the rotting

Blue jays create gouges in the sides of apples when feeding on the flesh of the fruit. (NH)

This windfallen apple was gutted by a blue jay, which fed on the ripened seeds within. (NH)

windfalls below the tree, feeding only on the seeds in the core. In this case, the brown pulp is left next to the sagging apples on the ground. Blue jays also feed on pears, with almost identical results. It is interesting to note that in the two orchards where I studied blue jays, they fed only on the apples that grew on trees that had not been sprayed with insecticides. The commercial crops, which received treatment, were never touched.

Crows feed on apple seeds while the fruit is still ripe and hanging on the tree. In such cases, their 3/4- to 1 1/4-inch (1.9- to 3.2-cm) round holes go straight to the cores from the sides of the apples.

According to Bang and Dahlstrom (1974), crossbills prefer to feed on smaller apple varieties and also are interested in the seeds, rather than the flesh. They slice off chunks of apple, working their way to the core; slices may remain hanging from the apple or be found below the feeding area.

American robins, mockingbirds, and cedar waxwings eat large quantities of fruit in their diet. The fruits of trees and shrubs, relished by these and other species, tend to be of a small size that can be swallowed in their entirety. The sign, therefore, becomes a lack of

Crossbills slice away at smaller apples to harvest the seeds. (Used with permission of Gads Forlag. Photo by Preben Bang.)

The skins and opened pits of bunchberries remain after a white-throated sparrow has fed. (NY)

These date palm fruits were opened by ravens, which ate only the seed—which is presented to one side. (CA)

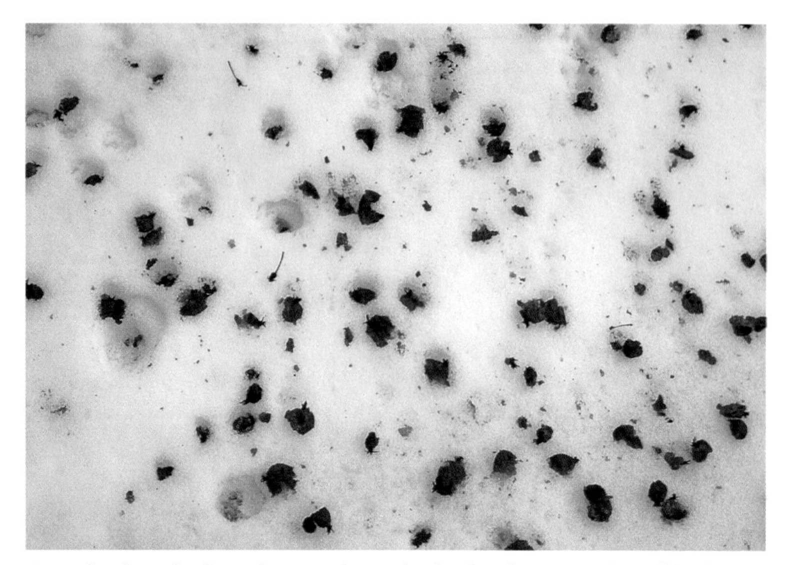

Gutted crab apples litter the snow beneath a bush, where returning robins feasted upon viable seeds in the spring. (NH)

sign. The fruit is missing. It will show up in the colorful droppings, though, having been transported by the birds and the seeds dispersed.

White-throated sparrows eat more fruit than most sparrows. They often feed upon the pulp of grapes and bunchberries, leaving behind the skins and seeds. If they feed on the bunchberry fruit earlier in its development, the pit can be cracked with ease, allowing them access to the seed as well. There is only a small window of time, though, when the fruit is ripe but the pit is still soft enough to penetrate.

In some instances, all a bird is interested in is the seeds the fruit is holding. Ravens feed heavily on date palms, which they jam into an available crack or hold under one foot, then peck a hole from which they remove the seed. The fruit is left or discarded.

Old fruits left hanging on trees are also used by birds. Winterberry fruits are a crucial late-winter survival food relished by wintering birds in the North. Any fruit that remains may be harvested in the spring with the early arrival of robins and waxwings, some of which may have wintered over, as well. Whether ornamental or wild, the sagging fruits of cherries and crab apples will not go to waste. As late as early spring, robins and waxwings will either eat them in their entirety or peck them out for their nutritious seeds within, as jays do on windfallen apples.

Browse

Browse refers to a woody diet of buds, stems, and barks. Many wildlife species graze on herbaceous vegetation from spring to early fall, and then browse until the green shoots reappear the following spring. Some birds are included in this group. Grouse and ptarmigan are heavy browsers of buds during the winter months.

Rabbits, rodents, and porcupines browse using their two upper incisors and their two lower incisors; the result is an easily recognizable, roughly 45-degree cut in the twig. Depending on the width of the stem, the cut may require more than one bite. This will be obvious if the cut is not completely smooth. Ungulates, members of the deer family, also browse in a unique fashion. As they have incisors only in the lower jaw, they grasp buds between the lower front teeth and upper hard palate, tearing them loose; the resulting sign is often a flat but ragged stem. On occasion, long strands may hang from the browsed ends.

Grouse and ptarmigan feeding signs are very similar; habitat and other clues will help differentiate them. Buds are often plucked without leaving any damage to the stem at all—the bud is just missing. If the bite is deeper on the stem, there may be some tearing to the end of the stem, and again the bud will be missing. Use your judgment

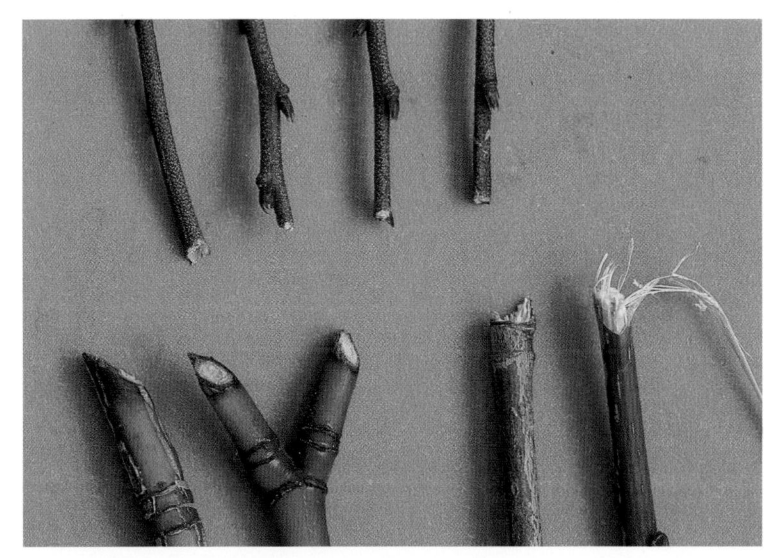

Compare the ruffed grouse browse along the top (low-bush blueberry) with snowshoe hare browse, bottom left, and white-tailed deer browse, bottom right. (NH)

in these scenarios. Is the stem of the size that a grouse or a deer would leave a ragged end?

Nip Twigs

Nip twigs are twigs of shrubs and trees that have been bitten off a branch by wildlife and may be found either hanging in the foliage or on the ground. Some species create nip twigs often. Porcupines create nip twigs during almost every season, feeding on acorns, maple buds, or hemlock foliage. Squirrel species and wood rats also nip twigs to collect fruit, eat buds, and build nests. In all of these cases, the ends of the branches will show the 45-degree cut characteristic of larger rodents. Woody debris, which superficially looks like nip twigs, is also a common sign of storms, high winds, icing, or insects. Be careful not to mistake storm sign for wildlife sign.

On several occasions, I have noted nip twigs in the roosting areas of wild turkeys. Jonathan Talbott and Michael Pewtherer, both excellent trackers, were the first to bring this to my attention. The twigs often show beak marks along the stem, in addition to the actual cut, which looks more similar to branches taken in storms than to mammal sign. But no foliage is eaten or altered. At this time, we believe the birds are opening areas for comfortable roosting. Note that Paul Rezendes, a friend and fellow tracker, has documented turkeys eating white pine needles under different circumstances.

White pine twigs carpet an area below a turkey roost, which the birds nipped and dropped for some reason. (NY)

This bristlecone pine shows separated cone scales where a nutcracker forced its bill inside the cone to obtain the seed. (CA)

Pine Cones and Pine Nuts

Pine nuts are an invaluable food resource for wildlife around the world. Regardless of the pine, pine cone, or pine nut, which all vary in size and shape tremendously, there will be a bird or mammal harvesting them. Larch and hemlock cones, which have a similar structure to pine cones, are utilized in much the same manner.

Depending on the maturation of the cone, different species are able to harvest the fat-rich pine nuts. At the start, cones are tightly closed, and the seeds slowly mature within. When the cone has fully matured, it begins to dry out, forcing the cone scales to separate, allowing the dried seeds to fall and be dispersed by wind or be visibly tempting to wildlife species. The winged seeds of certain pines catch the wind and disperse as far as the gusts allow. Pines seeds without wings are dependent on birds and mammals for dispersion.

Even when the cone is green and tightly closed, the seeds are harvested by three bird species: Clark's nutcrackers, red crossbills, and white-winged crossbills. These birds are competing with squirrels, which harvest and store entire cones at this time, too. Nutcrackers and crossbills work to attain pine nuts from green cones in very different ways. Nutcrackers have evolved long, thin beaks, ideally proportioned to reach pine nuts deep within the cones. They hang from branches, hammering the cones at just the right angle to force entry between scales. Reaching in, they remove the seeds, which sit in hollows at the base of the scales. They then wipe off excess sap from the green cones on the surrounding needles—which is a great sign to look for. After a nutcracker has fed from a "green" cone, the scales will be noticeably separated. (Not all maturing pine cones are actually the color green; for example, bristlecone pine cones are purple.)

Certain pines, such as lodgepoles, yield smaller pine nuts, and in these cases, nutcrackers feed in a different fashion. They tend to remove the entire cone from the branch and retire to a perch for feeding. They will work the side of the cone, hammering at just the right angle to obtain the pine nuts, as described above. When the bird has had enough or moves for some other reason, the cone is dropped to the ground. In the case of smaller pine nuts, nutcracker sign will be found around the base of the tree in which they fed, rather than still hanging from the branches.

As cones dry naturally, many pine tree species open from the end that is not attached to the branch, often the pointy end. If you find cones on the ground with the scales forced open on the sides of the cone, while the top remains tightly closed, you have most likely dis-

The three lodgepole pine cones on the right were partially opened by nutcrackers, while the two cones to the left have opened naturally to disperse seeds by wind. (CA)

White-winged crossbills plucked the cones and split the scales of the black spruce cones on the left to remove the seeds, and then dropped them to the ground. The cones on the right are untouched. (NH)

covered the feeding signs of birds. But study the tree the cones come from, as not all pine cones open in the same way.

I followed three nutcrackers in the high desert of the Sierras for an afternoon. They spent their time chasing each other, bark sloughing for insects, and plucking lodgepole pine cones to open at a perch. The cones were only partially harvested, as in every instance that a bird began working a pine cone, the others would fly in and harass it. In each case, the cone was dropped before it was finished.

The unique crossed bills of crossbills, when wedged into a tightly closed cone scale and manipulated in one of several ways, are able to pry open the scale enough so that the bird can use its lower mandible or tongue to scoop out the seed. It is incredible to watch. In the cases where the birds pluck the cone from the tree and retire to a perch, the sign will be found on the ground, but in many cases, the sign will be left hanging on the tree, as with the nutcrackers.

Crossbills tend to damage the cone scales as they harvest seeds. Evidence may be found in the form of scratches low down on the underside of the cone scales, the shredding of cone scales, or the splitting of cone scales. All three of these signs are common and may even be found together on the same cone. Shredded and split cone scales are conspicuous and sure signs of crossbill feeding.

There are some variables you also should consider. Nutcrackers, because they hammer to enter cone scales, on occasion break off cone scales, which is rare for crossbills. Crossbills are also much smaller than nutcrackers and feed on smaller seeds. White-winged crossbills tend to feed on the smallest of the cones, especially larch and hemlock, while red crossbills often feed on the slightly larger spruce cones and pine species. These are tendencies, however, not hard and fast rules.

Crossbills may also be found eating mud in the vicinity of a green cone crop. The mud coats and protects their stomachs, which can be upset by large intakes of sap. Look for scratching from awkward bills, which aren't designed to pick up anything from a flat surface.

The green cones that have not been harvested by crossbills, nutcrackers, or squirrels continue to mature. Eventually, cones begin to dry and brown, and in just the right conditions, scales separate with an audible *pop* or *snap*. At this point, the pine nuts become available to many more bird species, and a great harvest begins.

Left: ***This pitch pine shows the characteristic signs of red crossbill feeding. Note the scratches along the underside of cone scales, as well as split and frayed scales. (MA)*** **Right:** ***Jeffrey pine nuts opened by mountain chickadees. (CA)***

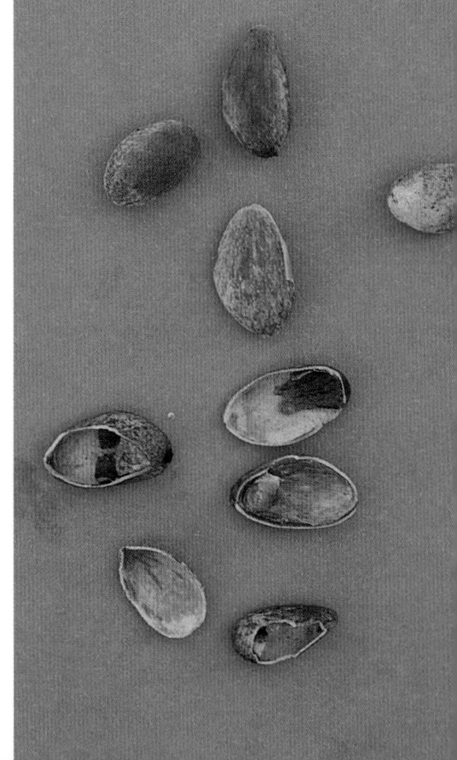

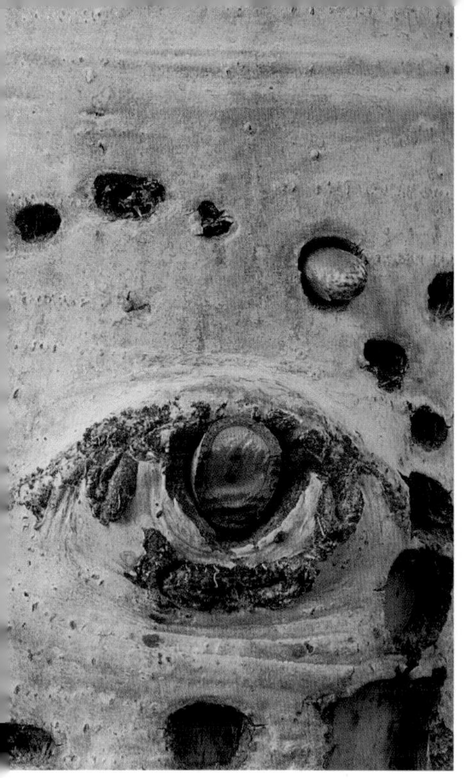

Left: *This old sapsucker well is the location of a pygmy nuthatch's workshop. The bird was spooked before this last ponderosa pine nut was opened. (CA)* Right: *Beech nuts wedged and opened in an old hemlock by a white-breasted nuthatch. (VT)*

Among those that feed at this time are chickadees, nuthatches, siskins, woodpeckers, and jays; the crossbills and nutcrackers continue to harvest as well.

When the cones open and the pine nuts become accessible to all, signs of feeding appear all over. To help differentiate among species, keep in mind where the different bird species tend to feed. Finches tend to feed at the source of the food. Chickadees tend to grab a seed and fly to a perch to open it, varying the perch with each seed or every few. Nuthatches and woodpeckers tend to choose one spot, taking seeds or nuts there to be opened over and over again. Pine nuts opened by chickadees will be found sprinkled under the tree in which they have been feeding. Nuthatch and woodpecker sign will be concentrated in specific locales. These areas, called *workshops,* accumulate sign with repetitious use. The actual workshop may be found at any height from the ground, although the sign tends to gather on the ground below, regardless of the height of the workshop.

The smaller ponderosa pine nuts were found at the base of a pygmy nuthatch's workshop. The larger pinyon pine nuts were found below a nearby downy woodpecker's workshop. (CA)

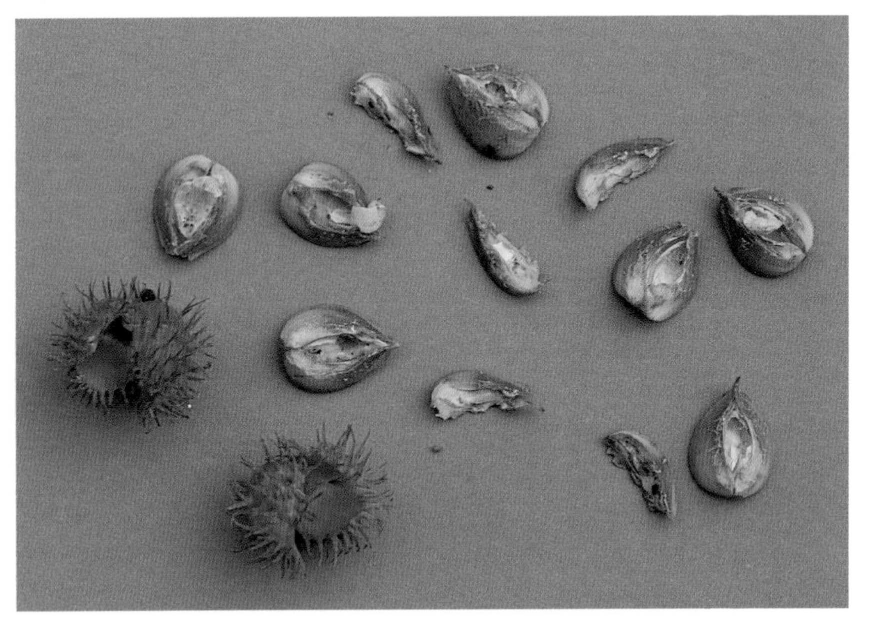

These beechnuts were found below the workshop of a white-breasted nuthatch. (NH)

Nuthatches and woodpeckers jam their pine nuts into bark crevices, snags, dead limbs, or whatever else will hold them still while they bang away to break into the nut. Brown-headed nuthatches are notorious users of longleaf pine nuts, which they jam into crevices of the bark of the trees and then hammer open. You can find the remains of these pine nuts stuck in the bark, but more often than not, the sign will surround the trunk, as an ideally sized crevice just right for pine nuts tends to be cleaned out and used again and again. Thus the signs of a nuthatch's or woodpecker's feeding are often found at the base of the tree or snag or under the dead branch stub it is using for its workshop, rather than in the wood where the seeds were originally jammed. Look for workshops very near the food source. In my field observations, downy woodpeckers varied their feeding sites much more than all the species of nuthatches I watched.

I once saw a pygmy nuthatch collect ponderosa pine nuts from a massive tree for several hours. When feeding on nuts on the lower right side of the tree, the bird used a hole in an old sapsucker well in an aspen to hold nuts still. A separate workshop, high up on a branch stub of an aspen, on the other side of the ponderosa pine, was used when feeding from the canopy and left side of the same tree. Flight means wasted time and energy, so it makes sense to keep the commuting to a minimum.

As time passes, the seeds also begin to fall to the ground. Certain bird species continue to feed. White-winged crossbills and juncos can be found en masse under hemlocks, larches, and spruce when the seeds fall, leaving a confusion of tracks and seed hulls. Chickadees and nuthatches may join them sporadically as well. This is the cycle of pine nuts. There are birds that take advantage of feeding at each level of cone maturation.

Birch, Maple, and Ash Seeds

There are several other notable relationships between tree seeds and bird species. In winter, redpolls can often be found knocking down birch catkins and feeding on the tiny seeds on the ground. The birds feed in flocks, and the feeding signs and track chaos can be quite impressive.

Grosbeaks are regular users of maple and ash seeds. Like redpolls, grosbeaks tend to feed in flocks, and their feeding signs can carpet the forest floor below trees in a mast crop year. Seeds are

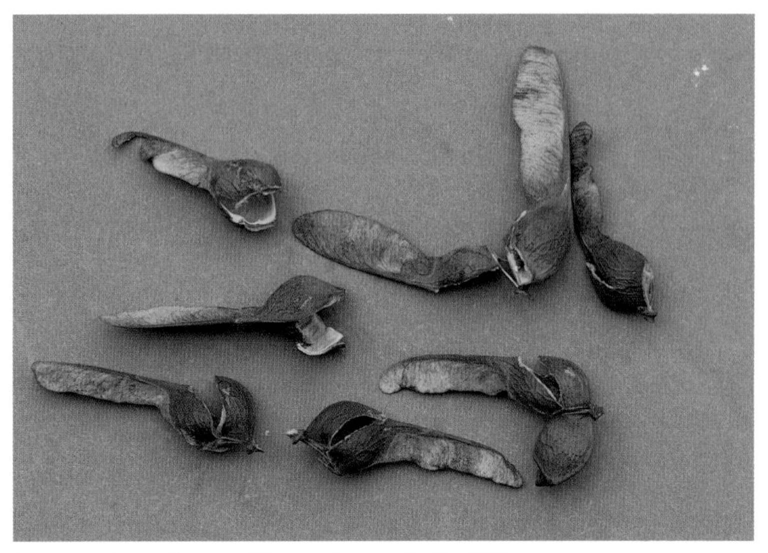

Sugar maple seeds opened by evening grosbeaks. (VT)

often split in half, and the winged hulls dropped after the seed has been removed. Seeds may also be removed by splitting the hull at one end, more often with maple than with ash.

Acorns

The incredible nutritional value of acorns has been acknowledged and well documented for hundreds of years. Where there are acorns, birds compete with mammals, including humans, in every harvest. The acorn, which varies tremendously in shape and size, offers a compact package of fats and proteins just in time to store for winter months ahead. The thickness of the acorn's shell also varies among oak species, and even among individual trees. This shell, thick or thin, must be opened to obtain the food, so the thickness of the shell is a major factor in how acorns are opened, and thus in interpreting feeding sign. With experience, you will learn the subtleties of feeding behavior.

One more important variable to consider when studying acorn sign is whether the acorn was dry or fresh when it was opened. A cached, and therefore dry, acorn splits more readily than a green acorn, which has a more flexible shell, and the nut meat will have shrunken and hardened and is easier to remove. The careful observer can learn to tell whether an acorn was opened green or dry, and this is useful to note with both birds and mammals. Also consider the implications in regard to behavior. Grackles love acorns and feast on them in the fall, but are they likely to be digging up acorn caches from under snow or retrieving an acorn cache from within a tree cavity? No. How about crows? Jays?

Crows and Jays

Crows are among the many that harvest and eat acorns. Generally, crows harvest an acorn and retire to a branch or nearby anvil, like a granite boulder or roof, where they open the nut. They either jam the acorn into a crevice to keep it still or, more often, hold it with one foot. They then hammer away at the acorn, creating one jagged entrance, and eat chunks of nut meat as they come loose. If the acorn is underfoot and the nut meat is fully removed, the shell may collapse under the weight of the bird, making the acorn remains appear as though they were run over. Beak marks will be obvious in and around the jagged entrance holes of the acorns. If you look on the side opposite the entrance hole, you will often find beak marks protruding from the inside outward, which is a good indicator that the perpetrator was indeed a bird. Little if any nut meat usually remains in the shell. These acorns will be conspicuous around trees where crows have been feeding or under and on nearby perches.

Fresh red oak acorns fed on by crows. (MA)

Wherever oaks and jays exist together, jays will be caching and eating acorns. Like crows, they generally hold the acorn in one foot and peck a hole in the shell from which to remove the nut meat. The entrance is almost always on the side of dried acorns, and the hole is made just large enough to remove the acorn meat. The size of the holes varies according to oak species and the size of

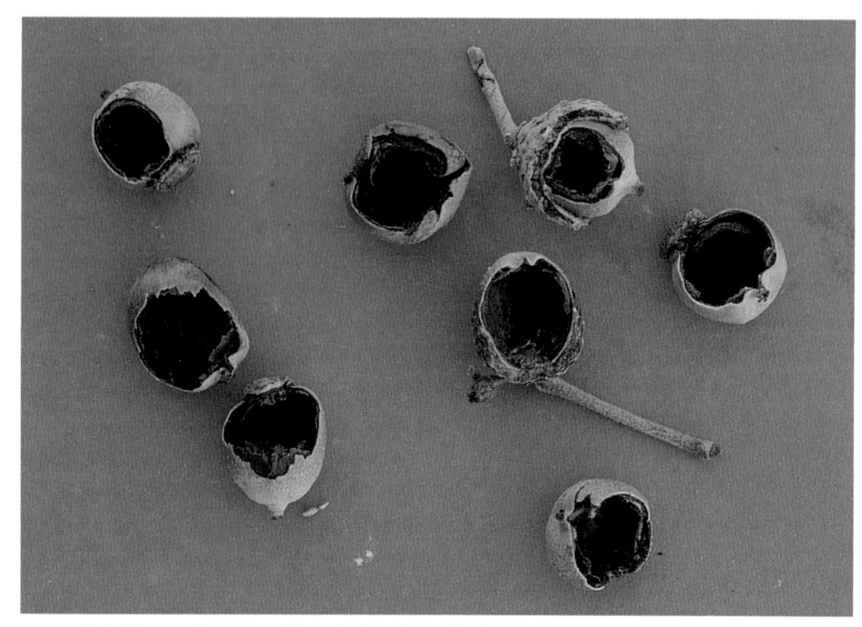

Dried acorns harvested by Mexican jays. (TX)

Jays often use anvils to open acorns and other nuts, such as seen here, where an acorn was opened by a Mexican jay on a convenient rock. (TX)

Pin oak acorns partially eaten by blue jays. (VT)

the acorn. The holes made by Mexican jays in these small acorns were quite small, although proportionate to the size of the acorn. Unlike the work of crows, the acorn shell is left intact other than a single entrance. This more efficient process results in a much cleaner opening. Look for acorn remains sprinkled under oaks and on nearby hard surfaces, such as rocks, that jays often use as anvils when hammering acorns. You may also find areas where caches have been unearthed and fed on.

Jays feed differently on green acorns. Fresh acorns are usually opened from the bottom, after the cuplike top has been removed and dropped. The hole is much more

These red oak acorns were opened by crows only to eat the weevil within. An acorn weevil—the larva—is pictured at the top. (MA)

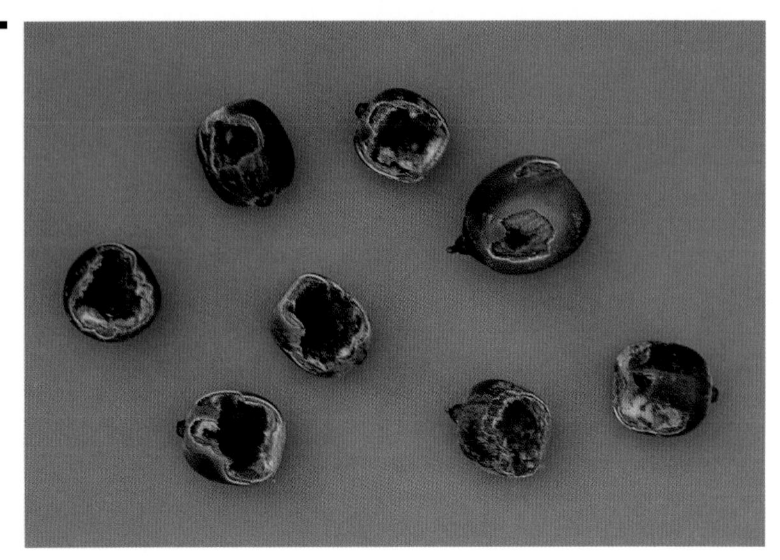

Compare these acorns, where it was blue jays that harvested the acorn weevils. (MA)

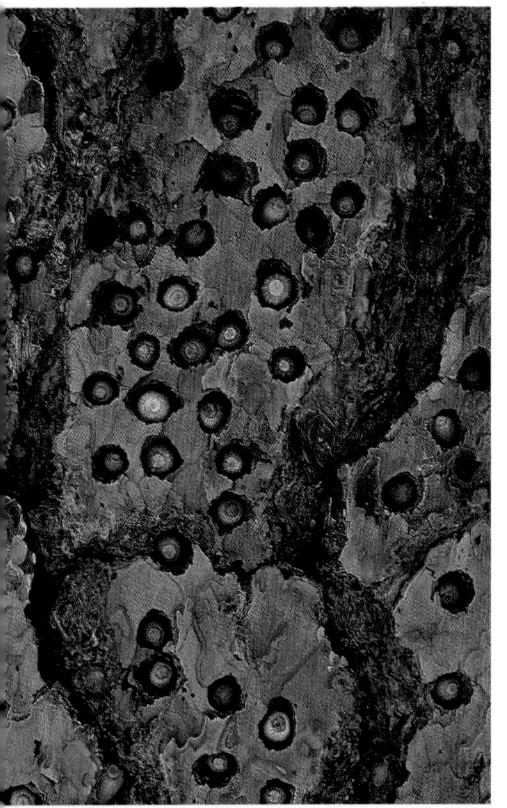

ragged, and the nut meat is eaten in hacked-up chunks. The remains of the shell are then dropped and likely contain some wasted nut meat.

Both crows and jays also harvest acorns to obtain the large, fatty acorn weevils that feed on the nut meat. The birds make jagged holes, smaller than those used to extract acorn meat, but just large enough to extract the weevils. The entrance holes made by jays tend to be smaller and more efficient than those of crows. Entrances for weevil eating, for the most part, are made at the base of the acorn—the area covered by the cap—but if the weevil has migrated within, the hole will be placed accordingly. The

The holes of an acorn woodpecker's larder are carefully sized. (CA)

An acorn woodpecker's larder in a shaved palm tree in an urban park. (CA)

inside of the acorn exposed by the bird will show signs of weevil activity—browning of the nut meat and dark, powdery material excreted by the insect—so look closely.

Woodpeckers

One of the most conspicuous users of acorns is the aptly named acorn woodpecker. These woodpeckers tend to work in family groups, collecting and storing thousands of acorns in one or several central locations. They drill holes in the

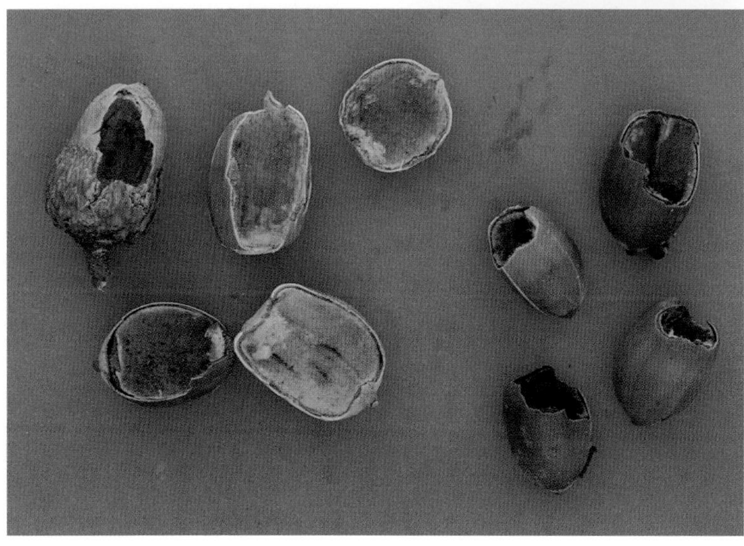

Acorns opened by acorn woodpeckers. On the right, the acorns were opened from the end while still in the hole, and on the left, the acorns were opened from the side after the bird removed them from the larder and retired to a perch. (CA)

Acorns opened by Lewis's woodpeckers. (CO)

bark of living trees, telephone poles, houses, shaved palm trees, snags, dead branch stubs, and other structures, in which they place acorns. These great mast collections go by several names, including *larders* and *granaries*. A substantial larder is an impressive sight to behold—an entire massive tree, base to crown, studded with acorns. Many smaller caches may be made in the softer dead branches of hardwood trees, such as those found in Arizona

Acorn remains strewn over an area 15 feet long and 4 feet wide, below a massive snag—evidence of a workshop and larder of a pair of Lewis's woodpeckers. (CO)

sycamores in the Southwest. The lives of acorn woodpeckers revolve around these larders. They are constantly moving acorns, which shrink as they dry, to make sure the fit is just right. If it's too tight, they may crack the shell, which leads to a rotten acorn. If it's too loose, pilferers may steal the crop. The family also works together to protect their great larders from other birds and mammals.

Acorn woodpeckers open acorns in two ways. They may eat an acorn directly from the hole in which it is stored, in which case they tend to enter the acorn from the base, which faces outward, and leave a ragged, circular hole. Or they may remove the acorn and retire to a perch to feed, holding it with one foot or jamming it in a convenient crevice, and entering the acorn from the side. This results in a ragged hole, often with beak marks on the inside of the shell. These beak marks may be found inside the acorn, pushing out, opposite the entry hole, just as with acorns opened by crows. Both kinds of remains are easy to find in and around larder sites.

Lewis's and red-headed woodpeckers are also regular users of acorns. Lewis's woodpeckers tend to have one or several central locations where they cache and later open acorns. The midden piles of acorn remains under such places, often dominant snags, can be impressive.

Grackles

The common grackle is particularly well equipped to eat acorns; it has a hard ridge inside its mouth that runs the length of the upper mandible. This ridge allows the bird to crack various nuts and to open acorns in a unique fashion. The grackle harvests an acorn and then rotates it in its bill, biting down and cutting the shell with each turn. A groove is cut around the entire acorn, which splits into perfect halves. No other bird or mammal opens acorns in this fashion. Grackles tend to feed in great flocks, leaving a conspicuous trail of opened acorns.

In the Pine Barrens of New Jersey, I was once surrounded by several hundred grackles eating acorns. It sounded as if I were in a great rainstorm, such was the noise made from acorn hulls and halves continuously dropping to the ground. The birds often dropped half of the acorn after swallowing the other half. When the crop is good, they don't bother to drop to the ground to find the missing half; they just reach for another acorn. (Refer to the illustration on the next page.)

Acorns opened by common grackles. (NJ)

Nuthatches and Chickadees

On occasion, nuthatches and chickadees feed on acorns. The acorns are left hanging on the tree and entered from the side. The birds make tiny holes and remove only a fraction of the nut meat.

Mammals

Many mammals also harvest and eat acorns. Acorns crunched by deer or porcupines can look like those opened by crows, but the surrounding sign will make identification easier. Deer pick up acorns from the ground, and there will likely be signs of shuffling debris or digging. Porcupines feed in the canopy before acorns drop, after which they'll continue to feed on the ground. If they've been eating acorns in the tree, you'll see the numerous nip twigs that have been cut to pull the outermost acorns within reach of their sharp incisors.

And rodents may leave tooth marks in the nut meat. Small mammals scoop with the upper incisors while gripping the outside of the nut with the lower incisors. These lower teeth leave tiny, white indentations along the outside edge of the hole, which trackers call *chatter*. Always check for chatter; acorns opened by birds will not have any.

Hazelnuts

Hazelnuts of the East and Midwest are relished by wildlife, though they are often not as available as acorns or other mast crops. Birds that eat hazelnuts are in competition with many mammal species, including humans, who horde the nuts in the New England area as much as any rodent. Birds, notably blue jays and red-bellied woodpeckers, harvest the nuts before they have fully matured and dropped from the tree. Mature hazelnuts are tough to open and require considerable energy expenditure to obtain the nut within. If the timing is right, the shell is easier to crack and the nut meat has developed enough to make the energy output worthwhile. Jays may be unable to crack the fully developed and dried shell of hazelnuts and thus hit the crop hard during the period when the shell is still moist and a bit softer, and when the nut meat has nearly reached full size. Woodpeckers are better equipped to handle tough nuts and use wedging techniques to hold the nut and then crack the shell with repeated blows. Therefore, signs of woodpeckers eating hazelnuts will appear long after the crop is gone.

Blue jays opened these beaked hazelnuts before the shells had completely hardened. Their sign will litter the ground below bushes. (VT)

Caches and Digs

Caching seeds, fruits, and food surpluses is common in many bird species. Nutcrackers and jays, both corvids, are notorious cachers, but others, such as chickadees, titmice, nuthatches, and ravens, also cache regularly. Seeds are cached in grass clumps, behind loose bark, or in forest litter. Most birds tend to wiggle their caches between and underneath leaves and other debris to make sure nothing can be seen. They may then pad the area to blend it with the surrounding area. I have walked over to caches of chickadees, jays, and ravens just after they were made and found no obvious visual evidence of their existence. Nutcrackers and jays make hundreds of such caches in the fall months, each nearly impossible to identify unless the bird was followed and carefully observed. On occasion, corvids may dig a bit more, creating an area to cache one or more items. Still, the digs tend to be within the debris layer, rarely entering the soil.

Other birds, such as raptors, tend to cache poorly, with the cache often out in the open and visible, or so poorly cached that half remains in view. Great horned owls use low-hanging branches or shrubs to cache partially eaten carcasses, returning when they are ready to eat again. They, along with eagles, also shove carcasses into the crotches of trees to keep ground scavengers away.

A nutcracker was spooked before it could cover its cache of lodgepole cones. (CA)

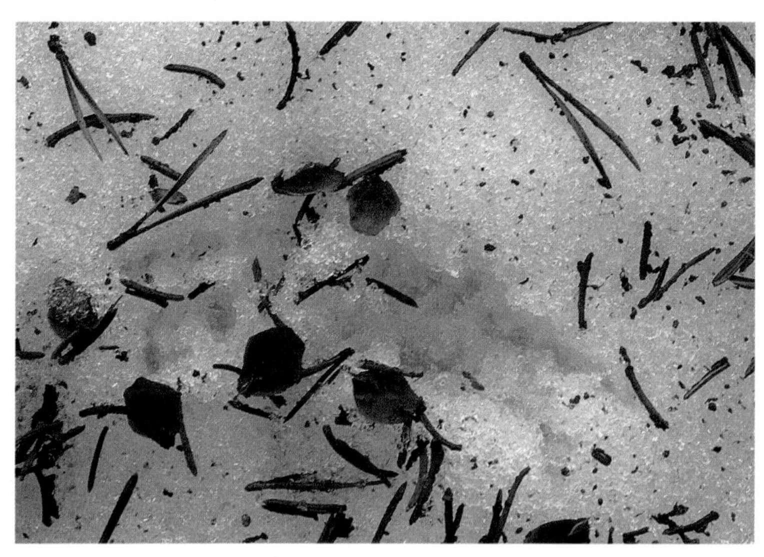

A nutcracker cache in snow. (CO)

Lawrence Kilham (1997) observed interesting caching behaviors in red-headed woodpeckers. Acorns were often placed in holes, which were then plugged with moist wood splinters. When the plugs dried, they became very hard to locate, which may have added an extra degree of safety to the cache.

The counterpart to caching is digging. The first questions to ask yourself when you encounter a dig of any kind are: What was the purpose of this dig? What was the bird or animal looking for?

A great deal of digging and scratching is nothing more than a shuffling of debris, which results

Just below this bush, a great horned owl had cached the remains of a desert cottontail. Not long before I took the photograph, the owl returned, finished its meal, produced a pellet, and flew off to its daytime roost. (UT)

in a miniature rolling landscape of piles and dips. Towhees, sparrows, pheasants, and thrashers use this strategy to stir up insects and uncover hidden prey. Jays and turkeys use similar movements to find acorns and other mast crops hidden within the current year's leaf drop. Deer, mice, squirrels, porcupines, and bears all rummage for mast crops as well. Each species leaves additional clues, as well as subtleties in how the leaves are moved. What length was the debris thrown? What is the width of each swipe of foot or beak? How far was the debris pushed in a single effort? Were both feet used simultaneously, as in towhees and sparrows?

Caches are dug up throughout the winter and often through snow cover in northern parts of the country. Snow, wind, and other kinds of weather compact the leaf litter around caches. This compacted litter can help you interpret the reason for the digs. Lightly feel with your fingers in the dig and try to figure out the shape of whatever was lodged in the litter. In the case of acorns dug up by jays, you will feel a flattened, circular impression in the walls of the debris where the acorn sat. For mammal trackers, this is very similar to the compression left when a porcupine or flying squirrel has removed a false truffle from beneath the soil.

Look for signs of feeding and other clues right around the holes. Jays tend to retire to a hard perch, which will become the anvil.

Leaf litter rummaging by Mexican jays in search of fallen acorns. (TX)

Wild turkey scratching can be dramatic, creating a rolling landscape, as the birds search for mast crops and insects. (NH)

This could be a decent-sized stick or log at ground level, a boulder, or a nearby branch of a tree. Gray squirrels tend to eat their caches of single acorns at the entrance to each hole and leave the remains there. Remember birds tend to cache within the debris layer, and thus their digs should be easy to differentiate from deeper digs into the soil, as when an animal was feeding on false truffles or the cache was made by a squirrel.

Feeding in Trees and Shrubs

A great variety of insects spend all or part of their lives on living and dead trees and shrubs. Birds have specialized over time to exploit specific areas of the trees and shrubs, feeding on particular groups of insects. Some birds pluck insects right from the leaves, called leaf gleaning; others search the outer bark for insects, called bark gleaning; still others remove bark or drill into the wood to feed. *Gleaning* describes any foraging behavior in which insects are plucked from some surface.

Leaf Gleaning

Numerous insects live along the leaves and buds of trees, feeding upon the foliage, fruits, and each other. A host of birds take advantage of this food source, including chickadees, warblers, and cuckoos. The sign of an insect plucked from a leaf is difficult to distinguish. It is easier to note the signs of feeding insects themselves, which often results in leaf browning, curled leaves, and holes in the foliage. Birds also use these visual cues, which help point the way to meals.

An exception to the lack of sign associated with leaf gleaning is sign left on cocoons, which may be found by leaf gleaners and often harbor a large, rich treat within.

Bark Gleaning

The outer bark layer of trees and shrubs provides a habitat for an entire universe of small insects. Some feed within the bark, others feed on lichens and mosses, and still others feed on each other. A few leaf gleaners, such as chickadees and warblers, also readily glean from the bark layer; other bird species depend on bark gleaning for survival. These birds include nuthatches and brown creepers. Downy woodpeckers and black-and-white warblers also engage in bark gleaning, and northern flickers occasionally feed in this manner.

Brown creepers usually begin at the base of a tree and move up, exploring every niche and crevice for insects. When they reach the lower branches of the crown, they then fly to the base of another tree and begin the process over again. Nuthatches, on the other hand, tend to hunt the bark layer while circling down, though they often move up trunks as well. By moving in different directions, these birds are able to share the same ecological niche, each finding different insects by taking unique perspectives of the same section of the trunk.

Signs of bark gleaning are hard to distinguish from other signs associated with tree bark. Creepers leave little behind to be interpreted; nuthatches occasionally pry off bark scales, enabling them to reach new layers of insects. This sign is hard to differentiate from scales of bark that have been removed by a passing creature, such as a climbing squirrel or raccoon, or by a falling branch. Bark gleaners may also encounter and open cocoons of various insects.

Cocoons are opened by many bird species. Downy and hairy woodpeckers peck small holes in larger cocoons, using their brushlike

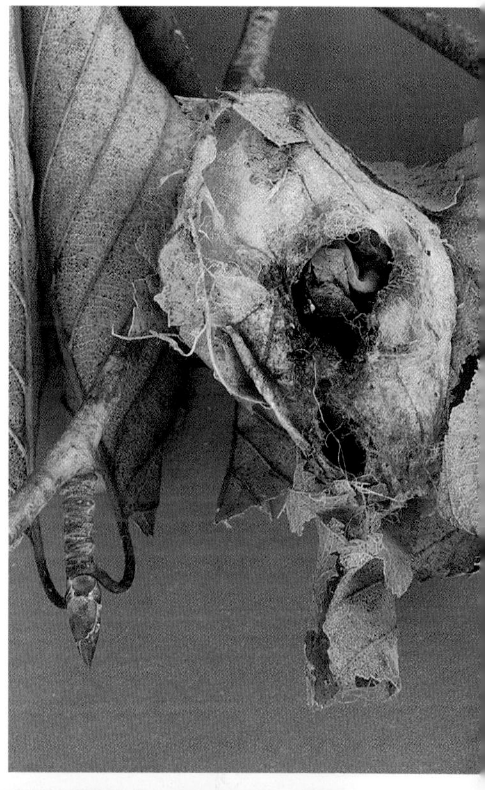

A cocoon opened by a black-capped chickadee. (NH)

tongues to remove the pupae within. Chickadees and titmice combine pecking with tearing; they make an opening, grab the sides, and pull back. I've watched both chickadees and titmice open cocoons in the field. The strength and vitality of these tiny birds always amaze me. They pull their heads way back, delivering incredible blows to create a hole, and then use their entire mass, albeit small, to pull the edge of the cocoon back to enlarge the hole. On one such occasion, illustrated here, a black-capped chickadee

Baltimore orioles follow tent caterpillar explosions, somehow managing to successfully wrestle and open the incredible webs. Blue jays also open tents and feed upon the caterpillars. (MA)

opened a massive cocoon on a cherry tree. It took the bird several minutes to create a large enough hole from which to remove the larva. It was huge, and the chickadee wrestled it into position and then swallowed it whole—which took several concerted swallows. The bird paused, wiped its beak on the branch, and then moved to an outer limb, where it burst into exuberant song.

Bark Sloughing and Scaling

As a tree matures, the lowest branches may die, as they are no longer useful in photosynthesis. As a tree or parts of a tree die, they go through many stages at which different insects invade and utilize the bark and wood. Certain bird species are adept at removing the bark layer on dead branches or tree trunks and feeding on the numerous insects, many of which are beetle larvae, underneath. Hairy and downy woodpeckers often work in this fashion. So do other woodpeckers, as well as Clark's nutcrackers. Prying off the entire dead layer of bark is referred to as *bark sloughing*. If only scales of bark are being removed, as done by nuthatches, this is called *bark scaling*.

A hairy woodpecker has used bark scaling to find insects on this spruce. (NY)

Certain bird species feed at specific heights of the trunk. Three-toed woodpeckers tend to feed low on dead conifer trunks, whether standing as snags or fallen as logs. Black-backed woodpeckers have been observed removing all the bark along standing trunks but leaving the bark along the branches. These are just tendencies, however, and the birds may feed in other places as well.

Signs of bark sloughing are clearly visible: bare branches or trunks, exposed trails of beetle larvae, and sawdust. But bark is also knocked off by passing and feeding mammals, and falls off with the passage of time. The easiest stage of tree degradation in which to read bark sloughing is shortly after a branch or trunk has died. At this point, beak marks are often clearly visible in the still-colored inner bark, made perpendicular to the branch or trunk. From a distance, this sign may look similar to areas where porcupines have fed on the inner bark of live trees. With time, though, the inner bark disintegrates, and beak marks are much more subtle.

This fir is a fine example of bark sloughing by a three-toed woodpecker. Note the horizontal lines created by the beak. (CO)

Entering the Wood

People commonly associate holes in tree trunks and branches with woodpeckers, and in many cases, they'd be right. Woodpeckers drill into the wood of trees, beyond the inner and outer barks. This depth of exploration reveals another layer of insects upon which to feed. Many beetle larvae exist just below the wood surface of decaying trees.

A greater understanding of the niche in which bird species live and find food aids in the identification of the species that has created the hole in question. The characteristics of the holes are very useful for identification, but before moving in for a closer look, stand back and absorb the overall feeding pattern created. Where is the sign occurring on the trunk, or at what level of the canopy of shrub? Are the holes arranged vertically, as is often the case with pileated woodpeckers? Are there horizontal rows of holes, as are commonly created by sapsuckers? Is the feeding concentrated on the lowest portion of the trunk, which is commonly how three-toed and black-backed woodpeckers tend to feed?

The S-shaped feeding pattern of hairy woodpeckers can be seen on this spruce. (NY)

Is there a long, vertical S-curve feeding pattern, which is common in hairy woodpeckers?

I spent several weeks one summer following hairy woodpeckers in the woods, which I'd pick up as they left their nesting cavity just outside my home. While working along a trunk, hairy woodpeckers rotate from a vertical position with the head up and tail propped below, to angles of 45 degrees to either side. As they descend or ascend, they continue rotating side to vertical, to the other side, creating an S-curve of sign along the trunk. Moving in this fashion may be a more efficient means of covering the trunk than a straight vertical route of horizontal rows. If the woodpecker returns to feed at the same source time and again, the S-curves overlap and the pattern is lost in myriad holes.

Sapsuckers

When assessing holes created by woodpeckers, first determine whether the bark surface is living or dead. Sapsuckers drill holes in living trees in order to feed on the sap flows and the insects that the sap attracts. These holes vary tremendously in shape and size, depending on the species of the tree; in each case, holes are created to maximize outward sap flow.

Sapsuckers most often create horizontal rows of similar-size holes, which they continue to use throughout a season. They return to these holes, feeding not only on the sap, but also on the numerous insects attracted by and then captured in the sticky flows. As the tree heals to prevent water loss and the potential of infectious invasion, sap flows slow and eventually stop altogether. More rows of holes are then added to create a larger and more permanent resource. Collectively, these holes are known as *sap wells*. Wells are often conspicuous and easy to find, especially when they are made in the thick bark of pines or junipers.

To make wells, the birds first remove a layer of outer bark, which creates a bright orange spot on the trunk, within which the well is made. These wells may be made at any height, but more often than not, they are made below eye level on the trunks of younger trees. Many tree species may be used for wells. Over two hundred tree and shrub species are used in North America. Some of the regional favorites are eastern hemlock, American linden, apples, aspens, cottonwoods, lodgepole pine, persimmon, willows, sycamores, mountain ash, and hickories.

A Williamson's sapsucker has created the classic sapsucker well in this quaking aspen tree—horizontal rows of holes in a young tree. (CO)

Sapsuckers tend to create evenly sized holes and neat horizontal rows in their wells, but this is not always the case. Sap wells on white birch are often in random patterns and involve different-size holes. Persimmon, a shrub in the Southwest, is another exception. When red-naped sapsuckers create wells on persimmon, they create very thin lines, about 1/16 inch (.2 cm) wide, running parallel to the ground. When these lines heal, they appear as black lines, more like lenticels or some natural part of the plant than the sign of birds feeding. In willows, red-naped sapsuckers often remove large sections of bark in addition to using small holes and rows. This looks similar to signs of mammals eating bark. Look for telltale beak marks or tooth marks, and note when the sign was created: Mammals tend to feed on bark during the winter months, whereas sapsuckers use wells during the warmer months. An exception is conifers, which store sap in special cells and may still be tapped by birds when temperatures dip below freezing.

While working on a persimmon, red-naped sapsuckers create fine lines rather than horizontal rows. Here the old taps have healed to form black scars. (TX)

Whether the wells are collections of holes or stripped bark, they can cause a tremendous amount of damage to the trees or shrubs. Entire young aspen clumps can be killed by annual sapsucker wells. Clumps of willow are also shaped tremendously by sapsuckers—in some cases, more so than by mammal browsing. Any puncture into the living tissue allows the possibility of fungal predators to move in. Sometimes sapsucker wells lead to the demise of trees; in other cases, the trees heal over

While tapping willows, red-naped sapsuckers remove large sections of bark as well as create the classic horizontal rows. (CO)

This aspen grove shows the damages of annual use by Williamson's sapsuckers. Note that some of the trees in the background have died, although likely all the trees are of the same clone. (CA)

and live on, none the worse off for the sign. Either way, well-making offers a lesson in field ecology. Birds are constantly shaping the forest and the environments in which they live.

Sapsuckers vigorously defend their wells against other species, many of which relish sweet sap. But while sapsuckers are away, squirrels, mice, hummingbirds, and other woodpeckers are among those that will slip in to feed. Over thirty species of birds have been documented feeding from sapsucker wells, as well as numerous mammals. An interesting relationship has developed between yel-

In this sap well, created by a ladder-backed woodpecker, the holes are placed in a chaotic fashion and were made in varying sizes. (TX) (Photo by Greg Levandoski.)

low-bellied sapsuckers and ruby-throated hummingbirds in the East. When hummingbirds arrive north in the spring, they follow sapsuckers around and depend on their wells for sustenance until flowers begin to bloom.

Other woodpeckers are known to create holes to collect sap from time to time. These holes tend to be haphazard in shape and size, and to occur singly or in small groups. An exception to this is ladder-backed woodpeckers, which on occasion will create a substantial well of holes. Unlike sapsuckers, though, their holes tend to be of varied shapes and in uneven rows.

Pileated Woodpeckers

Pileated woodpeckers also create holes in living trees. The work of pileated woodpeckers reveals the story of carpenter ants, which have invaded the dead core of a living tree. These insects feed within the trunks, leaving intricate collections of tubular cavities called *galleries,* while from the outside the tree may appear completely whole. Carpenter ants do not necessarily mean the end of the tree's existence. Ants feed on the interior wood, which provides structural support, rather than nutrient transport; many trees live long after becoming nearly completely hollow.

Pileated woodpecker holes are large and deep, and when fresh, as in this spruce, will be found in conjunction with masses of wood chips. (NH)

Left: ***Pileated woodpecker holes in eastern hemlock. (NH)*** **Right:** ***This pileated woodpecker hole has been plugged by white-footed mice, which create nests and larders within the hollowed core of the tree. (VT)***

The large, stout beaks of pileated woodpeckers are able to chisel off large chunks of wood; no other bird species in North America can do such work. The resulting holes are of such great depth and size that they cannot be confused with the holes created by other species. Holes are often several inches deep and square or oblong in shape. If you were to peer into one of these holes, you would likely see the galleries created by carpenter ants.

Pileated woodpeckers tend to work vertically, and what may begin as a vertical row of holes may be worked further and become one massive vertical stripe, several feet long. If the work is fresh, the site will be surrounded by wood chips and, likely, scat.

Whenever a large hole is made in a tree, there are many other animal and bird species that move in to take advantage of such a shelter. Those that move in are called *secondary users,* and the massive holes created by feeding pileated woodpeckers offer many of the same benefits as nest or roost cavities. White-footed mice are regular users of these holes, which are often jammed with wood chips and other debris to make the mice more comfortable. Mice also use these holes to gain entrance to the hollow core, where they can create a nest. Pileated woodpecker holes thus become exits and entrances for mice and other small mammals. Peer into these holes whenever you find one, and you are likely to find mouse droppings, feeding remains, or the mice themselves. Numerous times, we have peered in such holes only to come nose-to-nose with mice. On one such occasion, I found a tangle of mouse bodies, all wrestling to get a peek at the humans outside.

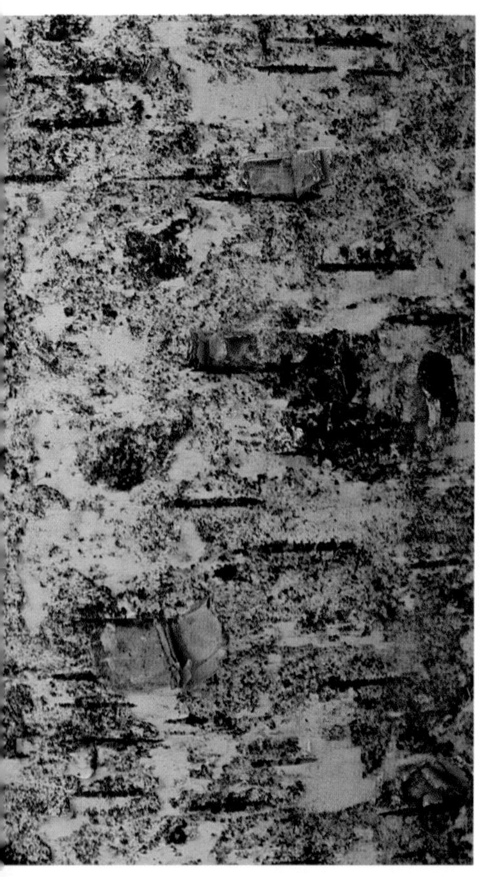

Trapdoors in a birch tree are a sure sign of downy woodpeckers feeding upon coccids. These doors have been folded back, making them more visible. (NH)

Downy Woodpeckers

In winter, a great place to look for downy woodpecker sign is on the stressed white and yellow birches on exposed ridges or in storm-hit areas. In these weakened trees, a coccid species, *Xylococcus betulae,* often spends the winter in the space just under the outer bark. Lawrence Kilham (1997) researched the phenomenon and called the sign resulting from downies feeding on coccids "trap doors," which the excavations resemble. The door is more often the result of the tree bark, rather than the woodpecker. Due to the horizontal peeling nature of birch, when a bark strip is cut vertically and pried to one side, the area just beyond the pry becomes a hinge upon which the bark bends back, creating the door. Downy woodpeckers peck and cut a vertical line in the outer bark next to the coccid, and then pry up the bark to expose the insect. On occasion, they will peck on three sides of the bug, thus truly creating a door effect.

Holes in Dead Wood

After a tree dies, different woodpecker species arrive to feed on insects from within the decaying wood. If carpenter ants persist, so will pileated woodpeckers. But these birds may also be joined by a great diversity of woodpeckers. Holes of these smaller species can be more confusing and overlap in size.

Hairy woodpeckers drill into wood far more often than downy woodpeckers. The holes of hairies tend to be 3/16 to 1/2 inch (.5 to 1.6 cm) in diameter and are fairly uniform in size regardless of tree species. Hairy woodpeckers also tend to drill into harder woods than downy woodpeckers. This makes sense when the bills of the two species are compared; the hairy woodpecker's bill is much longer and stouter. Downy woodpeckers are more likely to be found drilling holes into such things as sumac stems, cornstalks, phragmites, or insect galls. Individual holes of the two species are ordinarily fairly distinct, except in cases where an area is worked greatly, resulting in a larger, shallow excavation rather than a hole. Here, scat or associated sign may be more useful.

The pattern of the holes and their shapes and sizes are all great clues to what species created them. First take in the larger picture: Which species are present in the area around the holes in question? Then go deeper, closer, and note what stage of disintegration the wood is in.

After a tree dies, it goes through many stages as a standing snag before it falls, at which point it becomes a log and goes through additional stages before it is visible only as a hummock in the dynamic landscape. Jon Luoma (1999)

This staghorn sumac was drilled by a downy woodpecker. (MA)

writes about research in the Andrews Forest of Oregon, where it was discovered that there is actually more life, in biomass, in a dead tree than in a living tree. At each stage of disintegration, the tree becomes a different ecosystem, used by different insects, birds, and mammals.

Woodpeckers intensely utilize trees and snags at various levels of life and death. A sapsucker may drill holes into a young tree. A pileated may work the same tree many years later, while it is still alive. A three-toed woodpecker may remove chunks of bark just after the tree has died. A bit later, a hairy woodpecker may arrive and drill holes through the bark layer into the wood. In the next stage, nutcrackers and downy woodpeckers may slough off the bark. When the bark is gone, hairy and downy woodpeckers may continue to drill holes into the exposed wood to feed on larvae. After some time, while the tree still stands but the wood has softened, hairy woodpeckers and flickers may excavate cavities for nesting and roosting. Later, when the tree begins to crumble and stands at only a portion of its original height, a downy may excavate a winter roost. When the tree falls, it becomes a bridge for squirrels and fishers and cover for shrews and voles. Bears rip at its sides to expose insects, and seedlings begin to germinate along the top, held above the smothering leaf litter on the forest floor. This is a simplification of an intri-

A downed tree serves many purposes, including acting as a ruffed grouse's drumming log for courting a mate. (WY)

Left: ***Goldenrod galls opened by a downy woodpecker. (NY)*** **Right:** ***Aspen galls opened by a black-capped chickadee. (MA)***

cate and amazing process, but the importance of snags and logs in a healthy ecosystem cannot be overemphasized. Vital to birds and mammals alike, snags and logs enrich environments long after a tree has died.

Holes in Other Materials

In addition to wood, birds also drill holes in other things to find insects. Downy woodpeckers drill holes in cornstalks and phragmites to obtain larvae. Birds penetrate insect galls on an assortment of plants, shrubs, and trees to expose the insects, most often wasp and fly larvae. Certain galls are favored foods of specific birds. In the winter, downy woodpeckers can be counted on to find and open the large galls on goldenrods, a weed of roadsides and fields. The resulting hole generally enters straight from the side and is 1/4 to 3/8 inch (.7 to 1 cm) wide. In the spring, black-capped chickadees open willow galls, which at first glance appear more like flowers than insect galls. An assortment of other galls are used by woodpeckers and chickadees, as well as jays and other species. Blueberry galls, oak

Artist's conk, a bracket fungus, pecked open by a hairy woodpecker, which fed on the insect larvae found within. (NY)

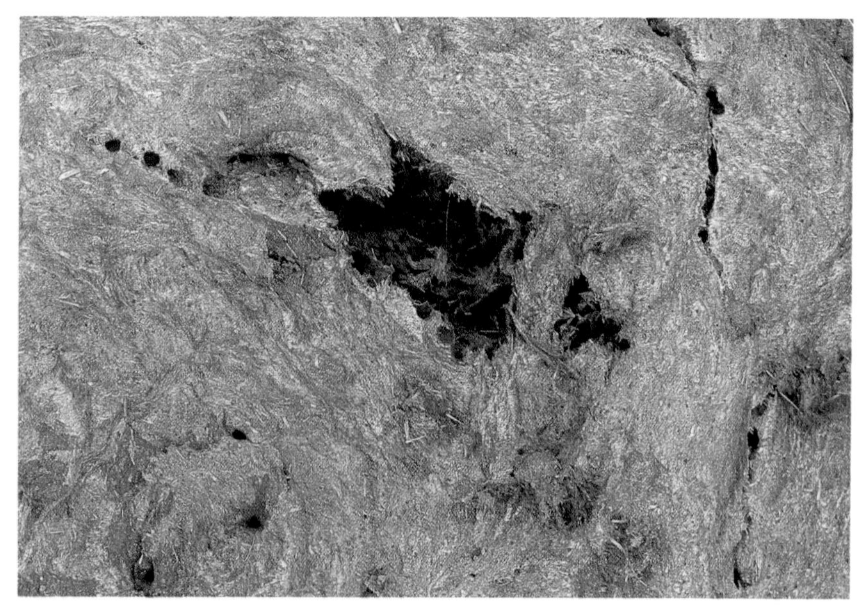

A black-billed magpie opened this bison scat to feed on the beetles that were feeding within. (WY)

galls, and aspen galls are among those commonly opened by birds.

Bracket fungi, softer fungi, and mushrooms are pecked open by birds to find the insects that reside within. If a tree is covered with bracket fungus, feeding patterns will be similar to those in wood. Hairy woodpeckers will peck open fungi in an S-shaped curve down the trunk, just as they do when drilling in wood. Artist's conk and tinder polypore are two species of bracket fungi that often show woodpecker sign, and others do as well. Softer mushrooms are pecked by a variety of birds, including woodpeckers, starlings, and jays. Birds tend to feed upon the invertebrates inside, but occasionally they will eat the mushroom itself.

Carpenter bee holes pecked open by a ladder-backed woodpecker in a dried lechuguillas stalk. One bee hole has not been disturbed. (TX)

Holes in wood made by insects might be confused with those drilled by birds. Look closely at all holes to make sure you are looking at bird sign and not insect sign. Insects such as beetle larvae or carpenter bees drill into living and dead trees to feed or lay eggs. Their holes are different from those made by birds in that they are very smooth around the edges. There are no beak marks or jagged edges at all. There may also be small wood fibers around or in the holes, which were excreted by the insects as they fed and moved deeper into the wood.

Probing, Gaping, and Anthill Opening

Probing and gaping are feeding methods used by many birds to find food hidden from view in earth, mosses, and vegetative tangles. *Probing* is a far more common feeding technique than gaping,

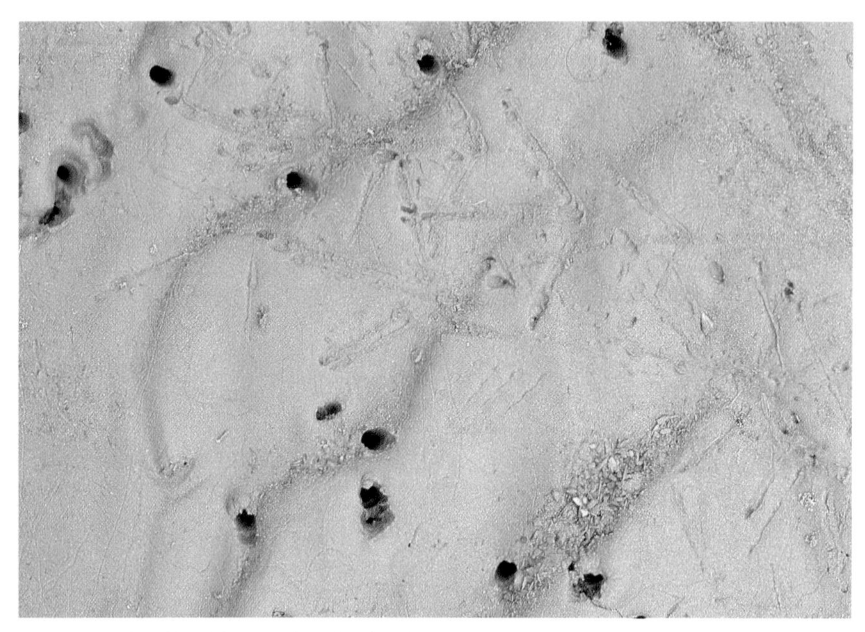

Woodcock probings and tracks. (MA)

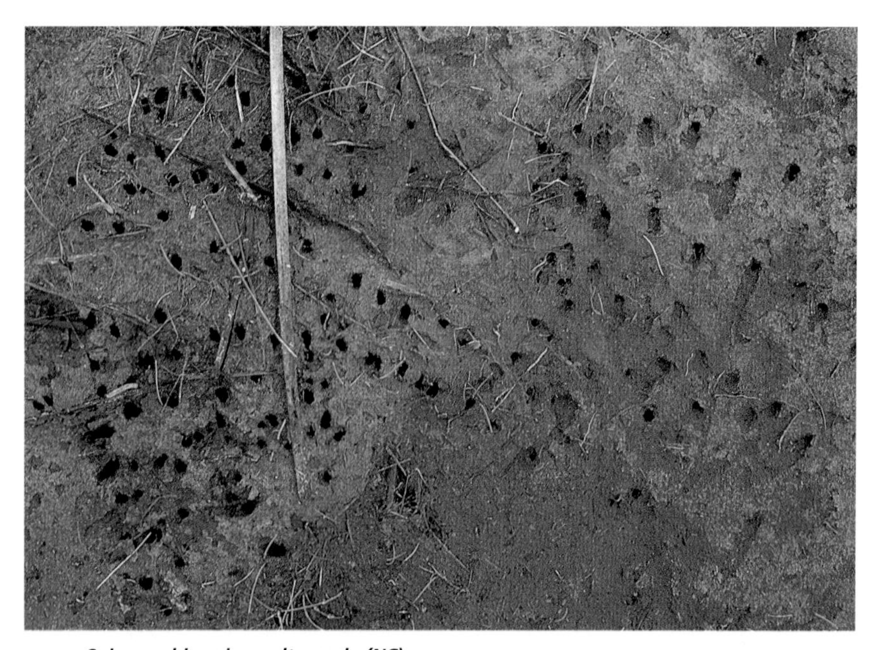

Snipe probings in a salt marsh. (NC)

involving the thrusting of the beak into earth and pulling out invertebrates. It is common among all sorts of shorebirds and easy to see at the beach. Bill lengths and widths vary, affecting the dimensions and characteristics of the resulting holes.

Gaping is a technique in which the beak is inserted into earth or tight vegetation and then opened, allowing invertebrates to be exposed for capture. The resulting sign looks very similar to probing in earth and mosses, but in dried grasses the holes look larger, making clear that the beak was spread open. Gaping is used most often by starlings, but also by blackbirds and some grackles.

Regardless of the hole, or whether gaping or probing occured, ask yourself what the reason for the hole was. What was removed? If you can determine the food source, then the identification of species is much easier. There are many well-known probers among the shorebirds. Snipes and woodcocks are both experts due to their long beaks, which help them feed on insects well below the surface. Woodcock probings are quite common in muddy, forested areas of the East. These birds live almost entirely on earthworms. Whimbrels turn pieces of tundra completely over, leaving conspicuous signs behind. Use habitat types to help differentiate among species.

Plovers, sanderlings, and sandpipers are all regular probers along our shores. Habitat and location in the mudflat are wonderful aids

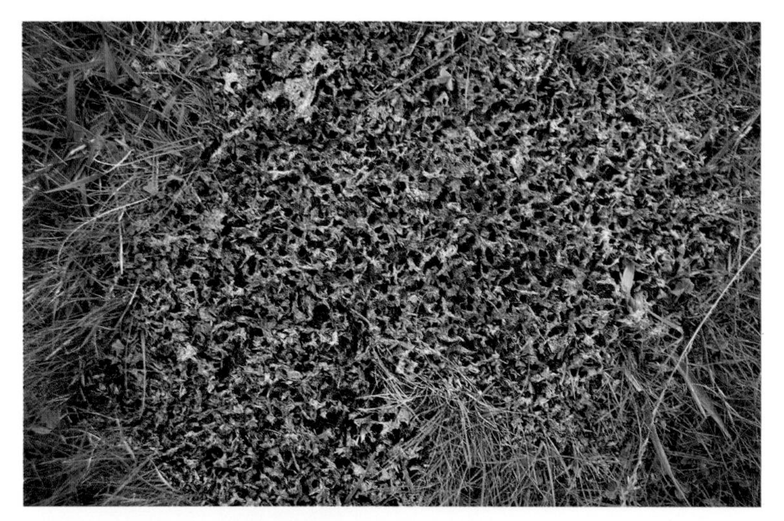

Gaping by European starlings is most easily seen in dried grass, where it becomes obvious that the mouth has been opened wide. (Used with permission from Gads Forlag. Photo by Mels Peter Holst Hansen.)

in identification, but the pattern of holes also offers the tracker information. Plovers hunt prey at the surface and thus move with incredible speed, faster than sanderlings, which also probe at speed, and much faster than sandpipers, which probe at a walk. Thus, plover probing is not concentrated, but spread in a thin line, and the

Probe feeding by a semipalmated plover (L); probe feeding by a sanderling (R).

amount of earth moved for such a shallow probe is considerable. Sanderlings, which probe at a bit slower pace, leave an incredibly regular stitching of holes. When they hit on a concentration of prey, they will intensively probe that area, which may be as small as 1 square inch (2.5 square cm). Sandpipers, on the other hand, continuously probe as they plod along, and their trails will show many more probes per step than those of either the sanderling or the plover. Sandpipers also tend toward a double-probe strategy—the head is fully raised every other probe, rather than every probe. Within the trail, this is often visible as paired holes or long holes, which are really two holes where the beak never made it to the surface before plunging down again.

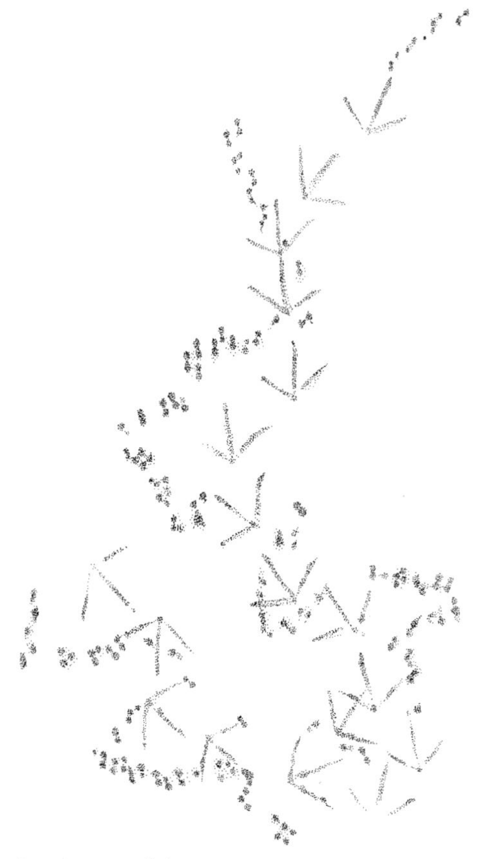

Probe feeding by a least sandpiper.

An anthill opened by a northern flicker. (NY)

A sign that looks similar to probing is the work of northern flickers opening anthills. Flickers feed on the ground often, making their sign easier to find than you might expect. I've watched flickers opening anthills in the field on countless occasions and often find their sign along roads and trails in open, dry areas. Look at the anthill's entrance for the impressions and marks of beaks; holes in smaller anthills generally measure 1/4 to 3/8 inch (.7 to 1 cm) wide. Also look for scat, which will always be present in an area filled with opened anthills. Tracks should also be found, as flickers often hop from anthill to anthill. If the beak marks are clear and clean, the sign is likely to be very fresh; ants begin rebuilding almost immediately. On occasion the sign will persist, as the remaining ants abandon their anthill to rebuild elsewhere.

Egg Remains

Eggs are eaten by both birds and mammals, humans included. Eggs are also damaged for reasons other than feeding. There are several important variables to consider when you find the remains of eggs. First, did the eggs naturally hatch? After the young have hatched, eggs may act as visual and olfactory attractants to the nest and therefore endanger the young, so parents will carry the eggs and drop them a suitable distance from the nest, or at least push them out of the nest. For this reason, egg remains appear in the strangest places, dropped by parent birds in streams, roads, trails, or the middle of large fields.

Eggs that have naturally hatched will be in two separate pieces or halves. The membrane that lines the inside of the egg will be folded

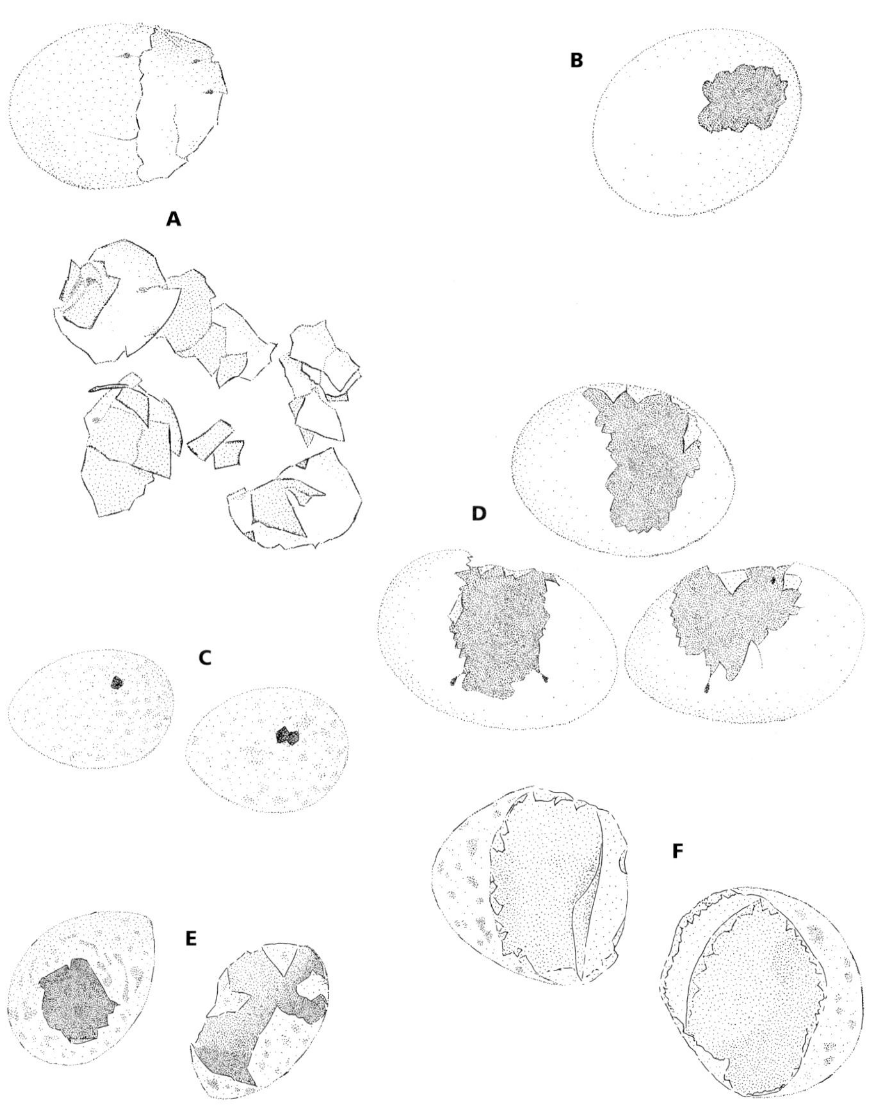

A Chicken eggs opened and smashed by a striped skunk
B Chicken egg opened by an American crow
C House sparrow eggs "beaked" by a house wren
D Mallard eggs predated by a mink—note the holes made by the canines.
E Sandwich tern eggs predated by a gull*
F Cliff swallow eggs which have naturally hatched—note the folding of the internal lining.
(*Used with permission of Gads Forlag.)

inward, which is a good indicator of successful hatching. Young birds also move in such a way that one half of an egg may end up within the other half; this is a very sure sign of a natural hatch.

Other variables to consider when assessing egg remains are the size of the egg and the number of eggs involved. The size of an egg influences how it is opened. Weasels tend to enter smaller eggs from the side, but larger eggs are entered from the smaller end, or tip. The quantity of eggs can also come into play. A single egg eaten by a crow may be cleaned of its contents, and then the shell may also be consumed. When there are many eggs, crows tend not to eat the shells, but leave them behind as evidence of the meal.

Bird Predation on Eggs

Most bird predators tend to open eggs from the side, leaving squarish, jagged holes. If the egg is small enough and the bird strong enough, the egg is lifted so that the yolk can be gulped like water. The size of the hole varies tremendously and can be fairly small or quite large.

House wrens are among the birds that will open an egg, but they do not feed upon it. A house wren will pierce the eggs of sparrows and bluebirds that nest in the same vicinity. The wren will enter a nest and poke the eggs with its beak, making just a single small hole, which is enough to kill the developing embryo. They may also go further and push the eggs out of the nest altogether. There are also other bird species that push bird eggs out of their nest, such as catbirds, which eject cowbird eggs.

Weasels and Skunks

Weasels and skunks are a constant threat to egg clutches. If an egg is large relative to the animal, it will be opened by the pointy end, as is the method when skunks open chicken eggs. If the egg is small in relation to the animal, it will likely be opened from the side, leaving roughly squarish, ragged openings, as when mink eat mallard eggs. The key indicator is evidence of the canines in the shell, which would be tiny holes near the entrance. If you can pair up a jaw's canines, then an estimation of the size of the animal is much easier.

Other Mammals

Many rodents, including mice, chipmunks, and squirrels, eat eggs and hatchlings. Foxes, raccoons, and opossums are also notorious nest raiders. Foxes and opossums, in my rather limited experience, eat the egg in its entirety—shell and all. I watched an opossum crack open an egg, slurp up the contents, and then return to eat the shell. I've heard that raccoons will do the same, although again, the number of eggs is a factor. A full turkey's nest would be a challenge for any animal to finish in one sitting, especially if it ingests the shells as well.

Dust Baths

Birds are known for bathing in puddles, ponds, streams, and even ocean tides. In areas where water is scarce, dust bathing becomes the norm. Bathing, whether in water or dust, is essential for plumage maintenance. The preen gland exudes oil, which birds then preen through their plumage. Dust bathing removes clots and excess amounts of oil, maintaining even oil distribution throughout all the feathers. Clots of oil result in feathers sticking together, which can lead to problems with flight, waterproofing, and regulating body heat. Dust bathing also may help eliminate parasites.

An impressive assortment of birds dust bathe, from wrens to raptors. As might be expected, there is greater diversity in desert regions. Throughout the country, the most common dust bathers are house sparrows and game species, such as turkey, grouse, quail, and pheasant.

The act of dust bathing is similar across species, and regardless of the bird, is tremendously entertaining to watch. Let's take a turkey dust bath as an example. A turkey approaches a suitable area, scratching from time to time, checking for ideal dust or sand consistency for optimum feather saturation. The bird finally selects a spot, and gives it a vigorous scratch to help prepare the loose earth. Then the turkey dives in, propping itself on one side or shoulder, using the other wing to scoop and throw dust up and over the body. To help spread dust throughout the feathers and skin, feathers are quickly raised and lowered, the entire plumage undulating in unison. The bird may roll around a bit and then switch sides, using the other wing to heave dust up onto the back and body. The turkey may even rub its head directly in the dust to help cover the scalp. This ritual can last from five to thirty minutes, at which point the turkey stands and proceeds a few steps. Then the entire plumage is shaken, feathers erected and lowered rapidly, which gives the bird the appearance of wearing a coat.

Billows of dust are released. It is hard to imagine the amount of dust that can be held by the plumage of one bird. It is truly impressive.

If a site remains as is, retaining the right requirements for wonderful dusting, it will be used again and again. In areas of really loose sand and dust, individual acts of bathing may be visible in the soil. If the ground is a bit harder, a spot may begin to wear into the earth, and these habitual bathing areas become more obvious. In addition, look for associated sign with dusting areas, such as feathers and scat. Feathers of the species may help in the identification of the bathing bird.

Turkeys

The dust baths of turkeys are often amoeba-shaped or oval, rather than perfect circles. The diameters of such baths range from 14 to 20 inches (35.4 to 51 cm), and they tend to be from 1 to 3 inches (2.5 to 7.7 cm) deep. We've found turkey dust baths in open fields, along field edges, along dirt roads and trails, as well as along the sandy shores of rivers and lakes.

Grouse vs. Cottontails

Grouse and cottontails create similar-size dust baths, and the two species do overlap in the same ecosystems as well. Grouse dust baths tend to be round, whereas those of cottontails are usually oblong or

A wild turkey dust bath. (NH)

A ruffed grouse created this dust bath along a deer trail. (MA)

A pheasant's dust bath. (Used with permission of Gads Forlag. Photo by Preben Bang.)

oval, longer than they are wide. Look closely for feathers or hair, which will help in the analysis.

Grouse dust baths are often found along sandy trails and roads, along field edges, and, on occasion, in the softer loam along deer runs in northern forests. Grouse will also bathe in the soft remains of rotted evergreen logs, as will cottontails. Grouse dust baths range in diameter from 7 to 13 inches (17.9 to 28.1 cm).

Quail

Quail species tend to create round dust baths, with diameters between 5 and 8 inches (12.8 and 20.4 cm). Quail dust baths can be found along sandy trails or roads, or in dried streambeds near cover.

House Sparrows

House sparrow dust baths are quite easy to find in urban areas. Look for small circular depressions in flower beds and other plantings near stores, as well as along footpaths and dirt roads. Round depressions between 3 and 6 inches (7.6 and 15.3 cm) in diameter can usually be attributed to dusting house sparrows. If dusting sparrows bother the gardener, a thick layer of mulch usually puts a stop to these social gatherings.

House sparrows regularly dust bathe in this spot behind the town hall of Cambridge, in Boston. (MA)

The remains of a feeding goshawk. (Used with permission of Gads Forlag. Photo by Stig Tronvold.)

Kill Sites

Within the realm of tracking, a *kill site* refers to an area where the signs of a predator taking prey are left behind. Some people feel that to be termed a kill site, it must have involved large prey and predators. But by our definition, for the purposes of this book, small prey, even insects, amphibians, or mollusks, could be involved. The victim's remains, along with any signs left by a food-getting technique, such as banging or dropping the prey, all constitute the kill site.

Signs along the Seacoast

There are numerous stories and adventures to be found along our seacoasts. Food is abundant in these areas, and both birds and mammals congregate to feast and live. Below are some examples of common signs of kill sites associated with birds along seacoasts.

Shelling Areas

Gulls and crows feed heavily on crabs and shellfish, such as mussels, along our shores. The birds often drop the crabs or shellfish from a suitable height to stun them and/or crack open the shells and gain access to the meat. This is best accomplished on harder ground,

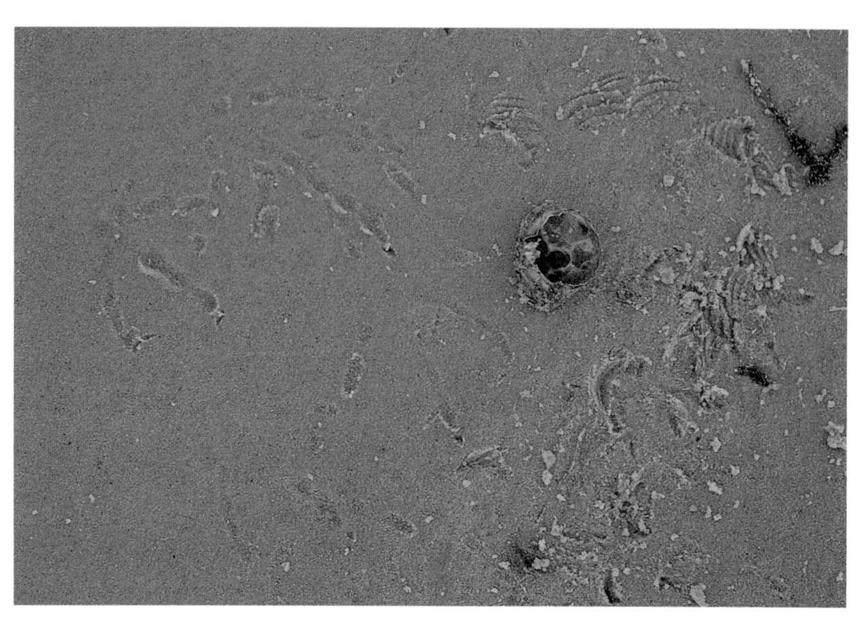

A crab kill site. A herring gull dropped the crab and then followed up by repeatedly slamming its victim into the hard-packed sand. Look for the clear imprints of the crab's carapace, or shell. (MA)

A shelling area used by gulls. (MA)

Shells dropped and successfully cracked by a herring gull. (MA)

and a spot may be used over and over, accumulating sign over time. Such a site is called a *shelling area,* and they are common along our shores. Another tactic, which gulls often use with smaller crabs, is to slam the creature against the ground in an effort to kill it and crack the shell. These tactics are often used in conjunction with one another: first the drop, with the gull hot on the heels of the crab in case of an escape attempt; then, if this hasn't produced the desired results, the crab is slammed against the ground repeatedly.

Depending on the history of a shelling area, it may contain a few dozen shells or hundreds of shell remains. As you begin to look for these areas, check flat salt pans, rocky areas, and parking lots. The areas will be as close to the source of the mussels and other shellfish as possible. At low tide, many more shelling areas will become apparent, some of which are hidden by high water.

Skate Egg Pouches and Sand Dollars

The egg pouches of skates and washed-up sand dollars are both taken by birds. Crows and gulls will both eat skate pouches, as will mammal predators. Sand dollars are pecked open by gulls. The ragged edges of peck marks by gulls are clearly evident when the

A gull opened the skate egg case on the left, while the chewed case remains on the right are evidence that it was opened by a raccoon. (MA)

Sand dollars opened by ring-billed gulls. (OR)

sign is fresh. Opened sand dollars may remain in the intertidal zone, washed back and forth in the sands, which naturally smooths the sharp edges, but the characteristic shape of the openings will remain the same.

Fish Remains

Ospreys and bald eagles are the great fishing raptors of America. Herons, terns, and pelicans are equally skilled at fishing, but they tend to swallow their prey whole, leaving little sign behind to tell the story. Gulls, corvids, turnstones, and egrets will also pick at dead fish.

Although bald eagles can pull live fish from the water, they are equally dependent on carrion for their survival. They are majestic scavengers. When salmon spawn and then die en masse, many birds and mammals appear to dine. Bald eagles often control the sandbars, pulling out massive salmon that float by or wash to shore.

When an eagle has fed upon a salmon, all that generally remains is the spinal column and the attached head. The head is often completely intact, except that the cheeks have been removed. (Any good fisherman knows that the cheeks of salmon and trout are quite a prize.) On occasion, the head may be skinned and left bare. In our studies, the tails were still attached on a small number of the salmon eaten by

This salmon was stripped clean by a bald eagle. Neither the salmon sign nor the eagle track are fresh. (WA)

In a time of plenty, alewife fins, spines, and tails carpet an area below an osprey's regular feeding perch. (MA) (Photo by Greg Levandoski.)

bald eagles. In a good area, a sandbar will be covered with skeletons in each of these conditions, with other variations as well.

When ospreys feed on fish, more often than not, there is nothing left of the fish to tell the story. I recently spent a lazy afternoon watching ospreys hunt and feed in a coastal bay off Texas. On several occasions, ospreys fed on the beach where I sat. Once an osprey landed on the beach right in front of me and began to feed. The bird ripped at the fish, swallowing stringy chunks for forty-five minutes without a break. All the while, a ring-billed gull attempted to steal portions of the kill. Twice the gull successfully stole fish straight from the beak of the osprey, just after a piece had been torn loose but before it had been swallowed. By the end of this session, the fish had been about half consumed, and the osprey took a break. The break lasted an hour, during which a ruddy turnstone unsuccessfully attempted to steal fish and the gull flew away. The osprey never let go of the fish for a second, but perched right atop it. When the hour had passed, the osprey took two more bites from the fish, just as a great blue heron flew in for the steal. But the osprey moved quickly, flying 15 feet down the beach; the heron walked away. The osprey took another break of forty-five minutes before beginning to feed again. A second feeding session of forty-five minutes completed the meal. A grand total of three hours and fifteen minutes after the process had begun, the meal was finished, and the bird walked into the shallows to wash its beak and talons. All that remained to tell the story were four squirts of uric acid the bird had excreted. Not even the fish tail remained.

In areas of fish abundance and easy fishing, ospreys become pickier about what they ingest. In times of plenty, look for discarded tails and fins under regular feeding perches near the food source.

Ospreys have also been documented carrying fish while migrating. If you watch an osprey fishing, you will come to realize how many failed attempts often occur before the prize meal is caught—and they seem to really hold on to this food once they've finally caught it, wasting absolutely nothing.

Impaling

A common practice among shrikes is to impale their victims upon thorns, barbed wire, or sharp sticks. Finding lizards, insects, small mammals, or small birds impaled is a clear indication of shrikes in your area. If thorns are unavailable, shrikes will stuff prey into the crotches of trees or shrubs. Although unrelated to raptors, the true

Beetles impaled on the thorns of velvet mesquite by a loggerhead shrike. (AZ)

A vole impaled by a northern shrike. (Used with permission of Gads Forlag. Photo by Torbjørn Moen.)

predatory birds, shrikes do make their living hunting smaller creatures. Lacking the strong-taloned feet of hawks or owls, they impale prey on sharp objects or wedge them into tight places to help hold them still. Once the prey is secured, they are able to tear off bite-size chunks with their hooked beaks.

According to the compilation study done by Norbert Lefranc (1997), there are four reasons why shrikes may impale prey. The first is to hold prey still while the beak tears off manageable chunks. The second is for food storage. Correlations have been found between bad weather and the use of stored food, indicating that shrikes use impaled food as resources in poor hunting conditions. It was also noted that shrikes that inhabit environments where weather patterns are more unpredictable create more food caches, or larders. Third, certain poisonous insects and amphibians lose their toxicity over time, which allows shrikes to eat food they would otherwise be unable to consume. And last, it is believed that impaling victims may play a part in courting and mate selection. Males impale prey in more conspicuous places at the start of the mating season. Entire bushes may be covered with prey in the hopes of attracting a female.

As in many species, individual shrikes may become particularly adept at catching one particular species—such as one particular type of beetle (pictured on the preceding page).

Carcasses

Magpies, crows, and ravens are common roadside scavengers and will take advantage of any dead creature, be it a wolf-killed caribou or a car-killed opossum. They will also hunt live prey when the opportunity presents itself or the urge arises. Without the strong-taloned feet or hooked beaks of raptors, they rely on other tactics to make the kill. They most often bludgeon their prey to death with repeated pecks to the skull. But without the sharp, hooked bills of a hawk or an owl, breaking the skin of a dead or living animal or bird is difficult. The answer is to use an existing entrance, pecking at the eyes of the prey, whether dead or alive, and eventually into the skull. Thus an injured creature that has been killed by a crow or raven will often show characteristic bloody eye sockets. Injured mammals may also show blood on their front paws, which they used to wipe and protect their stinging eyes. This is often the case when a corvid comes across an animal or bird that has been hit by a car but hasn't yet died. If the animal was dead before the bird arrived, the eye sockets will not bleed. Just the eye will be removed, and if the creature is small enough, the brain cavity will be broken into. These birds need someone or something to open a carcass for them.

The eye is often the only place a corvid can enter a carcass, unless it has been opened by another animal. In the case of this mule deer, both magpies and ravens were on the scene when I arrived. (CO)

This tuft of vegetation served suitably for beak cleaning after a magpie dined at a nearby carcass. (CO)

Research leads us to believe that ravens attract or lead wolves to carcasses in the North. This relationship may not seem to benefit the bird, but it does. The wolves open the carcass and leave enough scraps for the birds to feed. In areas without wolves or other large predators, cars open plenty of carcasses on which corvids feed.

Another sign to look for near carcass feeding is beak cleaning, which is commonly done by magpies, crows, and ravens. A nice clump of vegetation is ideal, although an appropriate piece of earth may be substituted, where the beak is inserted, rubbed, opened, and closed to remove any food remains from the beak.

Bones can be opened by corvids if the prey is small enough or the bones brittle enough. If this is possible, the birds will feed on the marrow as well.

Birds Eating Mammals

Many mammal species fall prey to birds, and they vary in size and shape tremendously. Therefore, characteristics of prey remains are difficult to describe.

From the details of the kill site and ecological clues, I'd surmise this gray squirrel was killed and eaten by a red-tailed hawk. (MA) (Photo by Paul Rezendes.)

You may occasionally find the place where the mammal was captured, and this spot may offer some clues to the identity of the predator. Certainly one of the tracks I most enjoy finding in snow is that of a predatory bird's entire body. A perfect body print shows wings and tail clearly, and occasionally a head and feet will show too. Some people believe that an inspection of the crispness of the feather marks will tell you whether the avian predator was an owl or hawk; owl feathers have small, ruffled edges, called *mufflers*. However, the crispness of the tracks of

A great horned owl has swooped and picked up a mouse. (UT)

The body print of an owl. Note that a vole tunnel enters the impression of the head. The kill was made with the beak. (NH)

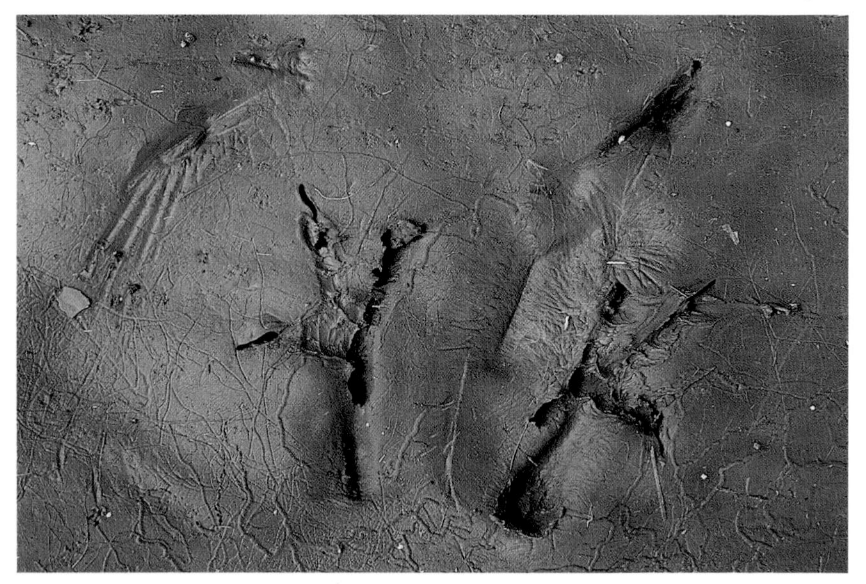

From the looks of things, this great horned owl came up with a beak full of mud, rather than the intended mouse. Look closely for mouse trails, which go back and forth along the top of the picture, as well as for impressions of the keel and feathers. (CO)

the feathers is more often a result of ideal snow conditions and perfect entry by the bird, rather than the structure of the feathers.

There are two features of a body print that do indicate an owl instead of a hawk. If there seems to be no apparent mammal trail on casual inspection, it is likely the work of an owl, although you may have simply found a spot where a bird came down to rest. Inspect the body print carefully, as a mammal trail may be visible entering into the print from beneath the surface of the snow. Owls hunt by sound and can detect small mammals below the snow surface.

Also consider whether the beak played a role in the capture. Owls often capture small prey with their beaks rather than their feet. Hawks, in general, do not use this hunting strategy.

Birds Eating Birds

I arrived an hour early for a meeting in Harvard Square, on a perfect spring day in the city. Being Saturday, the streets and sidewalks were mobbed. I aimed for a bench out in front of Grendel's Restaurant, where there would be ample opportunity for people watching. As I arrived, the pigeons rose in such a flock that I was temporarily blinded, and I wondered what could scare them in such a way. Just a bit farther down the walk, the answer was perched in absolutely plain view, 10 feet off the ground, on the large lower limb of a maple tree: A peregrine falcon gripped its prey with one talon and began plucking the victim's chest with incredible intensity. Clumps of the pigeon's feathers floated down on the slight breeze, landing amid the now peacefully feeding living flock. Standing in the middle of the path, which was the best position, I watched this bird for an entire hour. It tore into the chest, feeding continuously with powerful rips and swallows. I wanted to memorize the entire bird, the entire event, so that I could replay it over and over again in my imagination. The head and wings of the pigeon fell limp over the sides of the branch. As the chest was further thinned, the peregrine clipped the head from the carcass, and it fell to the other pigeons.

Out of the hundreds of people who continuously filed and pushed by me during that hour, only five saw the bird. Most people just complained under their breath about my being in the way as they passed. Of those who did see it, two thought it was an owl, two an eagle, and one asked me what I thought it was.

As the hour came to a close, so did the meal. The peregrine left the carcass, which now consisted only of skin and bones, wings still attached, draped over the branch where it had fed. It moved up the branch a bit, attempted some beak cleaning on a small branch, and then began to eye the flock below. Immediately the pigeons became nervous again, but none flew. And then it was time for my meeting. I pulled myself from the scene reluctantly and left the peregrine where I had found it, perched on the same branch of a maple tree in plain view in Harvard Square.

Birds do hunt other birds. Falcons are infamous hunters of birds, but herons and egrets also eat birds when possible, and owls include birds in their diet as often as they are easily obtained. Quite a number of bird predators, including great horned owls, screech-owls, and peregrine falcons, decapitate prey, whether it be another bird or a small mammal. And bird predators tend to eat the breast meat of bird prey first, leaving other tidbits for later. If the hunter is well fed, it may leave the carcass with only the breast meat removed and the plucked feathers nearby.

Peregrines and merlins eat prey in similar ways, and their kill sites are difficult to distinguish. I've watched both species tear off and

A yellow-billed cuckoo killed and eaten by a peregrine falcon. (ME) (Photo by Eleanor Marks)

A black guillemot killed and eaten by a peregrine. (ME) (Photo by Eleanor Marks)

This crow was killed and partially eaten by a red-tailed hawk. Red-tails do take bird prey from time to time. (NY)

Below the perch where a bald eagle fed hangs a clump of skin and feathers of a Canada goose. Wintering waterfowl are popular fare for bald eagles. (MD)

drop the wings of warblers and swallows, as well as pick clean the skeleton so well that the remains split in half—two wings, and a few bones. With larger prey, they tend to leave the wings attached to the carcass, and often all that remains are wings, the spinal column, and the pelvic bone. Terry Goodhue and John Drury, who have studied peregrines and merlins for many years in Maine, say that both species often decapitate their victims but that they are equally likely to leave the head attached or eat the brains. Carcasses in both conditions are found regularly in the field.

Evidence of a bird predator may also appear on the keel of the victim—the thin bone running down the centerline of the breast, to which the flight muscles attach. From time to time, triangular beak marks of the predator can be found on the keel.

Quill Damage on Feathers

If you find a kill site where feathers still linger, the arrangement and condition of the feathers as well as the environment around the kill site often aid in the identification of the predator. Is the kill site under a low-hanging branch, near a wetland, or out in the open? Have the feathers been dropped from a perch? Was there wind, or did the feathers drop straight downward? How high was the perch? Is there scat that can help you identify the predator? Are the feathers arranged in a ring on the ground, sometimes called a "fairy circle," as is often the case at hawk and falcon kill sites, or are they more randomly strewn around, as is often the case when the predator is a mammal? Has the carcass been moved, or was the entire bird consumed in one place? There is much to consider before moving in to look at the feathers themselves.

A long-tailed weasel ate this junco near a bird feeder. (NH)

Also consider the size of the bird that was eaten. The size of the prey says something about the size of the predator. If a sharp-shinned hawk takes a chickadee, there will be feather remains. If a fisher takes a chickadee, the bird will be eaten in its entirety. The remains of a larger bird, such as a grouse, would not rule out fishers, but it would rule out sharp-shinned hawks.

Moving in closer, an examination of the quills of individual feathers and clumps offers even more information. This is not an exact science, and there is variation on how quills are damaged even by a single animal or bird. Predatory birds, such as hawks and owls, tend to pluck their feath-

A house cat retired to the safety of a hay barn to eat this pigeon. Every feather was gripped with the teeth and pulled out, or plucked. (NY)

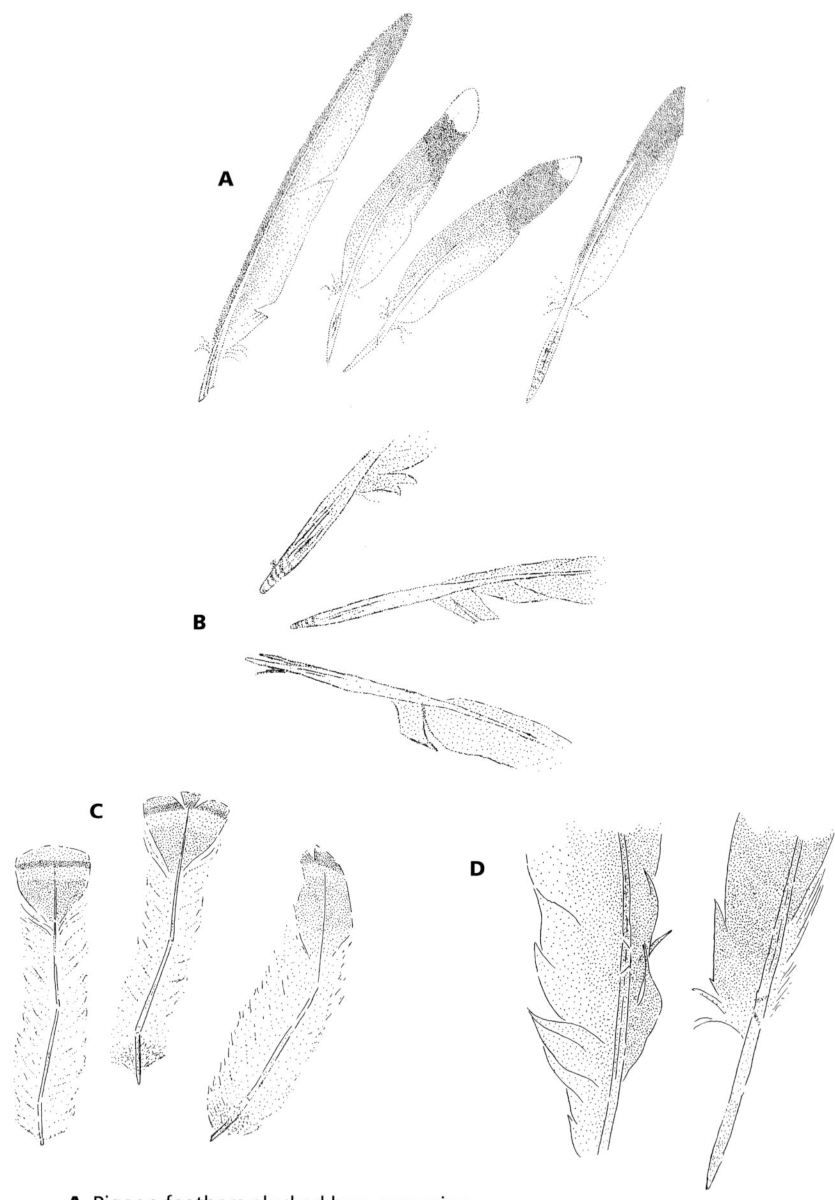

A Pigeon feathers plucked by a peregrine
B Ruffed grouse feathers plucked by a great horned owl
C Wild turkey feathers plucked by an owl, using the feet
D American crow feathers plucked by a red-tailed hawk—look for the V shape of the beak on the feather shaft

ered prey. The shaft of the quill is grasped firmly by the beak, and the feather is removed in its entirety, so the end of the feather extracted from the body will be intact. Occasionally birds will also pluck with the feet, grasping quills between toes. If the quill is wrapped under toe 2, over toe 3, and back under toe 4, the result will be two creases in the quill shaft. The area of the quill shaft that the raptor grasped will be damaged. In an extreme case, the shaft will be torn, but more often than not, the shaft will look relatively unscathed except for crease lines, which indicate the area where the shaft was squashed. If you squeeze the same area yourself, you will note a soft spot, which is exactly where the beak held the feather while plucking.

If the tip of the shaft next to the body has been cut, a mammal is responsible. The back teeth of such predators are carnassial teeth, which are used for shearing food into manageable chunks. In canines, these teeth are well developed, and this behavior is easily witnessed when domestic dogs are given a hunk of food. They slide the hunk to the back of the mouth and chew with only one side, using the back teeth. This is exactly what is found in the case of feathers. Canines, members of the dog family, will shear feathers off large, feathered prey. Large is relative among predators—what is large for a wolf and a gray fox differs. This is also true within other animal families, like the mustelids, or weasel family—large for a fisher is different from large for an ermine.

In the case of canines shearing feathers, you will often find clumps of feathers at a kill site. Sections of feathers are sheared off together, and the saliva of the mammal may cause these clumps to stick together. Look for the angle of the shear to be constant across the entire clump of feathers—this is crucial in distinguishing canines from other mammals. Weasels also shear off sections of feathers, but the length of the shear will be short, and if continued, it will vary in angle with each bite. This is why the junco feathers pictured on page 291 were sheared in a circular fashion—multiple bites were involved. Or the overall appearance of the sheared edge may be ragged. Bobcats, which have less developed carnassial teeth, leave much more ragged sign, and the clumps are much smaller than one would expect for such a large animal. The feathers appear to have been chewed through, rather than cleanly sheared.

House cats and foxes will also pluck feathers. Rather than damage to the quill, as on feathers plucked by raptors, the sign will be higher on the feather. The quill may show signs of handling near the

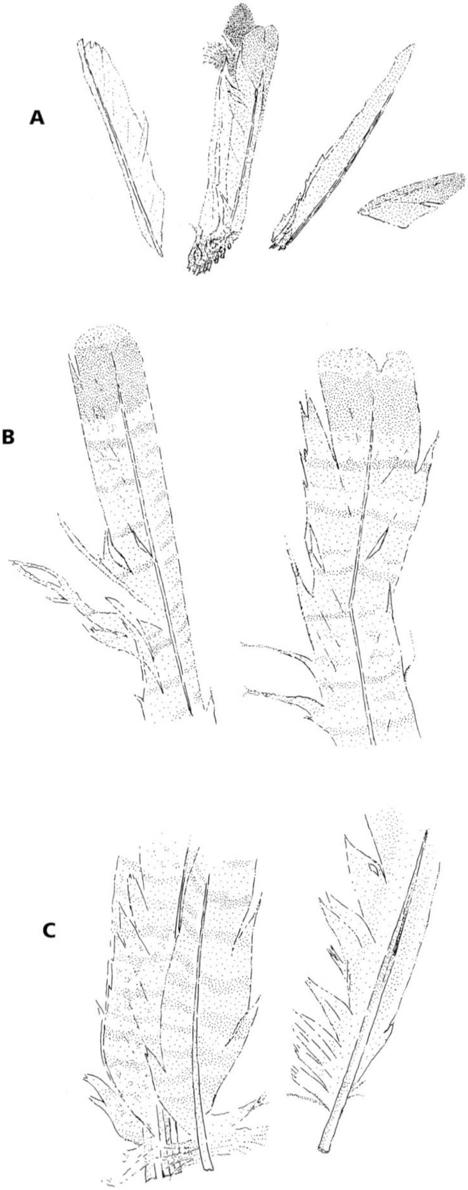

A

B

C

A Junco feathers sheared and cut by a long-tailed weasel

B Ruffed grouse feathers plucked by a red fox—note the holes in the plumage made by the canines

C Ruffed grouse feathers sheared by a red fox

Feather remains of a mourning dove eaten by a coyote. Note that several feathers have been plucked and others sheared. (NH)

top, but look for holes or impressions in the plumage itself, where canine teeth bit down and pulled the feather. I have found house cat kills where all the remaining feathers were pulled. At red fox kill sites, quite often you will find both bitten-off and plucked feathers. Tail feathers are often plucked by foxes, rather than bitten.

Harvesting seeds from the flower heads of burdock is a tricky business. Every year, goldfinches take the risk, and some become caught and remain there until they die. (Photo by Robert McCaw.)

Other Dead Birds

Birds are also found dead in the wild, without any signs of being fed on. They are often killed by impacts with human structures, poor weather, or in some cases, just bad luck. One such natural death is the slow starvation experienced by goldfinches that get their feathers caught and cannot escape from the Velcrolike structures of burdock plants, on which the birds feed.

Each of these birds was killed in the same night by the same 850-foot communication tower in Elmira, New York. (NY) (Photo by Bill Evans.)

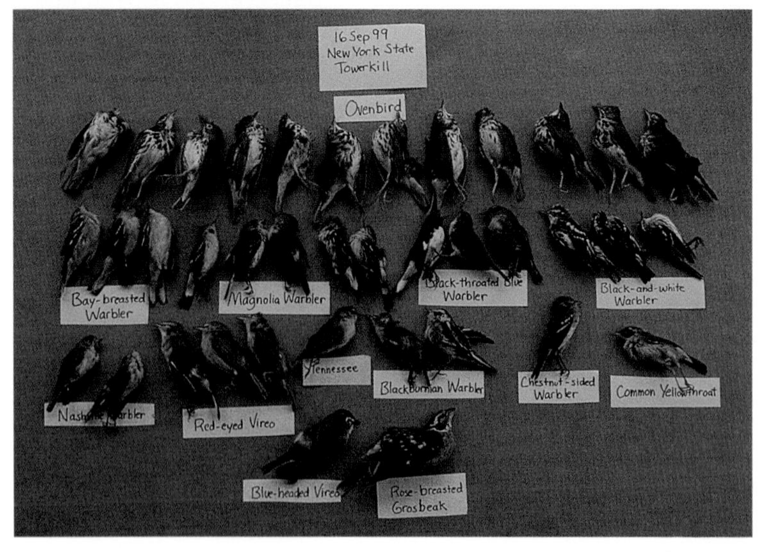

Some birds never leave the nest. (NH)

According to David Bird's compilation studies of bird mortalities in the United States, every year an estimated 500 million birds are killed by domestic cats, 80 million in collisions with windows, 57 million by motorists, and 1.2 million by television towers. Consider how many birds and small mammals could be saved by keeping cats indoors. Many birds are also strangled by fishing line or the plastic rings from six-packs, oiled, or choked by swallowing clear plastic bags.

In addition, untold numbers die of starvation, malnutrition, or cold temperatures in the winter. If a bird has not adequately fed before a cold night in the winter, it is likely to freeze to death, dropping from its roost without a sound. And many young birds never leave the nest or are bullied out by stronger siblings. The death toll for birds is naturally high—higher in some species than others. When we consider human-caused mortalities and loss of habitat as well, it is no wonder that populations of certain species are plummeting around the country. Awareness of the problem is the first step in an action plan of prevention and conservation. Ask yourself what you can do to help.

Nests and Roosts

5

I think of birds as feathered architects,
master builders and brilliant artists.
MARYJO KOCH, *The Nest*

E.M./M.E.

Probably one of the first nature treasures a child, whether growing up in the country, suburbs, or city, brings home to show his or her parents is a nest. I grew up deep in what would now be called the inner city and can still remember that at the age of three I was fascinated by the different nests at different heights in two enormous Seckel pear trees in our backyard. By the time I was four, I was spending endless—and fruitless, I might add—hours with a shoebox half propped up on a wobbly stick, which was tied to what I thought was a very long string. I myself was tied to the other end of the string, probably lying little more than 3 feet away in plain view on the concrete patch next to our back steps. Little pieces of white bread served as bait, carefully arranged under the box. I spent countless spring mornings trying to catch one of the seemingly hundreds of sparrows that swept through my yard throughout the day, momentarily a shadowy, racing canopy in front of the sun. I now know these brown city birds are commonly named house sparrows and are ornithologically called invasive and exotic, but back then I loved them with all my tomboy heart. In a place of endless concrete and asphalt, factories and rowhouses in blocks that stretched forever beyond the unseeable horizon, they seemed really free—they lived in the sky. My coauthor Mark, himself a country boy, also has wonderful stories of resolutely searching for warbler nests in fields in his native England while he was still very young.

Nests protect what is little—or at least try to. A tremendous variety of nests has slowly developed from the most ancient type, the original simple scrape form, which emerged roughly 135 to 70 million years ago. This wonderful diversity in nest building evolved over eons as a slow process of the accumulation of adaptations to promote particular species' survival in various habitats.

In the most likely scenario, as birds evolved from their cold-blooded reptilian ancestors to become warm-blooded creatures, they had to solve the problem of how to keep their vulnerable eggs warm. Probably the earliest solution was the simple scrape on the ground—a barely noticeable indentation, yet sufficient to keep eggs grouped together while the incubating parent's body provided the warmth.

The long process of cumulative adaptations called natural selection is driven by selective pressure because the survival of the species is slightly enhanced, generation by generation, through the selection of some characteristics rather than others to be passed on to succeeding generations of birds. The next evolutionary development of nest making was probably simply adding plant material to the bare scrape, an adaptation that over millennia evolved into raised platform nests, either on the ground or in trees, or, more

Young song sparrows stretch a nest to its limit. (NY)

rarely, floating nests in the waters of marshes with thick vegetation. Due to the pressure exerted by newly evolving mammalian predators on ground nesters, selective evolutionary pressure favored the survival of birds that created slightly more complex—and eventually, much more complex—nests. By adding height, materials such as grasses and reeds added to physical protection from predators due to increased inaccessibility and camouflaging, and the bulky nesting materials themselves also added insulating protection from the weather.

In all this unfolding evolutionary history, perhaps the most exciting part of the story for us has to do with biodiversity—the incredible array of different ways in which species of birds have solved the problem of "protecting what is little" to the best of their abilities, given all the complex opportunities and drawbacks of evolving environmental changes in myriad habitats that exist on the earth. And nests offer a story of personalities too, as some individuals are just better builders and more savvy in the selection of materials and site locations than others.

Nest Types

Based on Ehrlich et al.'s (1988) work, nest types can be broken down into nine basic categories defined by similarities in construction that relate as general patterns to specific bird families, although there are some exceptions to these generalizations. These categories will help you make sense in the field of the tremendous diversity of bird nests by organizing identification characteristics according to overall patterns exhibited by families of birds.

All of this stunning diversity represents how surviving species directly related to the challenges and possibilities presented throughout evolutionary history to birds. As birds evolved to respond to the challenges of predators, which were also evolving, as well as to sometimes dramatic changes in climate and to major changes in landforms and ecology, so too did their capacity for constructing astoundingly diverse forms of nests.

Note that these categories are adapted from Ehrlich (1988, p. xvi) and Dunning (1994). As trackers, be aware that there are considerable variations among these categorical lists within species, so that a particular species' nest may fall under two or more categories.

If you do find a nest, be aware that there are laws in the United States and Canada prohibiting the collecting of eggs, nests, and feathers, and protecting nestlings.

Scrape

Probably the oldest and simplest type, a scrape nest is often just a simple depression in the ground, so slight that it may resemble a natural indentation, yet deep enough to keep the eggs from rolling out of the area, either formed naturally or scratched out by the parent bird. A female least tern, for example, will settle herself into the sand and swivel her body around, using her breast to form the slight depression; the common tern prefers to "belly down" into the sand and kick it away with her feet. Usually, no nesting materials are added by the inhabiting bird, although some plovers may arrange a few pebbles or broken shell fragments at the bottom.

This nest type is typical of terns, plovers, and other shorebirds, some raptors, seabirds, and all nightjars. Some large owls will resort to this primitive nest form in the absence of large enough tree cavities or abandoned hawk nests; short-eared owls seem to specialize in this kind of "non-nest" nest. Dunning includes the marbled murrelet

A piping plover scrape. (MA) (Photo by Greg Levandoski.)

The platform nest of a mute swan. Note the manicured vegetation, all plucked for building. (MA)

in this category because they build nothing. Their nests are simple depressions in moss, but they are exclusively located high up off the ground, as high as 140 feet, and only on huge, old-growth Douglas fir branches in the Pacific Northwest. Murrelet nests have edges that are naturally built up by accumulations of scat in a rough circle around the nestlings.

Platform

Preferred by wetland birds, raptors, and most owls, the platform nest is an evolutionary development that, in a sense, moves the basic design of the ground-level egg-holding depression up in elevation and adds materials. In this type, certain bird families "relocate" the primitive scrape-type nest up either a little higher, as with ducks' "sideways throwing" and "sideways building" or pulling-in reed nests, or much higher into the trees to a better-protected location. This nest site is further enhanced, and the depression deepened by adding nesting materials such as grasses, sedges, reeds, or sticks and is built large enough for the bird to either clamber into or land on as it brings food for the nestlings.

Usually located in a marsh, a tall tree snag, or cliff, duck nests are built up by the pulling in of vegetation under and around birds' bodies, whereas the large raptors tangle sticks helter-skelter into what can become enormous and quite durable piles to form the nesting area, reusing and adding to the nest each year. Therefore, typically there is no uniform external form or shape of the platform; however, often there is an internal lining. Herons, for example, add small sticks and large leaves. Leahy (1982) notes that "a Bald Eagle's nest accumulated over decades could attain a diameter of 9 feet, a depth of 20 feet, and eventually weigh a ton or two." The enormousness of these nests makes sense if you visualize it having to support an 8- to 9-pound male and a 10- to 14-pound female eagle, as well as the nestlings.

Crevice

Crevice type nests include natural, unaltered openings in ledges, cracks in cliffs, narrow spaces between boulders, and even protected access points in human structures, such as gaps under the eaves of houses or barns. Taking advantage of this structurally safe location, the bird will either add nothing or use whatever material is necessary to complete the required nesting area.

Cup

The cup type nest is the classic bird nest we usually picture or have seen children bringing home to their parents. Probably the most advanced developmental form of the nest, from an evolutionary perspective, a cup nest is a deep half hemisphere, securely holding several eggs and closely embracing the female parent's body to add warmth during incubation. Made of twigs, grasses, rootlets, and other locally available natural and man-made materials, often reinforced with mud molded to the shape of the female's breast, it is usually securely woven to the branch or trunk on which it is mounted or suspended, or is so strongly woven that it can maintain its own shape if simply resting in a crotch or on a beam, supported from below. The inner layer is often made of fine grasses, moss, plant down, scraps of paper, and human or animal hair woven together to cradle the eggs and increase the insulation that protects the eggs during the incubation process.

The building of the classic robin's cup nest is in three stages: First a foundation of grass, straw, leaves, and even twine or string is woven, either on a tree limb or a man-made supporting shelf, and is

A common raven's cup nest. (ME) (Photo by Eleanor Marks)

continually molded by the female during construction to conform to her body's shape. Then pellets of mud are added to its inner wall, with the female rotating her body inside to plaster the walls, smoothing them like a potter working on the inside of a pot. Finally, fine grasses, moss, or feathers are pressed into the wet mud, creating a soft, ingeniously insulated lining. Because the robin's nest is statant, that is, supported from below and not suspended from rims attached to reinforcing branches, the mud layer adds crucial structural strength to prevent undue spreading.

As trackers and naturalists, all of us should be aware that while the form of the nest is quite consistent, materials selected for a cup nest will vary in response to what is available in the surrounding habitat. In her field sketchbook of nests, Maryjo Koch (1999) paints beautiful comparative examples of country versus city robin cups; her list of the "city" materials includes "Doublemint gum wrapper, Q-tip, straw paper, string, duct tape, paper, packing tab, Zigzag paper, business form, paper strips." Also, robins build anew each year rather than reusing the previous year's nest. Perhaps this fact along with the nests' adobelike buttressing for durability accounts for why they are so often found by children in the wild.

Saucer

The saucer type nest is basically just a shallower, simpler version of the cup shape, whose outer rim is usually approximately no higher than two times the diameter of the eggs it contains (the cup nest is typically at least several times higher than the eggs' diameter). Dunning includes the saucer nest within her cup category, as structurally the saucer is actually just a flattened or shrunken version of the classic woven cup. She illustrates this relationship, for example, by noting an amazing characteristic of hummingbird nests: The hummingbird's metabolism runs so high—its heart rate can be well over a thousand beats per minute while flying—that it has little tolerance for prolonged loss of body heat, due to either lack of food

Left: *A beautiful spherical nest built by a cactus wren. (TX)* Right: *The spherical nest of a black-billed magpie. (CO)*

calories or weather. Adults can go into a state of torpor to cope with these stresses for at least limited times (eight to fourteen hours or overnight) by deliberately slowing their heart rate to under fifty beats per minute and dropping their body temperature by 30 degrees to conserve energy. To protect their far more vulnerable eggs, which are not capable of torpor, the female hummingbird weaves an extremely fine cup that fits almost as tight as a snug woolen sock around herself and the eggs. The genius in her cup design is in its flexibility: It slowly expands as the eggs hatch and nestlings grow, until by the time her fledglings fly free, it is stretched out almost flat, transformed into a saucer.

A spherical dipper nest. (WY)

Spherical

Spherical nests are ball-shaped sphere versions of the typical cup nest, usually well roofed, with a small entrance at one end. In the deserts of Texas, it almost seems as though every chain cholla cactus that has grown higher than 2 1/2 feet (76.2 cm) harbors an active or abandoned cactus wren nest. Typically, cactus wren nests are found deep within the plant, an oblong construction of grasses protected by the cholla's spines. In the same desert areas, the spherical nests of verdins can be differentiated from those of cactus wrens by the fact that they are typically constructed on the outer branches of shrubs and are made with twigs rather than grasses.

One intriguing spherical nest is that of the black-billed magpie, whose nest can at first glance be easily mistaken for that of a wood rat. Just as wood rats do, magpies build nests concealed by outer branches in medium shrubs, or in short trees in sagebrush and desert country. I once made the wrong assumption about such a nest from a distance. A closer inspection, however, revealed the more neatly constructed nest of a magpie, made from twigs of similar diameters rather than a jumble of different materials, as is often the case with wood rats. The outside was clean, unadorned with cow dung, rocks, or trash, which is typical of rats. And there were no droppings below the nest, which is always the case with wood rats, which have noticeably littered front door stoops. Even if I hadn't been convinced after noting all the differentiating clues, a magpie helped confirm my identification by suddenly emerging from an unseen entrance. It flew 15 feet away and landed between sagebrush plants, eyeing me warily.

Although we categorize dippers and ovenbirds under spherical nests, some authors place them in a separate nest type termed oven.

Pendant

The pendant type nest is a tree nest, woven into an enclosed teardrop pouch that is suspended on the most extreme outer thin branches of a large shrub or tree, well elevated above the ground. This provides protection from most predators, which would likely be too heavy for the slight support structure. The insides are designed much like the cavity type of nest, with a small aperture entrance near the top leading to the lower chamber of the woven, udderlike nest. The sum effect is that of a cradle suspended and gently swaying high in a tree. The nest is woven out of grass, hair, string, or whatever other natural or man-made materials might be available, producing a light but well-insulated, double-walled pendulous nest.

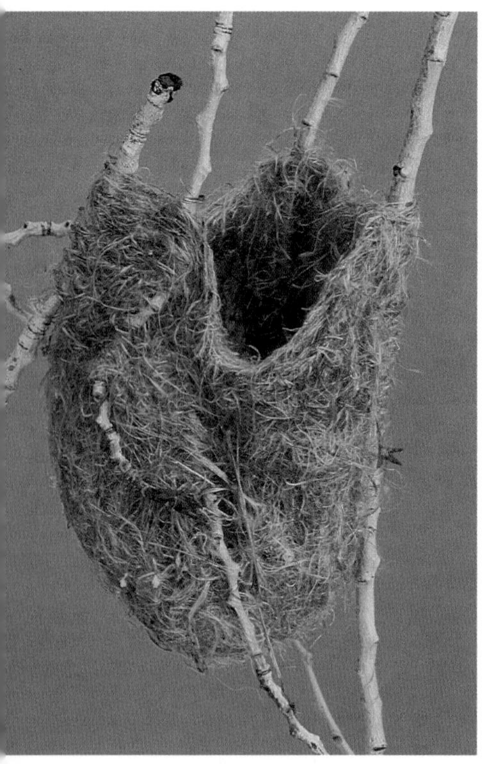

A Baltimore oriole nest. (CO)

Cavity

The tremendous variety of cavity nests and their importance to wildlife illustrate an absolutely essential survival requirement that

Left: ***The nest cavity of a hairy woodpecker in a red maple. Holes are generally a bit higher than wide, 2¼ to 2⅝ inches (5.7 to 6.7 cm) high and 2 to 2⅜ inches (5.1 to 6.1 cm) wide. (NY)*** **Right:** ***A ladder-backed woodpecker cavity in sabal palm. (TX)***

all these species of cavity builders must have in abundance: dead trees, which unfortunately many people thoughtlessly cut down, not realizing how precious they are. As trackers, we honor these aged specimens by calling them standing snags and revere them in all stages of their decomposition. Each stage and layer of their rot opens up more possibilities for more species to exploit this kind of weather-secured, many-floored residence hall and hammer their way in or, in the case of secondary users, chew their way in, or simply crawl up, fly in, move in, and take possession.

Cavities are either created, as by woodpeckers, or discovered within old trees. They have a small aperture for an entrance and a cup-shaped structure inside, either naturally formed or molded with

Left: ***A "candlestick" and cavity of a red-cockaded woodpecker. (SC)*** **Right:** ***The nest cavity of a red-bellied woodpecker in cabbage palm. Entrances tend to be between 1¾ and 2⅜ inches (4.5 and 6.1 cm) in diameter. (FL)***

mud or other materials by the inhabiting bird. Most cavities made by a bird are made for roosting purposes or mating rituals, rather than actual nesting. Many woodpeckers use cavities as part of the mating process. A male excavates a cavity and then drums from within the tree. The female, if she is satisfied with the location, may come and return pecks from the outside. She may still reject the cavity and force the male to go create another one. Quite a few cavities may be rejected until one is chosen.

An interesting cavity species is the red-cockaded woodpecker, which sadly is in great decline. Their cavities are unique in that not only do the birds excavate a nest, but they also peck smaller holes all around the main hole, creating a sap flow, which can drip all

A "saguaro hotel," with gila woodpecker cavities in the central area and gilded flicker holes near the top. Not all cavities are active simultaneously. (AZ)

about the entrance and down the tree, thus earning them the name candlesticks. Researchers do not agree on why red-cockaded woodpeckers create such sap flows, although a popular theory is that they deter nest predators.

Cavities in saguaro cactuses in the Southwest are common. Both gilded flickers and gila woodpeckers make these cavities for nesting, but they often choose different locations on the cactus. The stouter bills of the gilded flickers allow them to cut cavities through the wooden ribs near the top of the cactus where the ribs converge. Gila woodpeckers stay at midlevel on the cactus, where the ribs are separated enough to cut a cavity between them.

Cavities in saguaros are cut out by these birds the year before they are inhabited. The excavated cactus secretes a fluid that hardens into a scab, thus preventing water loss, which could kill the cactus, as well as waterproofing the inside of the nest cavity. This scab also creates a wonderful sign that a tracker can search for.

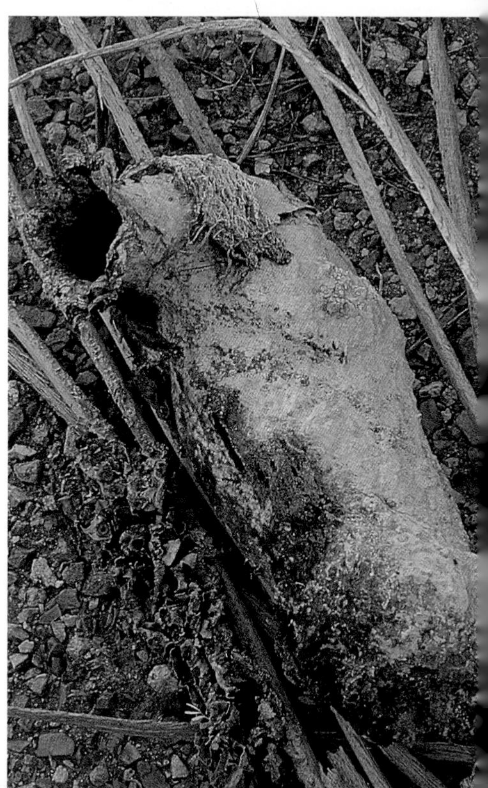

A saguaro boot, this one created by a gila woodpecker, as the entrance is small. Gila woodpecker cavities tend to be 1¾ to 2 inches (4.5 to 5.1 cm) in diameter. (AZ)

Saguaros live a long time, unless disturbed by bulldozers. Between two hundred and three hundred years old, a saguaro finally succumbs to age and dies. The standing cactus slowly begins to disintegrate and eventually snaps in half, loses limbs, and falls completely over, continuing the rotting process on the ground. The watery flesh rots first, and you can find the wooden skeletons of these cacti sprinkled throughout the desert areas within their range. Oddly, the scab created by the cavity lasts much longer than the flesh of the cactus and can be found lying within or next to the ribs of the long-dead cactus body. These cavities are called saguaro boots because they look like large shoes, with the cavity entrance forming the opening for the foot. Saguaro boots are difficult to find along trails in the Southwest, as they are a prized collector's item. But they are not difficult to find if you leave the trail and escape the majority of human traffic. The destruction and theft of saguaros are currently a threat to the species, so please be sensitive in your explorations.

Burrow

Similar in function to cavities but located in the ground, burrows are long, narrow tunnels leading to a larger nesting hollow at its end. Some are excavated by the birds themselves, such as kingfishers; oth-

The burrows of bank swallows, along with the larger entrance to a burrow of a belted kingfisher at the upper left. Entrances to swallow holes tend to be 1½ to 2½ inches (3.9 to 6.4 cm) in diameter. (NH)

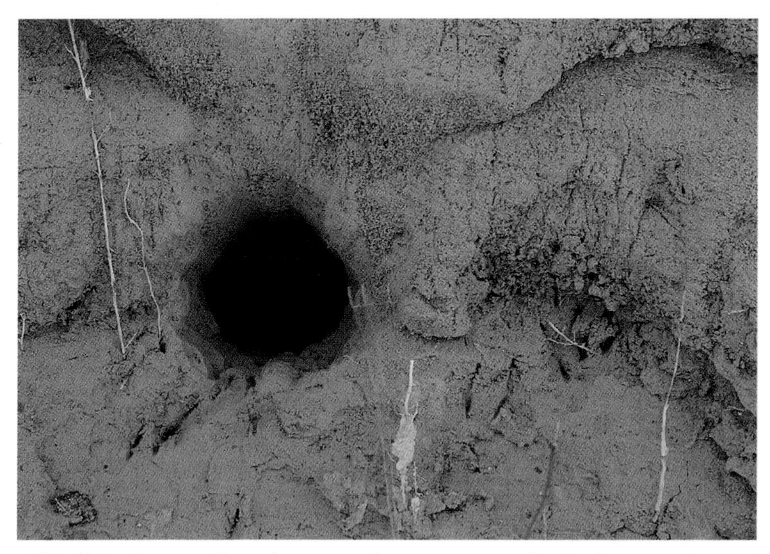

A kingfisher burrow. Note the two paths, or ruts, worn by the feet entering and exiting. Entrance holes are generally between 3 and 4 inches (7.6 and 10.2 cm) in diameter. (NH)

ers have been excavated and then abandoned by mammals. Northern rough-winged swallows actually steal burrows from bank swallows and kingfishers that worked hard to make them. Some tunnels, cut in soft dirt or sand, can be 15 feet long before opening out into the nest, which is a lot of digging for bills and narrow zygodactyl feet. This also means that birds that are born in burrows spend their first weeks in total darkness.

Burrowing owls often usurp and enlarge small mammal burrows. The authors had wonderful

The burrow of a burrowing owl, with pellets, prey remains, and droppings littered about the throw mound. Entrances range from 3 to 4 inches (7.6 to 10.2 cm) high by 4 to 7 inches (10.2 to 17.8 cm) wide. (CA)

looks at burrowing owls living amid the round-tailed ground squirrels of Southern California. The burrows of owls were easily differentiated from their mammal counterparts by the prey remains, scat, tracks, and pellets that covered the "throw mounds" just outside their burrow entrances. These mounds are often used for roosting.

Mud Use in Nests

Mud is a constant draw for the tracker, with the myriad mysteries that appear in mud puddles and along streambanks across our country. During the nesting season, many birds use mud. This is also the best time to find the tracks of swallows, which will be collecting mud for building purposes. The nests of barn swallows fall under the cup category, while the nests of cliff swallows are in their own category. In my studies, cliff swallows generally collected wetter mud than barn swallows. But this was not always the case, as limitations in supplies also governed what was collected. I've also seen blue jays collecting pine needles from mud puddles, and robins taking beaks full of mud to line their nests. The start of nesting season is an especially good time to check mud puddles.

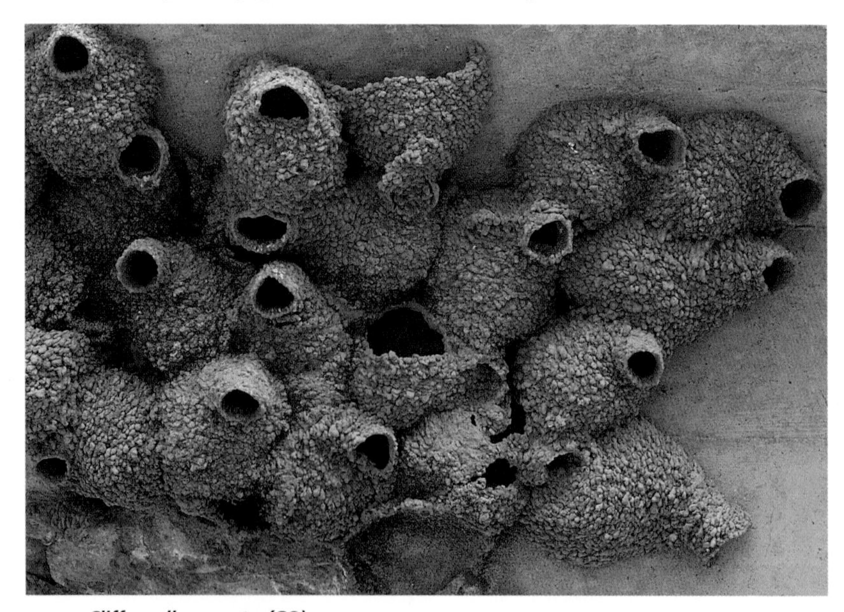

Cliff swallow nests. (CO)

Moist mud collected by cliff swallows. (CO)

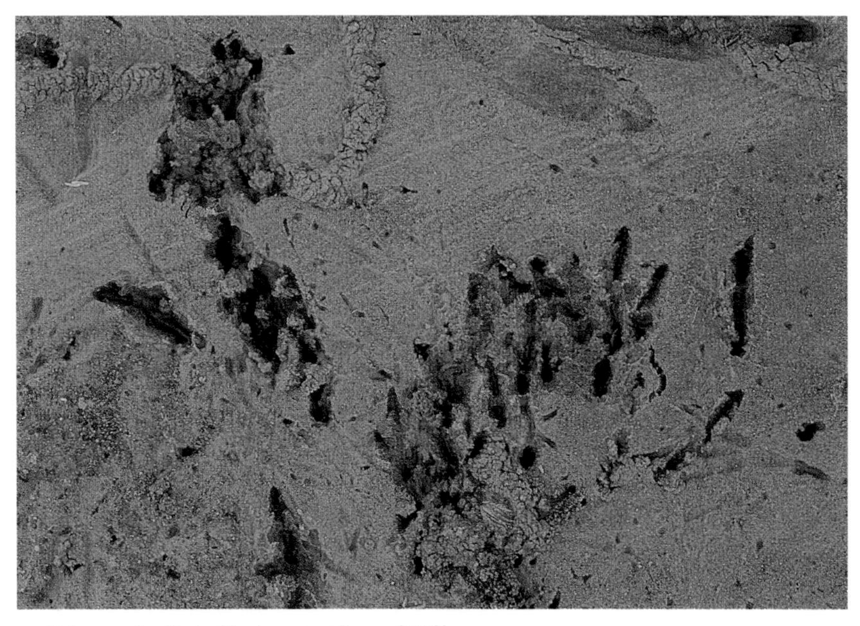

Drier mud collected by barn swallows. (WY)

Roosting

Roosting generally refers to a safe place on the ground or in trees where birds rest and sleep, sometimes in the day and sometimes at night, sometimes in large groups and sometimes individually. Owls, for example, roost individually while resting and digesting their latest morsels, leaving their ejected pellets beneath their favored roosts—which are often used repeatedly—before flying off to hunt again. Such sign, as well as accumulations of scat, called whitewash, can serve as clues to the location of roosts.

For some species, roosts are set locations used habitually for many decades; for others, such as migratory birds, they are essential rest and feeding areas at designated layover locations on their long flight paths, places where they must be able to safely refuel calories in order to complete their treks. On the other hand, birds may just perch alone on a branch or huddle in groups under drooping evergreen boughs in winter because heat is retained under thick evergreen branches. Many species also use cavities in snags for roosting.

A winter roost of a downy woodpecker. (NY)

Secure resting areas for gulls, terns, cormorants, shorebirds, and pelicans are called *loafing areas* and are always hidden from disturbance. They are often located in interdune swales, far from human traffic, and also in slightly elevated areas like salt pans far out in the middle of freshwater and saltwater marshes.

Some roosts look like nests. Downy woodpeckers, for example, create between four and eight winter roosts in rotting snags; from the outside, they look just like

A downy woodpecker winter roost, carved into a root wad of a fallen tree. Cavity holes range from 1¼ to 1⅝ inches (3.2 to 4.1 cm) in diameter. (NH)

nesting cavities. This is especially so in northern areas, where winter roosts may be excavated as well as lined for added protection against the elements.

Some birds actually roost in snow. Ptarmigan, grouse, and snow buntings are among the birds that are regularly found snow roosting. These birds either sit and wait patiently for a snowstorm to cover them or find a fluffy snow patch and fly straight into it, turning 90 degrees when just under the surface and settling down at a safe distance from the entrance. That extra turn is crucial; it helps buy the time, even just a second, needed to escape should a predator come in the entrance created by the bird. Birds always roost in the snow facing the wind during snowy gales.

Many years ago, on a clear morning after a heavy snowfall, I was walking in the woods of New York. The sun was bright and the mood was light as I strolled in search of the local fisher (a large member of the weasel family). Then, with a sudden flap of wings, a grouse exploded from the snow just where my foot would have fallen. Nigel, a young lad with me, fell backward in fright, and I'll admit I jumped too. When we stopped laughing and calmed down a bit, the sign of a long and patient night could be seen in the hole the grouse had created: A large pile of droppings rested at the bottom.

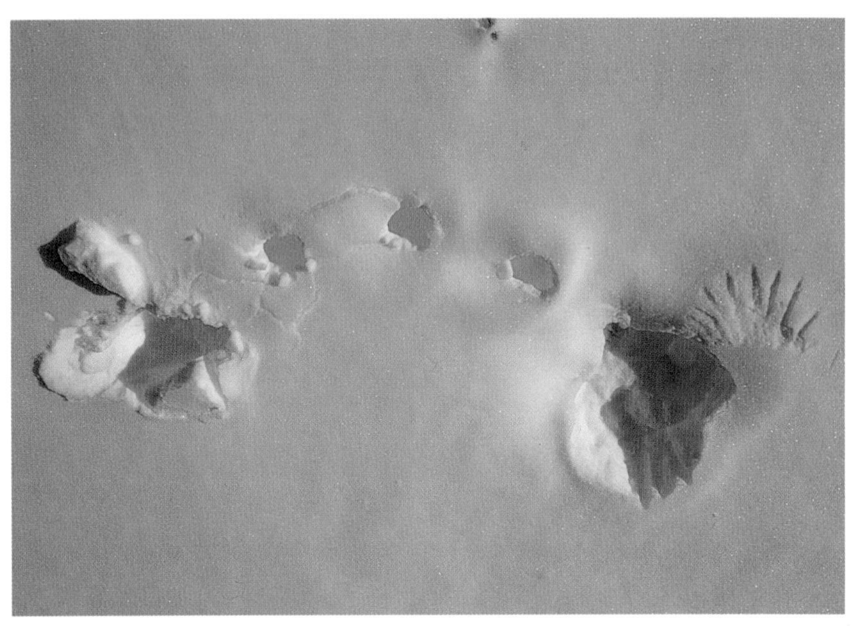

The snow roost of a ruffed grouse. (ME)

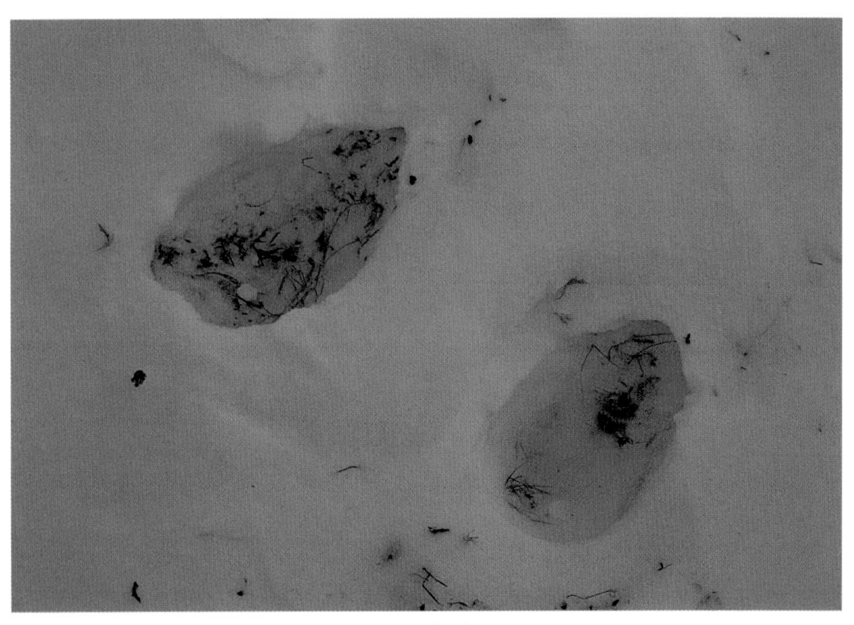

Snow roosts of white-tailed ptarmigan. (CO)

A great blue heron rookery in morning mist. (MA)

While it's obvious that nests are crucial to the reproductive life of birds, roosts also play an important role in the breeding phase of this cycle. When a roosting site is selected primarily for breeding purposes, however, it is usually given the more specific name of colony, rookery, or heronry. Birds that breed in close company with other birds of their own species are described as colonial, a pattern that provides group protection against predators. Rookery is an approximate synonym borrowed from Europe, where it refers to the colonial behavior of a corvid called a rook and is now applied in this broader sense. Heronry is a shortened form of heron rookery. Leahy (1982) observes that colonial behavior appears to offer a more successful form of security from predators than do the mating and nesting behaviors of individual birds, and "consequently many such species lay fewer eggs than do solitary nesters without diminishing breeding success."

Secondary Users

The moment the builder or excavator, also termed the *primary user,* abandons a nest or cavity, there is often another creature ready to take advantage of its hard work. These animals and birds are called *secondary users.*

White-footed mice are notorious users of old cup nests; mice are expert climbers and can be found nesting 30 feet above the ground. Nests are used by mice in two ways. Most often, nests in the Northeast are converted into storage bins for acorns, beechnuts, rose hips, or other storable food sources. These stores are often pilfered by other birds and animals, such as red and gray squirrels. Mice also may convert an abandoned nest into winter or nesting quarters. I have not found many such nests in my travels. The last such case was several years ago in a coastal area of southern Maine, where I was studying the effects of large deer herds in regard to browsing sign. As is often the case, I was distracted from the project at hand by a red fox trail, which I just had to follow. The fox looped a field, which at the far end held a barberry shrub. In the center of this shrub was a spherical bird's nest—or was it? No, upon closer inspection, it appeared to be a cup-shaped bird's nest, with a circular roof addition. Suddenly, before I had time to consider the implications, a white-footed mouse appeared from an entrance I hadn't noticed at the bottom of the nest. We stared at each other for only a moment before the mouse deftly ran straight down the ladder of crisscrossed branches, without ever breaking stride, and disappeared down a small hole in the snow.

This nest is filled with bittersweet fruits, which were harvested and stored by a deer mouse. (ME)

The winter quarters of a deer mouse, which added the roof to enclose a bird nest. (ME)

The birds that create cavities do a tremendous service for ecological communities, as so many birds and mammals rely on cavities for nesting, raising young, storing food, and surviving cold and stormy nights; this is why snags are so important. Flying squirrels and white-footed mice often take up residence in abandoned woodpecker cavities, and many birds also use abandoned cavities for nesting, including bluebirds, wrens, tree swallows, kestrels, screech-owls, and starlings.

Ehrlich and Daily (1998) noted that seven different bird species raised young in the abandoned cavities made by red-naped sapsuckers alone: "tree swallows, violet-green swallows, house wrens, mountain bluebirds, mountain chickadees, northern flickers, and a Williamson's sapsucker pair." They also said that the red-naped sapsucker produced many more cavities than other woodpeckers in the area of Colorado they studied and therefore should be considered a keystone species, as so many others depend on these cavities for their existence.

In saguaro country, the abandoned cavities of gila woodpeckers and gilded flickers allow elf owls, screech-owls, kestrels, tree swallows, purple martins, starlings, and wrens to have secure places to raise young.

Feathers

6

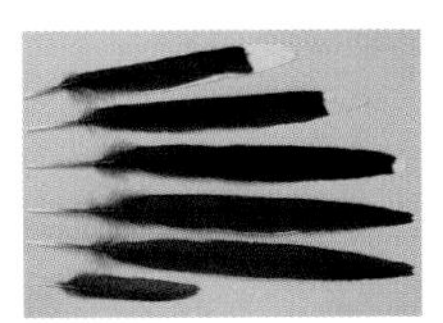

E.M.

Birds are the only living or modern animal with feathers. Although there is still debate among experts, many believe that birds as a class evolved from reptiles. Reptiles' scales and birds' feathers are both composed of keratin, an insoluble protein that also makes up their bills and nail sheaths, as well as our hair and nails. And there are other similarities. Scales and feathers both serve as an outer layer to protect the body from external damage, rain, snow, sunburn, and temperature extremes of heat and cold. Scientists have not yet found fossils of all the intermediate stages of adaptations that would document tiny, incremental physical changes as they unfolded over time from a pure scale to a fully formed feather, changes that could definitively show us how this evolutionary leap from a completely scaled reptile to a completely feathered bird happened. Up until very recently, it was largely agreed that the earliest fossil of a bird species was discovered in 1857 in Germany. Dated from approximately 138 million years ago, during the Jurassic period, and called Archeopteryx, meaning roughly "beginning wing," this specimen already had fully developed feathers, but they were embedded over a skeleton that was still very reptilian.

Evolutionary theory currently speculates that the missing links, if we ever find them, would probably tell the story of how a tree-climbing reptile gradually developed the capacity for passive gliding, in a form anatomically similar to flying squirrels, and that somewhere during this process of evolution, its reptilian scales gradually elongated into insulating feathers to protect it as it became increasingly warm-blooded and more of a flyer than a glider.

Though their evolutionary origin remains a mystery, the usefulness of feathers to the tracker is clear. Often, based on feather type, tip shape, and color, combined with your knowledge of local species, their preferred habitats, and life cycles of breeding, molting, and migration, you will be able to use feathers found in the field as important signs for species identification. Be aware, however, that there are laws forbidding the collection of feathers, unless you are a state-licensed educator or a member of a Native American tribe.

The possible variations among feathers are enormous, given different species, different types of feathers, different colorations, the age and sex of the bird, and the particular season in its life cycle. For example, winter, or basic, plumage often differs from breeding, or alternate, plumage. Therefore, positively identifying a species from a single feather you discover in the field can be a challenge. However, some knowledge of each of these categories, combined with what you have learned through tracking about the specific birds in your area and the information provided in this field guide, will help you to successfully identify many of your finds. The plates in this chapter illustrate the feathers of many North American species, but it's best to begin by narrowing your study to the bird species commonly found in your area in a particular season and focusing on family characteristics.

Structure and Types of Feathers

Although some books refer to all feathers as contour feathers, we will use the term to refer to the feathers that hug the bird's body, covering the downy layer closest to the skin. The contour feathers give the bird its characteristic shape and provide insulation and aerodynamic effect. Other feathers extend beyond the body into the wings and tail, serving as flight feathers.

At the lower end of each feather is a hornlike shaft called the inner quill, or calamus. This shaft has a tiny hole at its base, called the inferior umbilicus, and a second hole farther out where the wispy branches begin, called the superior umbilicus. While growing at a rate of almost 1/4 inch a day, and embedded in a follicle similar to the root of human hair, the feather is fed via blood vessels through this lower hole. Once the feather is fully grown, the hole closes and the feather technically dies, meaning that it is no longer made up of living cells.

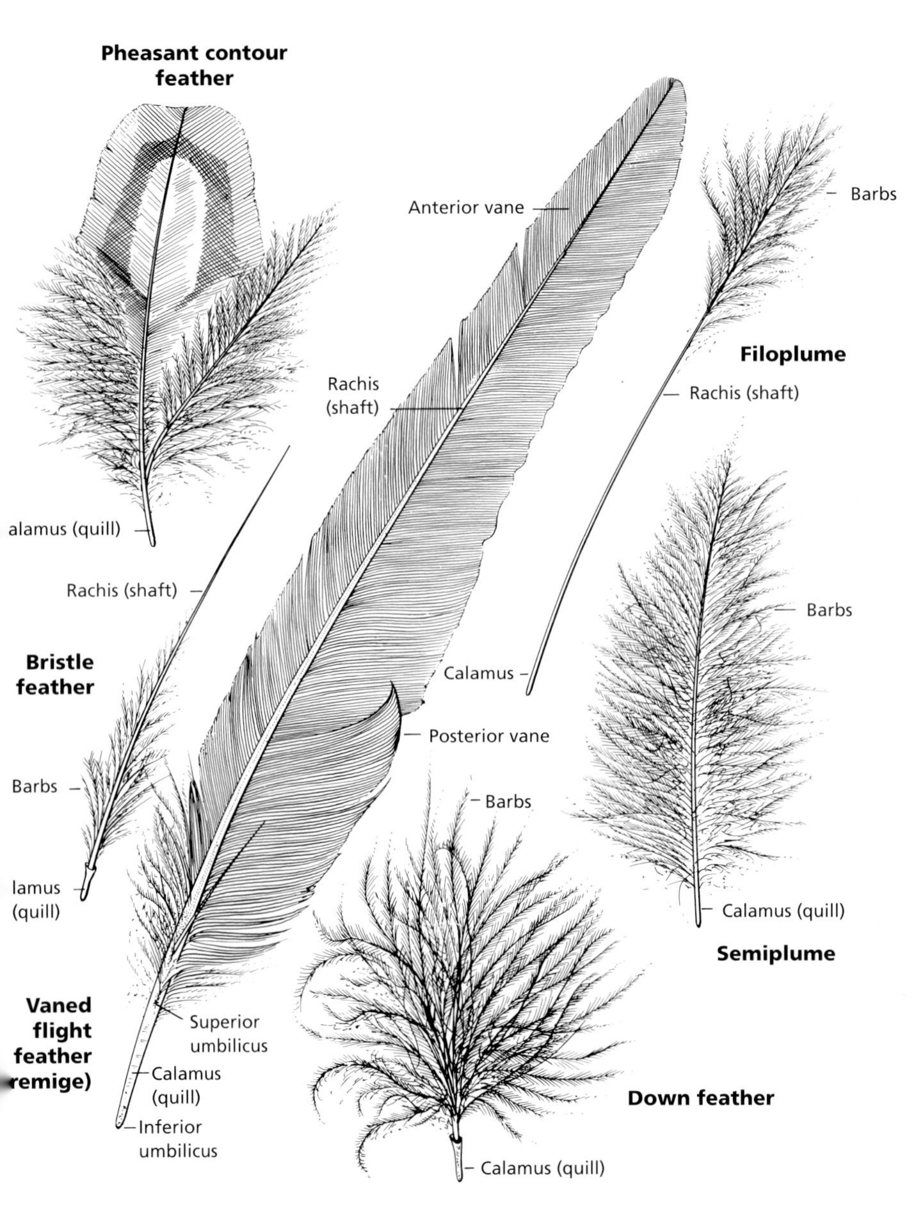

The major types of feathers. (Used by permission of Yale University Press.)

The mid to outer horny quill, or rachis, provides structural strength, running from the superior umbilicus all the way to the outermost, or distal, tip. Although it is round at the base closest, or proximal, to the body, it develops a groove on its underside farther out where the feathery vane, or web, begins; there it becomes flat on its sides, thus assuming a hollow, extremely strong, yet flexible boxlike structure. The initial wispy branches are fluffy barbs and barbules, called plumaceous, meaning plumelike. The farther, solidly connected vanes, called pennaceous (think of a quill writing pen), are stiff, flat, and interconnected, divided by the quill into the two flat, strong, yet flexible webs so essential to flight. If you viewed a feather under an electron microscope, you would see tiny hooklets on each strand of the thousands of barbules. This structure is so amazingly complex that a single 6-inch pigeon's flight feather has 600 barbs on either side of its quill; from each of these extend 275 barbules, and on these approximately 660,000 barbules are millions of microscopic interlocking hooklets—all this on every feather.

The hooklets reach out and grasp the adjacent barbules on the next strand, which is why when a feather vane becomes split, a bird usually can easily preen it back together again by drawing the feather through its bill and rehooking the barbules. You also can do this with most damaged feathers by simply pulling them repeatedly through your fingers while gently squeezing; the vane should almost automatically zip back together again. This seemingly magical ability of feathers to be reconstructed is essential to birds' survival. They are constantly preening damaged feathers, cleaning them in a variety of ways, from water baths to dust baths; using oil from the uropygial, or preen, gland, located on top of the rump where the tail begins, to preserve feathers, to reduce abrasion, and thus promote feather life; and diligently rearranging feathers that have become twisted out of place—all aspects of an unceasing task that guarantees their protection from the elements and their ability to float. This fundamental insulating capacity is destroyed by pollutants such as oil slicks, which mat the feathers, seeping through and coating the bird to the skin, thus destroying vital air pockets and possibly leading to hypothermia and drowning.

These basic elements are common to all feathers; different particular types are simply variations or specializations of these general fundamental structures. Contour feathers are usually identifiable by their shape. The webbed vanes are most often symmetrical, of equal width on each side of the quill, and usually broad at the outer tips. Flight feathers are asymmetrical and tapered in varying degrees toward the outer tip, depending on the bird's family, wing type, and preferred form of flight. (See the illustration on page 328.) The leading, or anterior, edge of a flight feather, which cuts through the wind during flight, is usually the narrower side, and the trailing, or posterior, edge vane is usually the wider side. The flight feathers of owls are easily identifiable by their unique soundproofing on the leading edge, surface, and trailing edge—soft, velvety serrations that reduce wind resistance to muffle the sound of wingbeats and produce the silent flight needed for successful hunting.

The innermost down feathers have a very flexible and thin central shaft, which often does not extend beyond the calamus, the part of the quill closest to the body. They are short and fluffy and serve exclusively for heat conservation and flotation. When fluffed out, these tiny feathers are remarkably efficient at holding trapped air to thoroughly insulate the bird from lower temperatures and provide buoyancy to waterbirds.

There are other, more specialized, types of feathers as well. Semiplumes are an intermediate form between contour and down feathers, having a central, thin quill but only fluffy barbs. They lie beneath the contour feathers, providing insulation and enhancing smooth, aerodynamic body contours. Filoplumes are like long thick hairs with tufts of feathers at the outer ends. Rich in nerve endings at the base, they enable a bird to sense when it is being touched, which down and contour feathers can't detect, and when flight feathers are out of position or damaged and need to be preened. They may also possibly aid in controlling movements of feathers during flight. Bristle feathers, the reverse of filoplumes, have sparse tufts located at the base and stiff, bald stems extending upward. These may serve a function similar to that of cats' whiskers. They are found around the eyes, as in northern harriers, or nose and mouth, as in nightjars and flycatchers, where they may enhance insect sensing, and in woodpeckers, where they protect the nostrils from sawdust. Powder feathers have barbs that slowly disintegrate at their outer tips into a fine, scaly talc distributed by preening, which aids in waterproofing and protecting other feathers.

The major types of feathers you are most likely to find in the field are wing flight feathers, also called remiges, which include the outer wing's primaries and the inner wing's secondaries and tertials, which

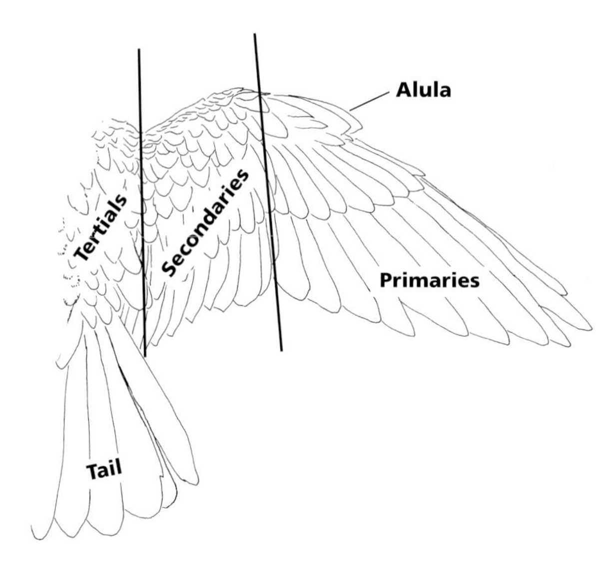

Prairie falcon—right side schematic of feathers.

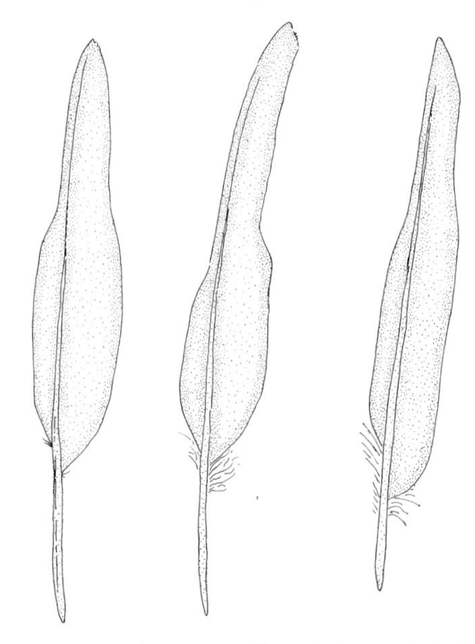

Feather shapes (from left to right) of a Canada goose, sharp-shinned hawk, and common raven. Note the different distal shapes—the formation of "slots" to aerodynamically streamline the upstroke.

fill the gap between the wing and the body when the wing is fully outstretched, as well as tail flight feathers, or rectrices. It becomes easy to distinguish primaries, secondaries, and tail feathers from one another as you gain field experience. A primary tends to have a much thinner leading edge than a secondary, and therefore the webs on each side of the quill are noticeably unequal, with the trailing edge usually much wider, and for most of their length the quill is generally much straighter. Primaries are often angled at the outermost tip so that despite being arranged side by side, they can be positioned to create gaps, called wind slots, to promote aerodynamic flight. They are located from the midwing outward and numbered in that direc-

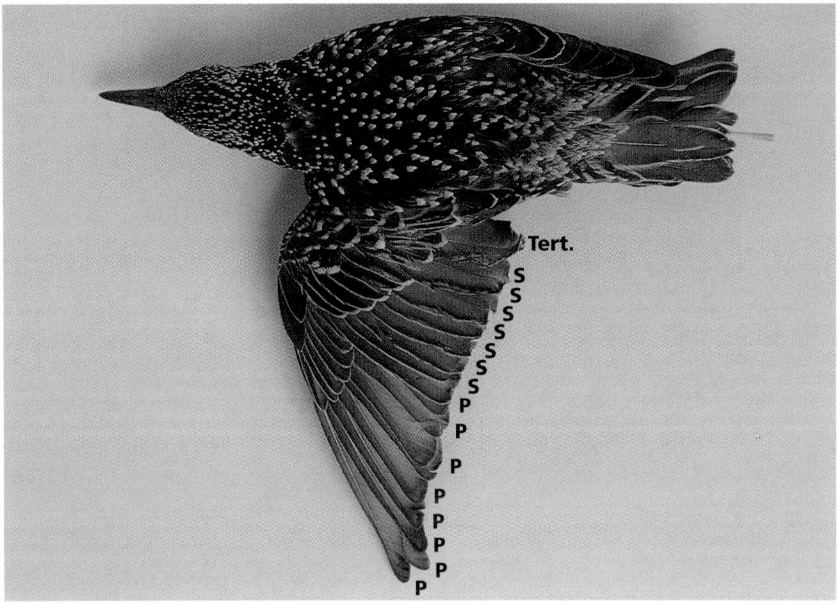

European starling skin and wing. (A complete set includes ten primaries and nine secondaries.)

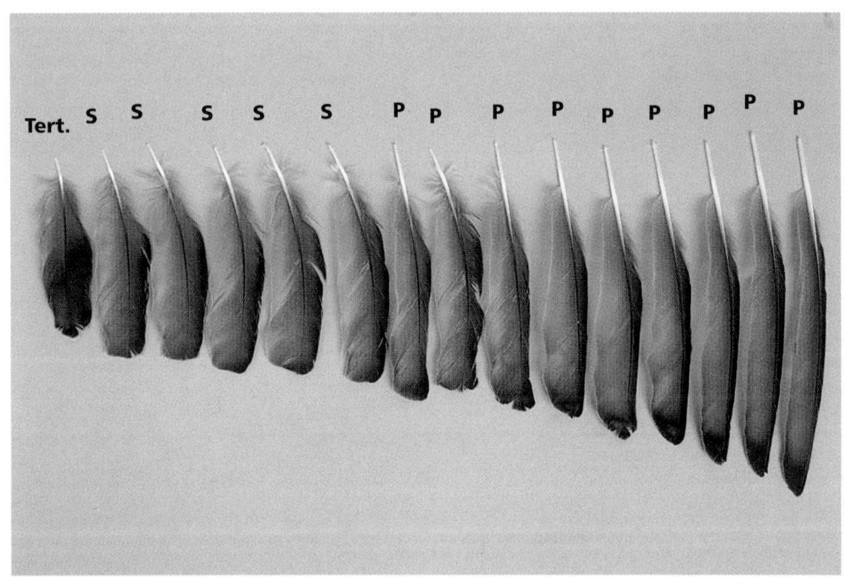

Individual feathers of a starling wing. From left to right: *tertial, secondaries, and primaries.*

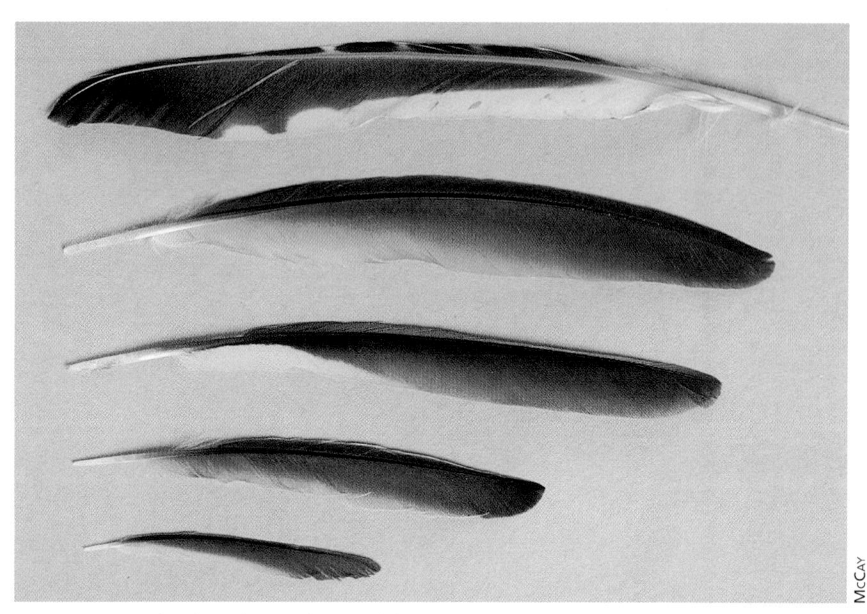

Primary feathers (top to bottom) from a northern flicker (red-shafted), yellow-billed cuckoo, blue jay, cedar waxwing, and yellow warbler.

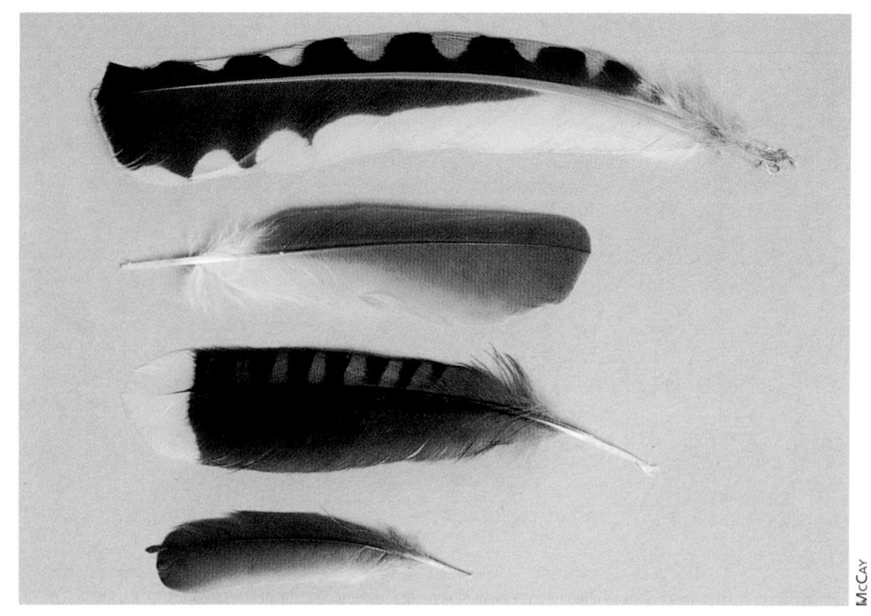

Secondary feathers (top to bottom) from a northern flicker (red-shafted), yellow-billed cuckoo, blue jay, and cedar waxwing.

tion. Most perching, or passerine, birds have ten; some songbirds, such as wood warblers and tanagers, have nine. The maximum is twelve.

In contrast, the webs on each side of the quill of a secondary are usually equal in width, and the quill has a noticeable curve. Secondaries also tend to have broader tips, whereas the outer primaries' tips are sharper. They are located from the midwing in toward the body, where tertials may close the gap, and are numbered in that direction. The number of secondaries is more variable depending on the length of the wing: six for hummingbirds, nine for passerines, thirty for the albatross, which needs a very long inner wing to sustain long, gliding flight at sea.

The quills of tail feathers have a unique characteristic that is very reliable for identification: When viewed from the side, the calamus is bent where the tip enters the body, almost like the angle on a gasoline pump nozzle. The number of tail feathers is not necessarily related to family characteristics, and can vary from eight for the groove-billed ani to thirty-two on a common snipe, with the average being twelve, numbered as pairs from the center of the tail outward. Most of the entire length of the quill of a tail feather, except for the proximal end, tends to be straight and strong, although there are several major differences in tail types, based on patterns of placement and shape of feathers. These

McCay

Tertials (left to right) from a cedar waxwing, blue jay, yellow-billed cuckoo, and northern flicker (red-shafted).

Red-bellied woodpecker wing: Compare to schematic of wing on page 328.

Red-bellied woodpecker tail.

McCay

Barn swallow skin—note tail.

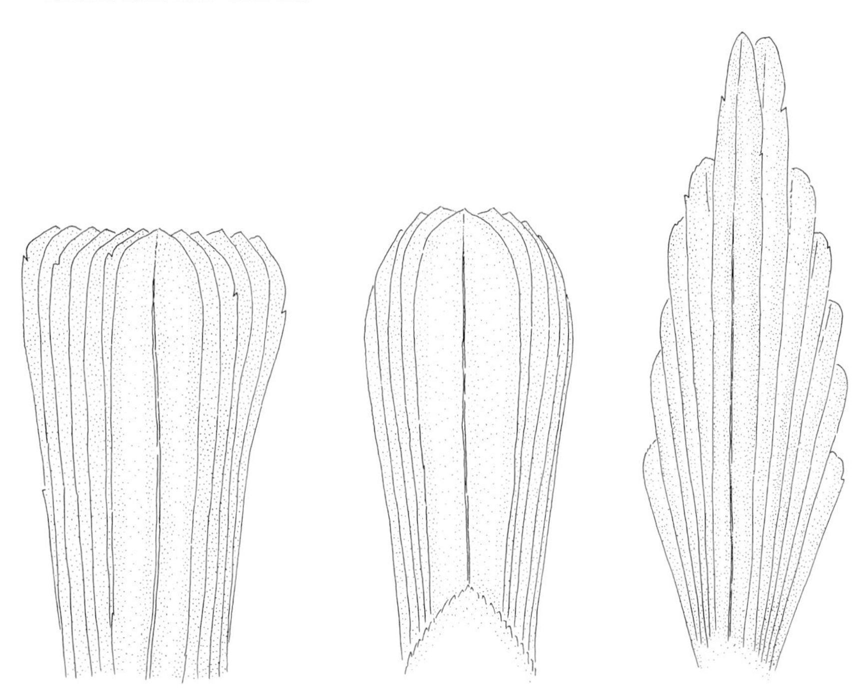

Left: *square nuthatch tail;* Center: *rounded crow tail;* Right: *acute mourning dove tail.*

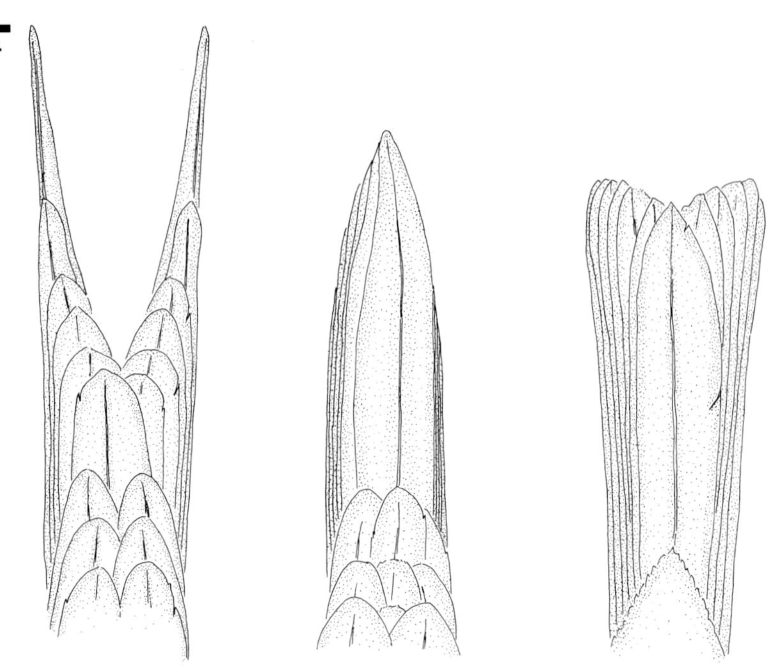

Left: ***forked swallow tail;*** **Center:** ***acute mourning dove tail;*** **Right:** ***emarginate snow bunting tail.***

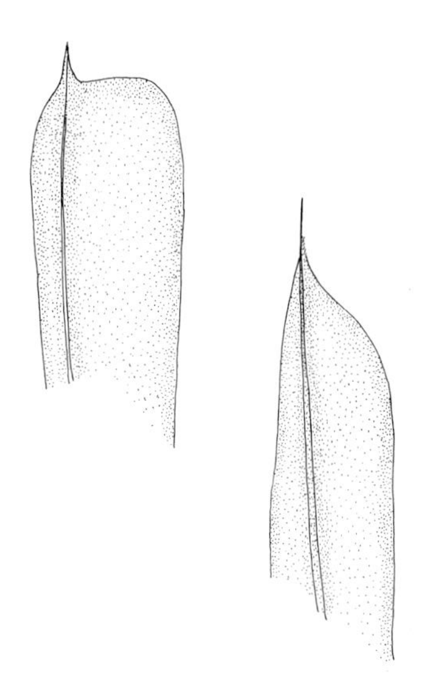

Left: ***woodpecker tail;*** **Right:** ***spinose chimney swift tail.***

are illustrated here and in the photograph section of this chapter.

Differences in feather shape, or morphology, and arrangement on the rump to form the tail, termed topography, combine to create the various major tail types, with each type linked to particular behaviors and the performance of specific types of preferred flight—the kind of flight a bird specializes in but is not limited to. The square, strong tail of the nuthatch props the bird against the tree as a fulcrum in an emergency or as a brace to intensify the power of its beak while gleaning. The rounded, strong tail of the crow distinguishes it from the raven, which has a wedged tail.

McCAY

Compare the tail feathers (top to bottom) from a northern flicker (red-shafted), yellow-billed cuckoo, blue jay, and cedar waxwing.

McCAY

Barn swallow tail vs. flycatcher tail feathers.

The magnificent flight maneuvers of the magpie are aided by the bird's very long, graduated tail feathers. When there is an abrupt difference in the length of tail feathers, as in the mourning dove, where only the middle rectrices are much longer than the others, the tail is termed *pointed* or *acute*. The snow bunting's tail, called *emarginate,* is the opposite, with the middle rectrices being the shortest, producing a slight convex curve. Terns and swallows are such aerial acrobats due to both their aerodynamic wings and extremely forked tails; from the middle pair outward, the rectrices become successively longer.

Even individual tail feathers have unique designs that aid tail function. Woodpeckers' very strong and sharply pointed, acuminate tail feathers are used to brace the bird against tree bark and help support its weight from below while it chisels away at tree trunks. This characteristic is carried to an extreme in the chimney swift's spinose feather. Its quill is pointed also, but it extends even beyond the barbs at the end of the feather, perhaps so it can lock in to small cracks in chimney mortar and bricks without snagging the webs of the feather itself.

Functions

Feathers serve two functions: protective insulation and flight. As warm-blooded animals, birds must maintain a reliable internal temperature, a more challenging task for them than for humans because their bodies are so small. Their downy feathers enable them to accomplish this to an extraordinary degree; by fluffing their body feathers to entrap pockets of insulating air, tiny chickadees are able to maintain their body temperature even in subzero weather.

But perhaps the most dramatic accomplishment of feathers is flight. The curvature of the wing, feather placement, muscle contraction, and skeletal structure all work together to provide the gravity-defying miracle of lift for the bird, enabling it to fly. In a normal, outstretched position, the wing's convex shape causes the air to pass faster over its top than across its hollow underside, creating a vacuum greater than the pull of gravity and lifting the bird as it glides through the air.

The wing of a bird has a strong, blunt, but smooth leading edge; a smooth, convex surface; and a tapered trailing edge. Sleek, aerodynamic contour feathers, called wing coverts, contribute to lift by eliminating rough surfaces that would produce resistance or drag, and the birds' mostly hollow bones reduce weight. (One notable exception is the loon, whose solid bones provide the weight it needs to accomplish astoundingly deep dives, to 200 feet or more.)

But maximizing lift alone is not enough to produce takeoff and then maintain flight. Some form of propulsion is required. In comparing the wing of a bird to the human arm, the necessary lift is produced mainly by the fore and upper "arm" of the wing with its secondary and tertial feathers. The "wrist" and "fingers" of the wing, which bear and manipulate the primary feathers through the downstroke and upstroke, create propulsion. The combined shape and

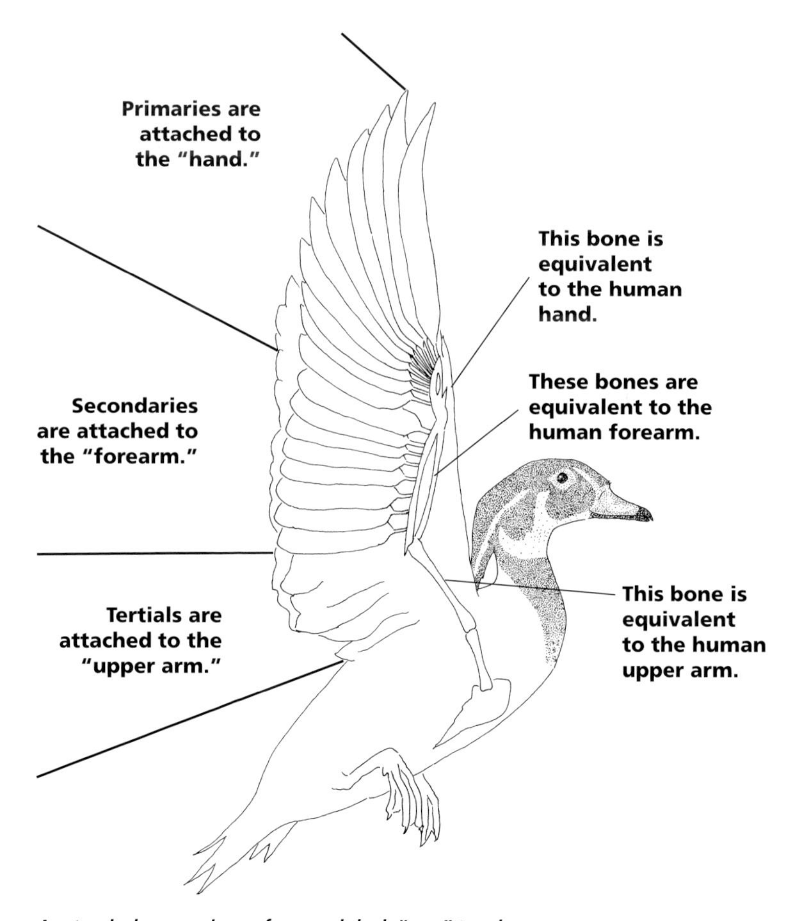

Anatomical comparison of a wood duck "arm" to a human arm.

twisting motion of these primaries function like the angled propeller blades of an airplane, strongly pulling the bird through the air.

With propulsion also comes turbulence, unless the wing and body of the bird are continually readjusted to meet the oncoming air in a streamlined way. Otherwise, increasing turbulence would build resistance, causing the bird to lose lift and therefore stall, with a complete loss of forward motion. A bird not only is able to selectively adjust its body feathers to eliminate the rough spots that create turbulence, but it can also manipulate the alula feathers, at the leading edge of its "wrist," to correct and ventilate the flow of air over the top of the wing and thus fine-tune and stabilize that airstream.

Flight Patterns and Wing Shape

The shape of any flight feather you find in the field is directly related to its function. Different types of feathers can be linked to different overall wing shapes, which determine the preferred form of flight of various species. According to Paul Kerlinger in *How Birds Migrate* (1995): "Birds with long, pointed wings fly faster and live in open country . . . birds with short, rounded wings fly more slowly and live in forested areas. The long, narrow wings . . . enable [birds] to glide rapidly over the water or engage in long-distance powered flight; short, rounded wings are [better] . . . for maneuvering."

Kerlinger also lists five major types of flight patterns that birds use during migration: "powered, bounding, undulating (also called

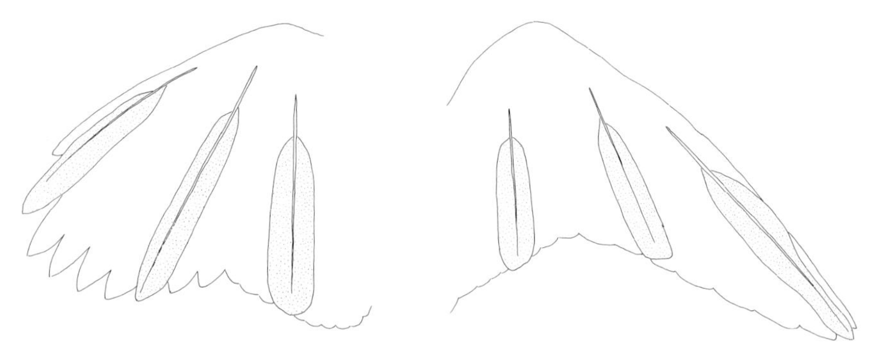

Comparison between ruffed grouse at left (short, powerful wings for bursts of speed in the forest) and Canada goose wing shapes (long, powerful wings for the powered flight of long migration in the open).

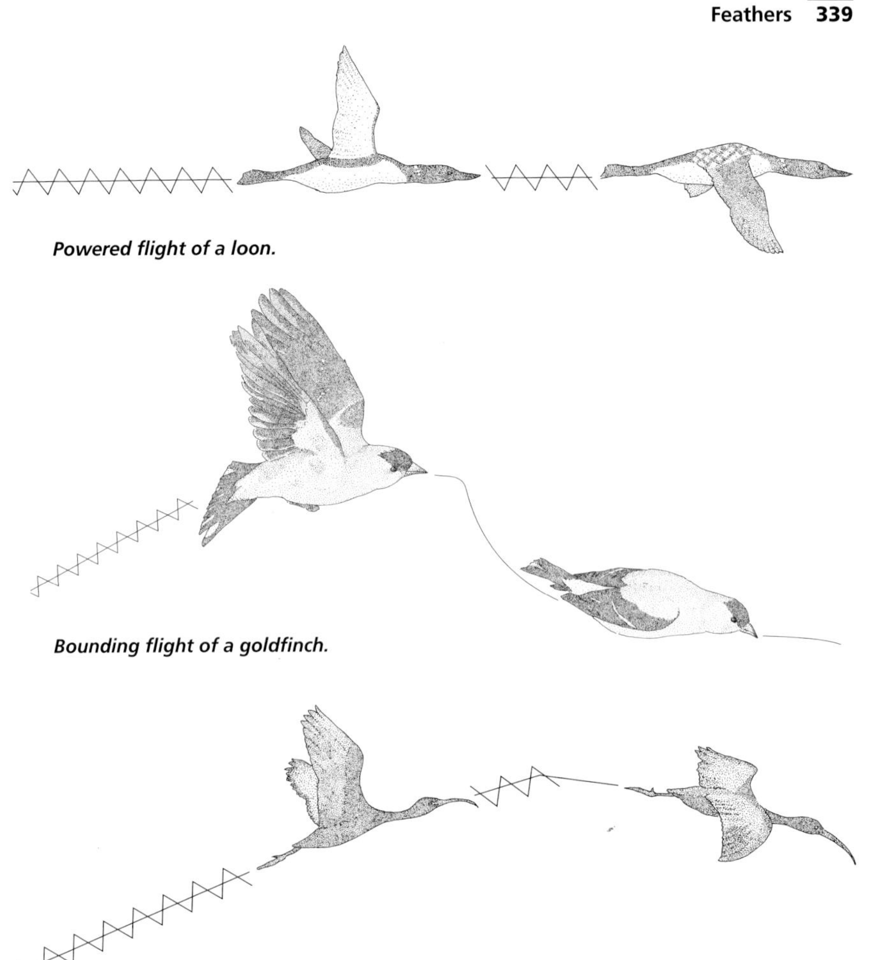

Powered flight of a loon.

Bounding flight of a goldfinch.

Undulating flight of an ibis.

flap and glide), partially powered gliding, and gliding." Depending on weather or stress factors, such as the sudden appearance of a predator, some species are capable of more than one type.

According to Kerlinger, the wings of the birds using powered flight never stop moving except for when landing. At that time, the wings will stop flapping, allowing the bird to glide downward. A characteristic of powered flight is its even course in the air.

Another type of flight related to powered flight is called "bounding." Birds bound upward in short but powerful bursts of speed and then tuck their wings in a brief, steep descent in which the accelerated forward movement uses the momentum of free-fall. Then they begin climbing again via powerful flapping. The equal rhythm of climb and descent averages out to a steady level of progress.

Undulating flight is similar to bounding with one crucial modification—between steep climbs the bird maintains forward movement with outstretched, flat wings in order to maximize a glide. Partially powered flight copies the flapping sequence of undulating flight, but with little gain in altitude in the power stroke, thereby allowing the bird to maintain a mostly level course without dramatic climbs.

Gliding is "flight with fixed wings and no flapping." It usually entails a gradual loss of altitude unless a bird is able to take advantage of thermal updrafts in order to soar in slow circles upward.

A Canada goose wing (see the illustration on page 338) is ideally designed for long, powered flights in which the bird flaps continuously: The wing is pointed, the outermost primaries are extremely long, and the broad secondaries are arranged to provide a lot of lift on a wide wing, which therefore has a long chord—the measurement from the leading edge of the wing at the "wrist" to the back of the wing at the end of its trailing edge.

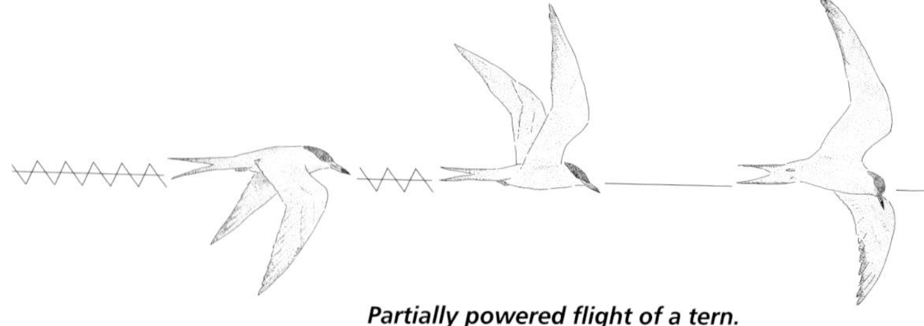

Partially powered flight of a tern.

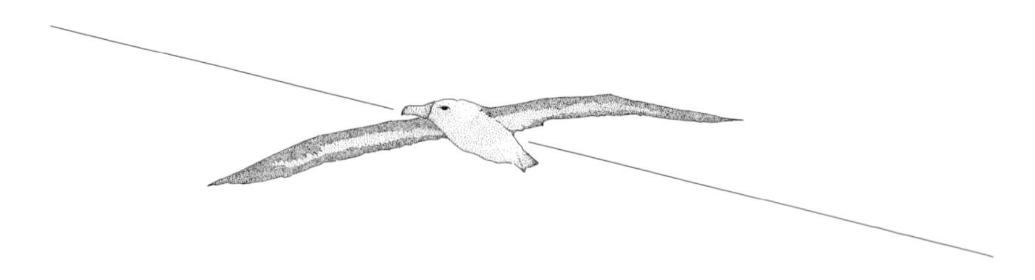

Gliding flight of an albatross.

Comparison beween tern at left (long, narrow wings for manueverability) and broad-winged hawk wing shapes (broad, hefty wings for soaring and maintaining glides).

A goldfinch wing is relatively short, compared with the length of the body, and compact yet has very flexible primaries with angled tips. The secondaries seem strong, suggesting that this little bird could easily power itself up, getting good lift for the drive stroke of bounding flight, but the short wings probably could not maintain a long soar afterward, so the bird conserves energy by tucking its wings against its body and swooping down, gaining momentum like a roller coaster at the bottom of a hill for its next flapping sweep upward.

An ibis wing is long and narrow and seems designed for maneuverability. The primaries are deeply emarginated, and the secondaries relatively short, suggesting long, undulating patterns created by the combination of short bursts of flapping speed to gain altitude and banking, descending glides.

The tern's wing has very long primaries and very short secondaries, which contributes to a partially powered glide. (Much of the tern's extraordinary ability for sharp turns and swooping, twisting dives for fish also comes from its doubly ruddered forked tail.) The albatross wing has the perfect anatomical structure for a long-gliding seabird that endlessly trolls just above the waves.

The broad-winged hawk's hefty wing accurately fits the name of the bird and is broad due to the relatively long length of both primaries and secondaries. It can maintain a glide for quite a while and can soar even longer on outstretched, fixed wings when it finds a thermal updraft to spiral up in. During peak migration times on the eastern flyway, broad-wings are well known for moving ahead of northwest fronts in waves of thousands of birds, with dozens of them sharing the upcircling elevator of a thermal, soaring and gaining altitude for thousands of feet, in a formation referred to by ornithologists as a "kettle."

The grouse's rounded, extremely wide, almost chunky wing suggests the enormous strength it needs to suddenly explode from the forest floor for short bursts of flight when threatened by a predator, and many other ground-dwelling game birds have similar rounded wings. The longest primaries are the middle ones, and they decrease in length toward the outer edges. Indeed, the outermost feather is extremely short. This powerful, tightly feathered design also helps the grouse produce its courtship drumming sounds, as it flaps its partially tucked deeply convex wings against its flanks.

The hummingbird's flat wing, combined with the bird's minuscule weight, may explain the bird's unique ability to sustain lift during both the downstroke and the upstroke of its wingbeats. Wing strokes can reach 80 beats per second.

Feather Colors

Given their greater mobility and consequently a somewhat less pressing survival need to camouflage themselves from predators, birds are among the most colorful of the vertebrates. Colors are produced by one of two pigments. Black, browns, and grays are produced by melanin, tiny granules found within or on the surface of the feathers. Carotenoids produce the reds and yellows, as well as the modification of red into pink. White is caused by albinism, a total lack of pigmentation in the keratin protein of the feathers. Pigment is derived from the bird's diet of seeds, grains, and vegetables. Flamingos in captivity turn white if they do not receive their natural diet of shrimp, and if a yellow canary is fed red peppers, its feathers will change to bright orange with successive molts.

Another source of green, red, and brown is porphyrins produced by the breakdown of hemoglobin by the liver. Some researchers believe that blue and green are not produced by pigments at all, but rather are the result of a peculiar refraction of light on the rough surface of the barbs, which reflects only blue wavelengths to the human eye. Greens are a combination of this blue-reflecting phenomenon over feathers with yellow pigment. Brown et al. (1987) observe that the seasonal intensification of color in some birds occurs not due to pigmentation, but to wear. The bright outermost feather tips gradually wear away, leaving only the main color of the feather. In the snow bunting, the white-tipped feathers wear off over winter, leaving bolder black feathers behind for spring breeding plumage.

Snow bunting feather wear. (Used with permission of Yale University Press.)

Molts are another reason for color change in birds. Since feathers are not continually restored from within, as are the living cells of the bird's body, when they are sufficiently damaged, they require total replacement. Sometimes this replacement of individual feathers occurs as needed during the year, especially with tail feathers; this takes place simply by a new feather pushing out the old, much as an adult tooth pushes out a baby tooth in humans. A total replacement of all feathers occurs either annually or semiannually for most species through molting, providing the bird with alternate (breeding) and basic (winter) plumages. Seasonal, environmental, and metabolic changes signal this complete change in plumage, which takes place in late summer in the northern hemisphere, with a second, more limited molt occurring in some species in spring, just prior

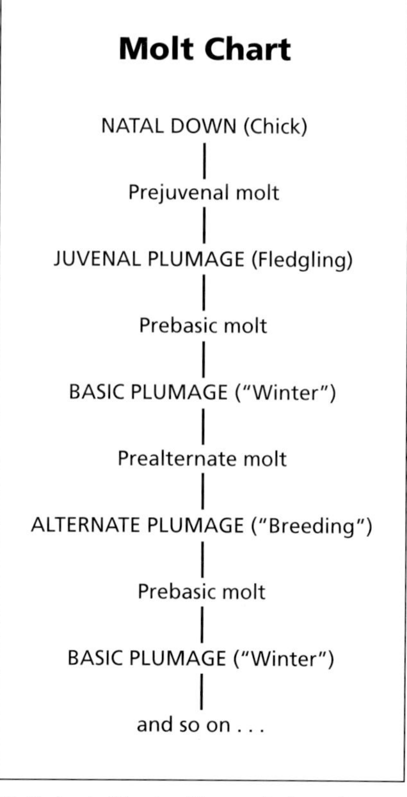

Molt chart. (Used with permission of Yale University Press.)

Roxie Laybourne and the Smithsonian's Feather File

Roxie Laybourne received her master's degree in 1932 and still works every day with great accomplishment and delight in the science she created—the identification of individual bird feathers using microscopic and whole-feather characteristics. Most of the feather plates that appear in this chapter exist because of her decades of research on feathers. In a laboratory on the sixth floor of the Smithsonian Institution's Natural History Museum, Roxie and her assistant Carla Dove have established the crucial reference source for this chapter: an international compilation that is simply referred to as the "Feather File." This treasure trove is the product of decades of research on birds.

Roxie has been identifying feathers since the early 1960s when a plane crashed in Boston after hitting a flock of starlings and claimed the lives of sixty people. Annually, the Federal Aviation Administration and Air Force report at least 5,000 bird strikes (with the actual number likely much higher). Sometimes the evidence left from these cases is just a feather shard that has passed through a jet engine (though Roxie notes that feathers are astoundingly strong, and therefore those durable fragments are often large). Each positive species identification is crucial to building databases that can then be used to analyze hits based on patterns such as the incidence of flocking versus individual birds or the involvement of migratory pathways.

The field ecology approach of tracking—learning birds and mammals, as well as trees, plants, insects, and ecosystem dynamics—fits naturally with Roxie's practical and gifted approach to solving serious identification mysteries. Whether helping airports determine the specific species involved in collisions with airplanes in order to help aeronautical engineers design safer jets, engines, runway placement, and traffic patterns or giving expert testimony in law enforcement cases involving illegal trafficking in birds, mammals, and animal parts, Roxie (as she is affectionately known throughout the world of ornithology) has revolutionized feather identification by her use of light microscopy.

What is Roxie and Carla's method?

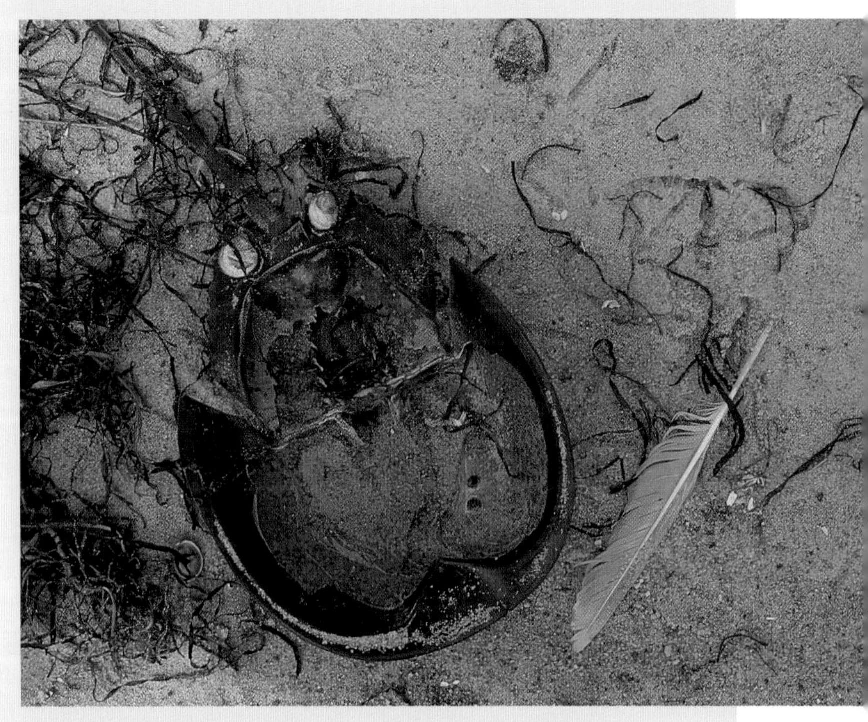

Horseshoe crab shell with gull tracks and feather; scavenging site.

"Most cases are identified by analyzing whole feather fragments. Careful comparison with bird specimens from the museum's extensive collection, consideration of species ecology, and biogeography are all considered. Frequently, feather remains are too small for positive identification. These cases require using light microscopy to examine the downy barbules, which often vary in morphological characteristics that can guide us to the family of birds," explains Carla.

Carla emphasizes that when trackers have questions about feather identification they should always feel free to approach local museums, whether large or small, and ask museum collection curators research questions. This helps build community among all those interested in studying birds and promoting their conservation.

to breeding season. Since a molt involves a major drain of energy and protein, it must not coincide with the other two major annual energy demands of breeding and migration.

Because of the molt, feathers are most likely to be found in the greatest abundance and variety in late summer. Another bonus for trackers is that these feathers will probably be molted from breeding plumage; therefore, despite wear and tear, they will be brighter and more colorful than feathers found at other times of the year.

For most species, two flight feathers are usually dropped at the same time in balanced pairs, from the same numbered position on each wing, beginning from the inner primaries outward. Then the tertials drop, triggering the secondaries to begin from the first mid-wing secondary inward, so the bird can maintain its capacity to fly by having only small, balanced gaps in its wings. Similarly, tail feathers for most species (except woodpeckers and brown creepers, which need to retain their central pair for support until a new pair has grown in) begin dropping from the central pair outward. Leahy (1982) notes the other major exceptions to this flight-preserving pattern: "Loons, grebes, anhingas, flamingos, ducks, geese, and swans, most rails, many alcids, and dippers lose their wing feathers simultaneously and so are rendered flightless for a period ranging from a few days (dippers) to 7 weeks (swans). Ducks, geese, and swans also lose all their tail feathers at once during this molt, and smaller owls lose just their tail feathers in this precipitous manner." Canada goose parents take turns going through complete molts of wing feathers so that one parent will always be capable of defending their goslings.

For trackers who have not spent a lot of time birding or learning the field marks of male versus female birds and adults versus juveniles, it's easiest, and most enjoyable, to start with adult males in their bright spring breeding plumage. While you are observing the males' plumage, as well as bill and tail shape, and perhaps leg color where appropriate, begin to take note of the females accompanying them.

Tracking birds in the spring will whet your appetite to learn more about them the rest of the year.

Editor's Note: In the feather plates P = primary, S = secondary, Tert. = tertial, and T = tail. Measurements are given in inches, followed by centimeters.

Loons *(family Gaviidae)*

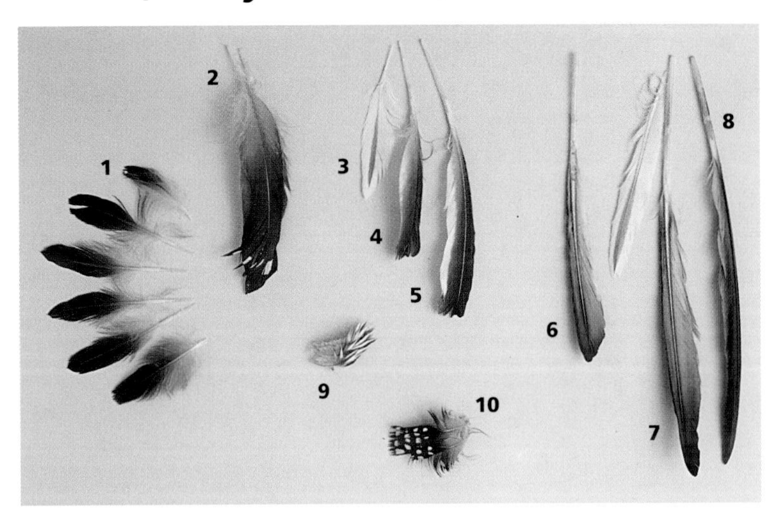

Common Loon *(Gavia immer)*

1 outer T: 2 (5.1) to center T: $3\frac{3}{8}$ (8.6)
2 Terts.: $5\frac{3}{8}$ (13.7)
3 S covert: $3\frac{1}{2}$ (8.9)
4 inner S: $4\frac{3}{4}$ (12.1)
5 outer S: $6\frac{1}{8}$ (15.6)
6 inner P: $6\frac{3}{4}$ (17.2)
7 outer P w/ coverts: $9\frac{1}{8}$ (23.2)
8 outer P: $8\frac{7}{8}$ (22.5)
9 neck clump: n/a
10 back clump: n/a

Storm-Petrels *(family Hydrobatidae)*

Leach's Storm-Petrel *(Oceanodroma leucorhoa)*

1 partial wing (outer Ps): $5\frac{3}{8}$ (13.7)
2, 3 P: $4\frac{7}{8}$ (12.4)
4 S coverts clump: $3\frac{1}{2}$ (8.9)
5, 6 S: 3 (7.6)
7 T (2): $3\frac{5}{8}$ (9.2)
8 contour: $1\frac{1}{8}$ (2.8)

Pelicans *(family Pelecanidae)*

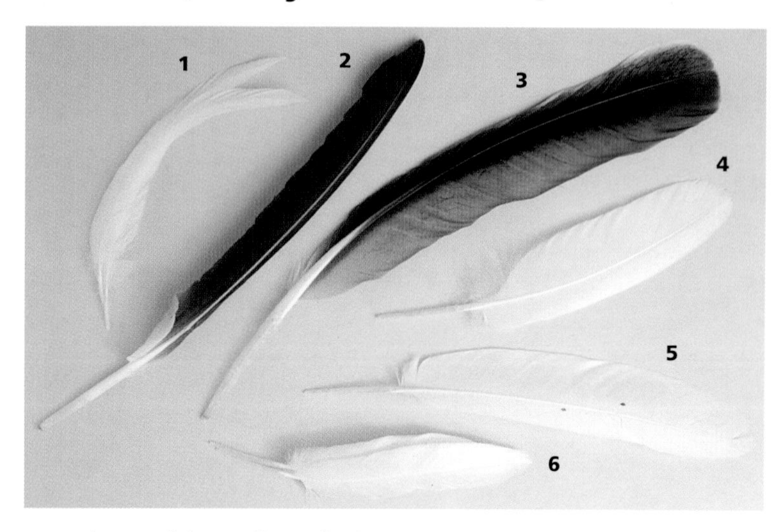

American White Pelican *(Pelecanus erythrorhynchos)*

1 wing coverts: 8 1/2 (21.6)
2 outer P: 13 1/8 (33.3)
3 outer S: 15 (38.1)
4 inner S: 9 1/2 (24.1)
5 inner T: 10 7/8 (27.7)
6 outer T: 7 3/4 (19.7)

Note: The longest midprimaries may reach 22 inches.

Darters *(family Anhingidae)*

Anhinga—female *(Anhinga anhinga)*

1 throat to chest "necklace": n/a
2 one plume of crest (3): 1 1/2 (3.8)

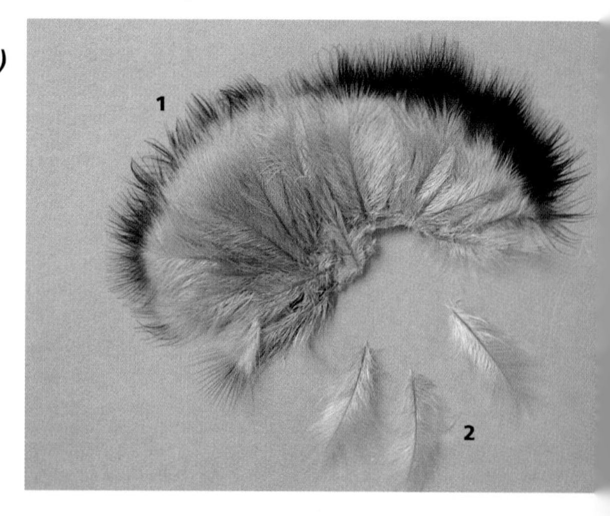

Cormorants *(family Phalacrocoracidae)*

Double-crested Cormorant *(Phalacrocorax auritus)*

1 wing: n/a
2 P: 7 3/4 (19.7)
3 S: 6 3/8 (16.2)
4 Tert.: 4 5/8 (11.7)

Herons, Bitterns *(family Ardeidae)*

American Bittern *(Botaurus lentiginosus)*

1 P covert: 4 1/2 (11.4)
2 S: 5 7/8 (14.9)
3, 4 S: 6 (15.2)
5 S: 6 3/8 (16.2)
6 inner P: 7 (17.8)

Great Blue Heron *(Ardea herodias)*

1 leading edge shoulder covert clump: n/a
2 T: 6 7/8 (17.4)
3 T: 6 3/4 (17.1)
4 T: 6 7/8 (17.4)
5 one feather of covert clump: 1 (2.54)
6 outer P: 11 (28.0)
7 mid P: 13 1/8 (33.3)
8 S: 10 1/4 (25.9)
9 S covert: 5 3/4 (14.6)

Great Blue Heron

breeding plumes:
up to 13 1/2 (34.3)

Storks *(family Ciconiidae)*

Wood Stork *(Mycteria americana)*

1 P (calamus damaged): 10 1/8 (25.7)
2 outer S (damaged): 8 3/4 (22.2)
3 outer S (damaged): 9 (22.9)
4 contour: 2 5/8 (6.6)

Ducks, Geese, Swans *(family Anatidae)*

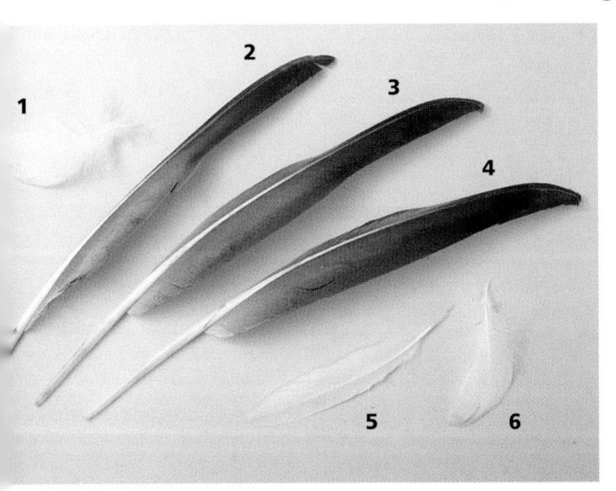

Snow Goose *(Chen caerulescens)*

1 contour: 4 (10.2)
2 P: 12 3/4 (32.4)
3 P: 13 1/2 (34.3)
4 P: 13 7/8 (35.3)
5 wing covert: 6 (15.2)
6 contour: 3 3/4 (9.5)

Canada Goose *(Branta canadensis)*

1 P: 13 (33.0)
2 S: 8 3/4 (22.2)
3 inner S: 8 5/8 (21.9)
4 breast contour: 1 3/8 (3.5)
5 breast clump: n/a
6 undertail covert: 3 5/8 (9.2)
7 center T: 6 1/2 (16.5)
8 mid T: 6 7/8 (17.5)
9 outer T: 6 3/4 (17.1)

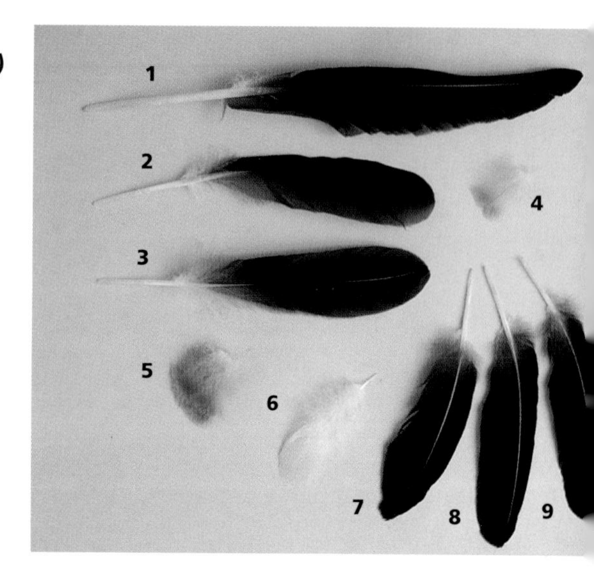

Perching Ducks

Wood Duck *(Aix sponsa)*

1 flank contour: 2 (5.1)
2 flank contour: 2 3/4 (7.0)
3 T: 4 5/8 (11.7)
4 flank contour: 3 1/8 (7.9)
5 Tert. clump: 3 1/4 (8.3)
6 Ss: 3 5/8 (9.2)
7 P: 5 1/4 (13.3)
8 P: 6 5/8 (16.8)
9 underwing covert: 2 1/4 (5.7)
10 breast clump: n/a

Dabbling Ducks

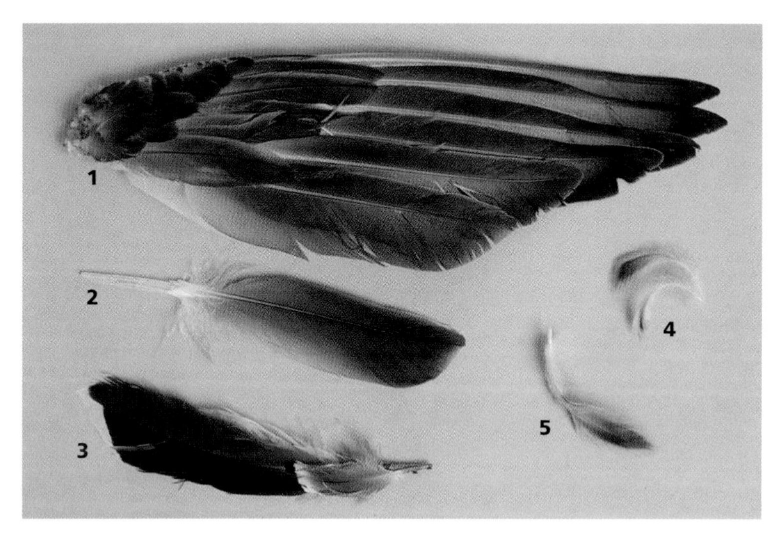

Mallard *(Anas platyrhynchos)*

1 partial wing (outer P): up to 8¾ (22.2)
2 inner P: 5½ (14.0)
3 (speculum) S w/ covert: 5 (12.7)
4 breast contours: 1½ (3.8)
5 flank contour: 2½ (6.4)

Gadwall *(Anas strepera)*

1 P (damaged): 4¾ (12.0)
2 P: 4⅝ (11.7)
3 S w/ coverts: 4⅜ (11.1)

Note white greater secondary covert.

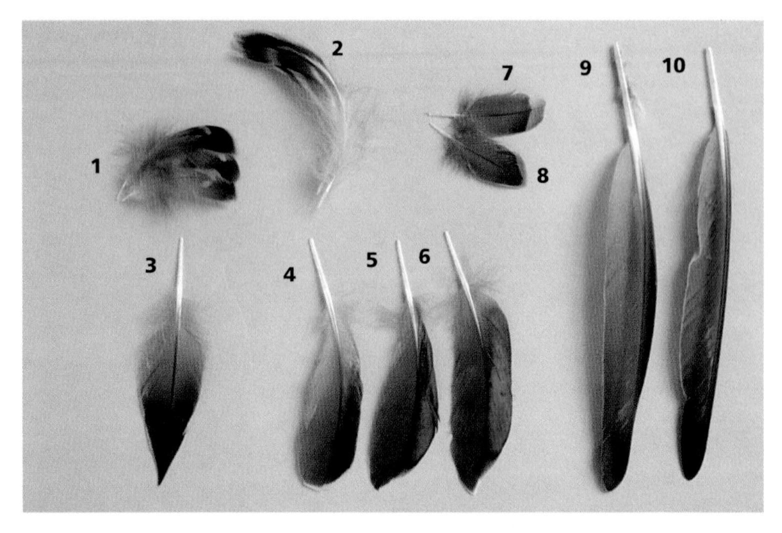

Green-winged Teal *(Anas crecca)*

1 breast clump: n/a
2 back contour (2): 2 3/4 (7.0)
3 T: 2 7/8 (7.3)
4 S: 3 1/8 (7.9)
5 S: 3 (7.6)
6 S: 2 7/8 (7.3)
7 crown contour: 1 1/2 (3.8)
8 crown contour: 1 3/8 (3.5)
9 outer P: 5 1/8 (13.0)
10 outer P: 5 (12.7)

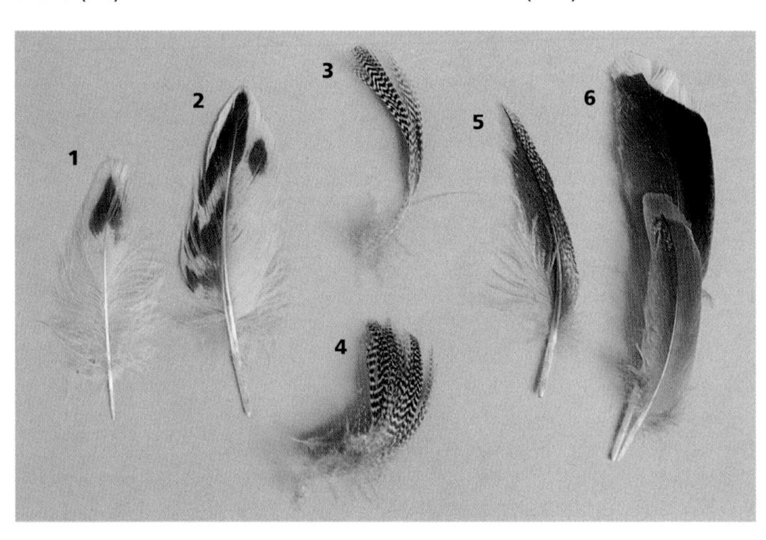

Northern Pintail *(Anas acuta)*

1 back contour: 2 1/2 (6.4)
2 back contour: 3 1/8 (7.9)
3 flank contour: 2 1/4 (5.7)
4 flank clump: n/a
5 flank (in pin): n/a
6 S w/ covert: 4 (10.2)

Northern Shoveler *(Anas clypeata)*

1 T: 4 3/4 (12.1)
2 S w/ covert: 4 (10.2)
3 covert: 3/4 (1.9)
4 P: 6 1/2 (16.5)
5 outer P: 5 5/8 (14.3)

Blue-winged Teal *(Anas discors)*

1 marginal covert: 1 (2.5)
2 S covert: 1 3/8 (3.5)
3, 4 S covert: 2 (5.1)
5 S: 3 3/8 (8.6)
6 S: 3 1/8 (7.9)
7 S w/ covert: 3 1/8 (7.9)

Mergansers

Red-breasted Merganser *(Mergus serrator)*

1 wing covert: 2 (5.1)
2 flank clump: 3 (8.0)
3 wing covert: 2 (5.1)
4 S covert clump: n/a
5 inner P: 4 1/8 (10.5)

Vultures *(family Cathartidae)*

Turkey Vulture *(Cathartes aura)*

1 S (in pin): 6 3/8 (16.2)
2 wing section (outer to inner Ps): 18 1/2 (47.0)
3 T: 11 1/2 (29.2)

Note: Damage to Ps and T due to collision with power lines.

Black Vulture *(Coragyps atratus)*

1 neck contour: 1 3/4 (4.4)
2 shoulder contour: 2 5/8 (6.7)
3 outer P: 14 1/8 (35.9)
4 S: 9 1/2 (24.1)
5 S w/ covert: 8 1/4 (21.0)

Hawks, Kites, Eagles *(family Accipitridae)*

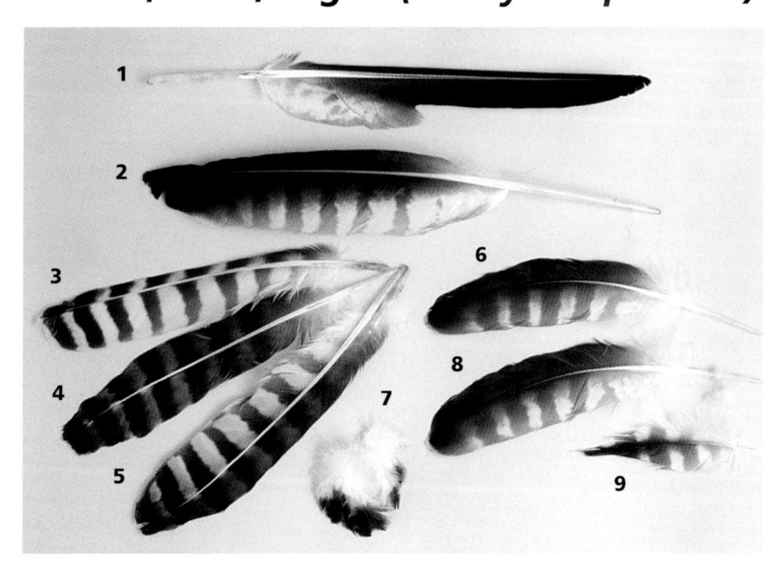

Osprey *(Pandion haliaetus)*

1 outer P: 13 (33.0)
2 outer P: 13 3/8 (34.0)
3 T: 9 1/4 (23.5)
4 T: 9 1/2 (24.1)
5 T: 9 5/8 (24.4)
6, 8 S: 8 5/8 (21.9)
7 breast clump: n/a
9 Tert.: 4 1/2 (11.4)

Golden Eagle *(Aquila chrysaetos)*

1 outer P: $12\frac{3}{4}$ (32.4)
2 partial wing (Ps): 21 (53.3)
3 S w/ covert: $11\frac{5}{8}$ (29.5)

Bald Eagle *(Haliaeetus leucocephalus)*

1 nape: $5\frac{3}{4}$ (14.6)
2 P w/ coverts: $15\frac{7}{8}$ (40.3)
3 S covert: $7\frac{1}{4}$ (18.4)
4 nape: 4 (10.1)

Bald Eagle

1 nape: 4 3/4 (12.1)
2 nape: 4 (10.2)
3 P w/ coverts: 15 7/8 (40.3)
4 S covert: 7 1/4 (18.4)

Bald Eagle—immature

1 undertail covert: 6 (15.2)
2 center T: 15 1/4 (38.7)
3 T: 14 1/2 (36.8)
4 outer T: 13 1/4 (33.7)

Bald Eagle—immature

1 inner P: 14¼ (36.2)
2 outer P: 19⅞ (50.5)
3 outer P: 17⅜ (44.1)
4 S covert: 5⅛ (13.0)

Bald Eagle—immature

1, 2 S: 13 (33.0)
3 S w/ covert: 13⅜ (34.0)

Note: Feather damage due to electrocution from power lines.

Accipiters

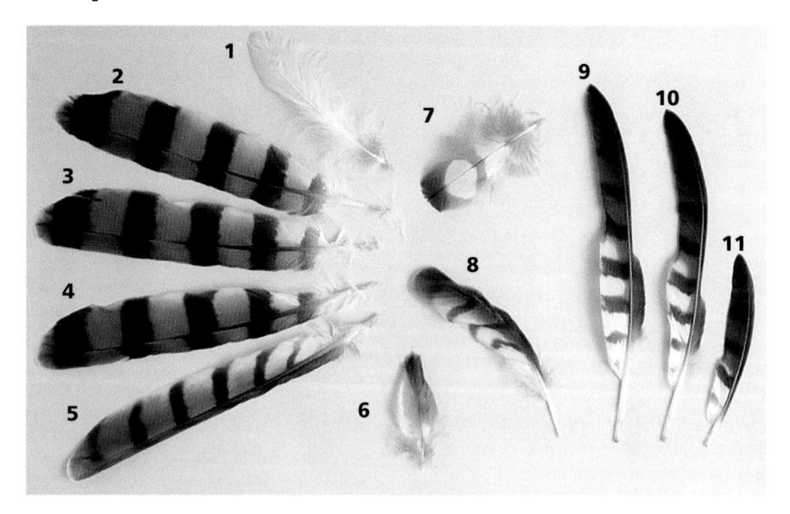

Sharp-shinned Hawk *(Accipiter striatus)*

1 undertail covert: 3 3/4 (9.5)
2–5 T: 6 5/8 (16.8)
6 Tert.: 2 1/4 (5.7)
7 underwing covert: 2 7/8 (7.3)
8 S: 4 (10.2)
9 P: 6 1/2 (16.5)
10 P: 6 (15.2)
11 outer P: 3 5/8 (9.2)

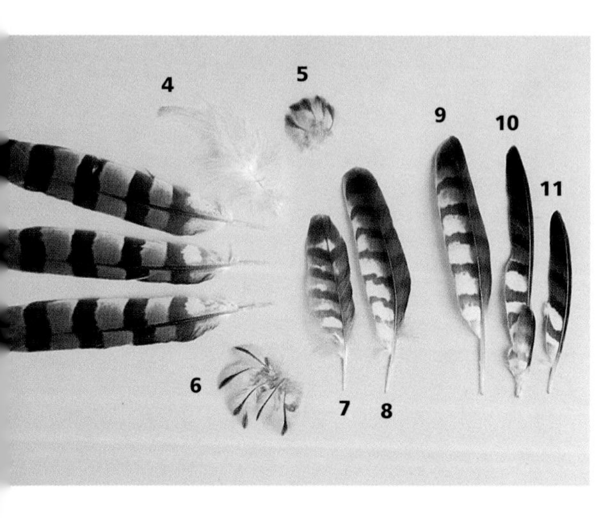

Cooper's Hawk *(Accipiter cooperii)*

1 center T: 9 5/8 (24.5)
2 outer T: 9 1/2 (24.1)
3 outer T: 9 3/8 (23.8)
4 undertail covert: 4 1/2 (11.4)
5 nape clump: n/a
6 breast clump: n/a
7 inner S: 5 1/4 (13.3)
8 mid S: 6 3/4 (17.1)
9 mid P: 7 5/8 (19.4)
10 outer P: 7 1/4 (18.4)
11 outer P: 5 1/4 (13.3)

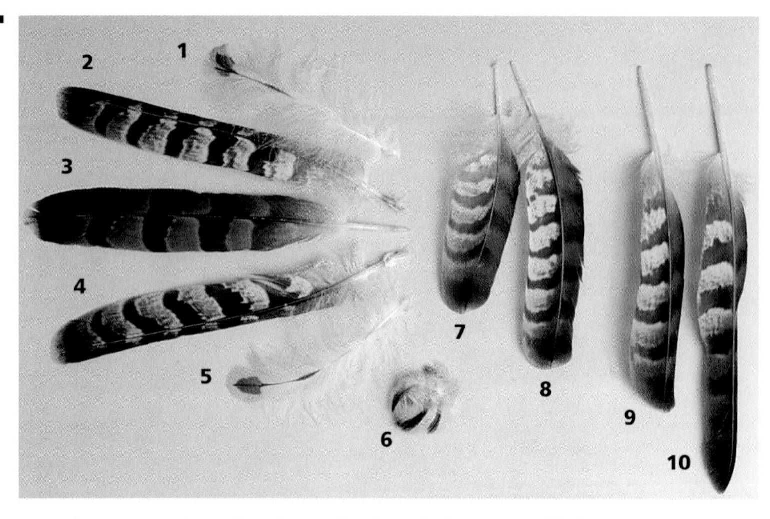

Northern Goshawk—female *(Accipiter gentilis)*

1 undertail covert: 5½ (14.0)
2 outer T: 9½ (24.1)
3 inner T: 10 (25.4)
4 outer T: 9½ (24.1)
5 undertail covert: 5½ (14.0)
6 neck clump: n/a
7 inner S: 6⅛ (15.5)
8 outer S: 7½ (19.0)
9 inner P: 8½ (21.6)
10 outer P: 10⅜ (26.4)

Buteos

Broad-winged Hawk *(Buteo platypterus)*

1 outer T (damaged): 6¾ (17.2)
2 inner T: 7 (17.8)
3, 4 S: 6 (15.3)
5 inner P: 6½ (16.5)
6 inner P: 7½ (19.1)
7 outer P: 8¼ (21.0)
8 throat contour: 1¾ (4.5)
9 throat clump: n/a
10 breast clump: n/a
11 breast contour: 1¼ (3.2)

Red-shouldered Hawk *(Buteo lineatus)*

1, 2 wing covert: $3\frac{7}{8}$ (9.9)
3 Tert.: $6\frac{7}{8}$ (17.5)
4 S: $8\frac{1}{4}$ (21.0)
5 P: $10\frac{7}{8}$ (27.6)
6 alula: $4\frac{1}{4}$ (10.8)

Red-tailed Hawk *(Buteo jamaicensis)*

1 outer T: $8\frac{1}{4}$ (21.0)
2 inner T: 8 (20.3)
3 outer T: $8\frac{1}{4}$ (21.0)
4 inner S: $7\frac{1}{4}$ (18.4)
5 outer S: $8\frac{1}{2}$ (21.6)
6 mid P: $12\frac{1}{8}$ (30.8)
7 outer P: $7\frac{7}{8}$ (20.0)
8 flank clump: n/a
9 neck clump: n/a

Swainson's Hawk *(Buteo swainsoni)*

1 partial wing: n/a
2 outer P: 10 1/2 (26.7)
3 outer P: 12 7/8 (32.7)
4 outer P: 14 (35.6)
5 S: 6 3/8 (16.2)
6 S: 6 1/2 (16.5)
7 underwing covert: 3 3/4 (9.5)

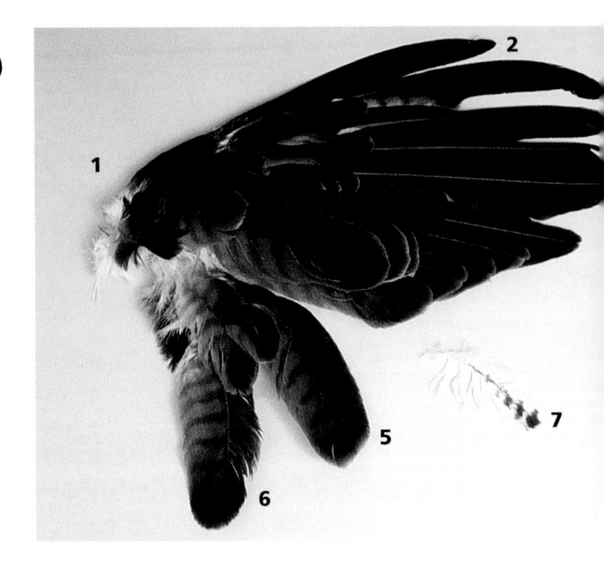

Caracaras, Falcons *(family Falconidae)*

American Kestrel *(Falco sparverius)*

1 female center T clump: 5 3/8 (13.7)
2 female mid T: 5 1/8 (13.0)
3 female outer T: 4 5/8 (11.7)
4 male mid T clump: 4 7/8 (12.4)
5 male outer T: 4 1/2 (11.4)
6 partial wing (inner Ps, Ss): 5 3/8 (13.7)
7 S w/ covert: 3 1/8 (7.9)
8 partial wing (shoulder to Ps w/ Ss): 7 3/8 (18.7)
9 outer P (2): 6 1/4 (15.9)
10 breast down clump: n/a

Note: Sex comparison of tail feather.

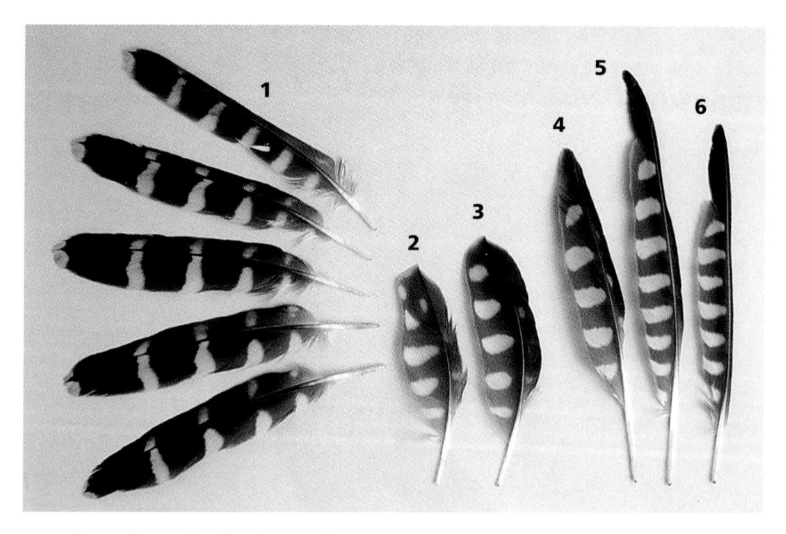

Merlin—female *(Falco columbarius)*

1 outer T (5): 5¼ (13.3)
2 inner S: 3½ (8.9)
3 outer S: 3⅞ (9.8)
4 inner P: 5¼ (13.3)
5 outer P: 6½ (16.5)
6 outer P: 5½ (14.0)

Peregrine Falcon *(Falco peregrinus)*

1 T (5): 8¼ (21.0)
2 S: 6 (15.2)
3 S: 6⅝ (16.8)
4 inner P: 6¾ (17.1)
5 outer P: 10⅞ (27.6)
6 outer P: 8⅛ (20.7)

Partridges, Grouse, Turkeys (family Phasianidae)

Ring-necked Pheasant—male (*Phasianus colchicus*)

1 P: 6 7/8 (17.5)
2 P: 7 7/8 (20.0)
3 S: 7 (17.8)
4 Tert.: 6 1/2 (16.5)
5 lower back: 4 1/2 (11.4)
6 Tert. covert: 4 1/4 (10.8)
7 Tert. covert: 4 (10.2)
8 center T: 17 1/8 (43.5)
9 outer T: 7 7/8 (20.0)
10 throat (3): n/a
11 side neck clump: n/a
12 flank clump: n/a

Wild Turkey (*Meleagris gallopavo*)

1 S covert: 7 7/8 (20.0)
2 T: 14 3/4 (37.5)
3 T covert: 10 1/2 (26.7)

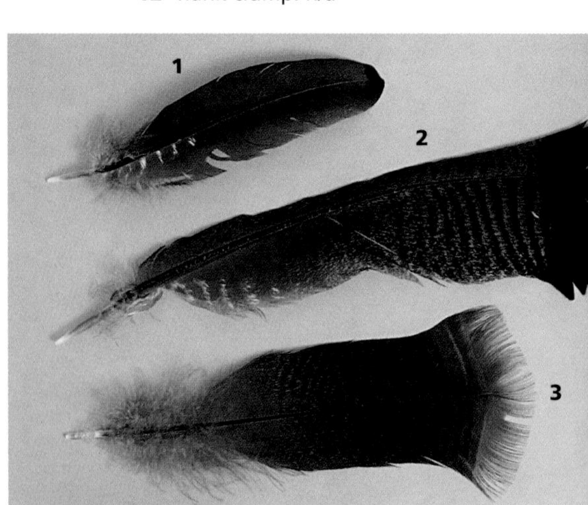

Ruffed Grouse *(Bonasa umbellus)*

1 outer T: $6\frac{7}{8}$ (17.5)
2 T: $7\frac{1}{8}$ (18.1)
3 Ss w/ coverts: $4\frac{7}{8}$ (12.4)
4 S (damaged): $4\frac{5}{8}$ (11.7)
5 Ps w/ coverts: $6\frac{3}{8}$ (16.2)
6 P: $5\frac{3}{4}$ (14.6)
7 P: $4\frac{1}{2}$ (11.4)
8 breast contour: $2\frac{3}{4}$ (7.0)
9 shoulder contour: $2\frac{1}{8}$ (5.4)

New World Quail *(family Odontophoridae)*

Montezuma Quail *(Cyrtonyx montezumae)*

varied contours, range: $\frac{3}{8}$–$3\frac{1}{4}$ (0.1–9.5)

Limpkins *(family Aramidae)*

Limpkin *(Aramus guarauna)*

1 wing coverts: 3⁷⁄₈ (9.9)
2 Tert.: 6³⁄₈ (16.2)
3 S: 7⁵⁄₈ (19.4)
4 P: 9¹⁄₂ (24.1)

Rails, Gallinules, Coots *(family Rallidae)*

Virginia Rail *(Rallus limicola)*

1 P: 4³⁄₈ (11.1)
2 wing: 5¹⁄₄ (13.3)
3 S: 4 (10.2)
4 underwing covert: 1⁷⁄₈ (4.8)

Cranes *(family Gruidae)*

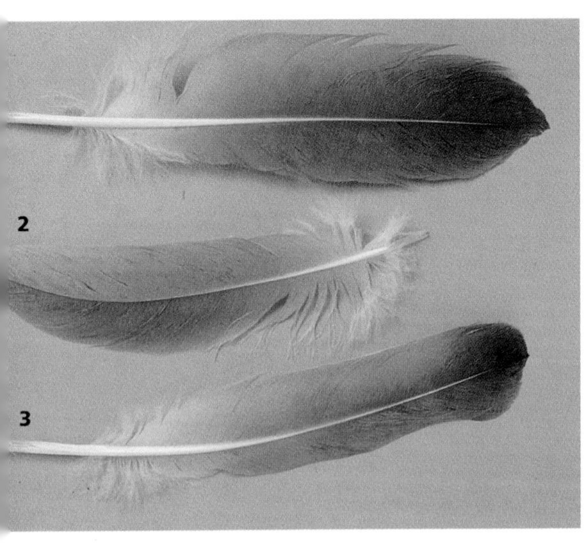

Sandhill Crane *(Grus canadensis)*

1 Tert.: $11\frac{1}{2}$ (29.2)
2 S covert: $8\frac{3}{4}$ (22.2)
3 S: $10\frac{7}{8}$ (27.6)

Whooping Crane—immature *(Grus americana)*

1 T: $9\frac{1}{8}$ (23.2)
2 T covert: $7\frac{1}{2}$ (19.0)
3 Tert.: $14\frac{3}{4}$ (37.4)
4 S: $12\frac{3}{8}$ (31.4)
5 P: $12\frac{7}{8}$ (32.7)
6 P: 13 (33.0)

Note: Species is highly endangered.

Lapwings, Plovers *(family Charadriidae)*

Black-bellied Plover *(Pluvialis squatarola)*

1 partial wing: 6 1/8 (15.5)
2, 3 outer P: 4 3/4 (12.1)
4 outer P: 4 3/8 (11.1)
5 inner P: 3 7/8 (9.9)
6 wing covert: 2 (5.1)
7 S covert: 2 3/4 (7.0)
8 back clump: n/a
9 breast clump: n/a

American Golden-Plover *(Pluvialis dominica)*

1, 2 T: 2 3/4 (7.0)
3 crissum: 2 1/2 (6.4)
4, 5 Tert.: 3 (7.6)
6 S: 2 3/8 (6.0)
7 P: 2 1/2 (6.4)
8 P: 5 1/8 (13.0)
9 back clump: n/a

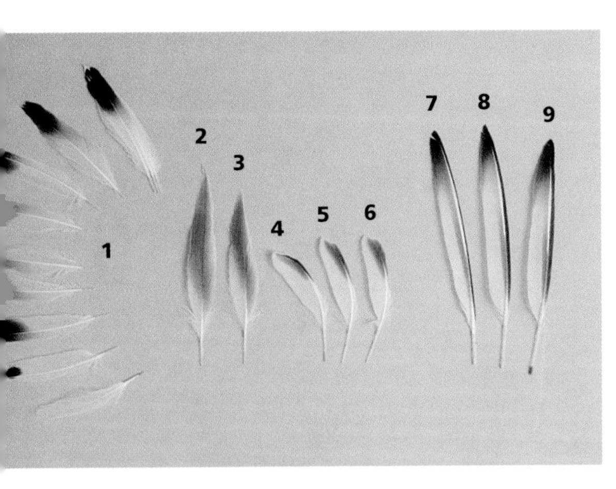

Piping Plover *(Charadrius melodus)*

1 T (9): 2 (5.1) avg.
2 Tert.: 3¼ (8.3)
3 Tert.: 2¾ (7.0)
4–6 inner S: 2 (5.1)
7–9 outer P: 3¾ (9.5)
Note long tertials.

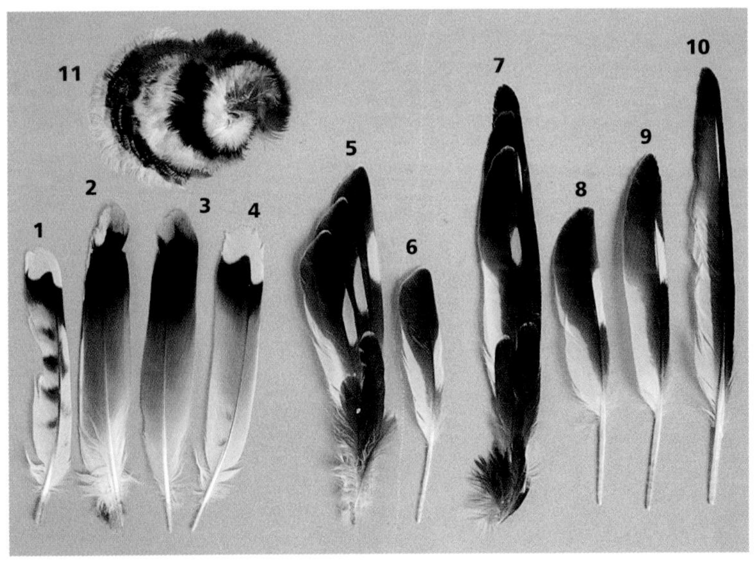

Killdeer *(Charadrius vociferus)*

1 undertail covert: 3⅜ (8.6)
2, 3 T: 3⅞ (9.9)
4 T: 3¾ (9.5)
5 partial wing (Ps): 4⅛ (10.5)
6 S: 3 (7.7)
7 partial wing (Ps, coverts): 5¼ (13.3)
8 inner P: 3⅝ (9.2)
9 mid P: 4¼ (10.8)
10 outer P: 5¼ (13.3)
11 breast clump: n/a

Northern Lapwing *(Vanellus vanellus)*

1 T (5): 4 7/8 (12.4)
2 undertail covert: 3 3/4 (9.5)
3 back contour: 2 1/2 (6.4)
4 back contour: 1 1/2 (3.8)
5 S: 4 1/4 (10.8)
6 P: 7 1/8 (18.1)
7 P: 7 3/8 (18.7)
8 partial wing (outer Ps): 8 1/8 (20.7)

Sandpipers, Phalaropes *(family Scolopacidae)*

Black Turnstone *(Arenaria melanocephala)*

1 outer P: 4 1/2 (11.4)
2 outer P: 4 5/8 (11.7)
3 partial wing (Ps, Ss, coverts): 5 (12.7)
4 underwing coverts: 2 (5.1)
5 S: 2 5/8 (6.7)
6 S clump: 2 5/8 (6.7)
7 partial wing (shoulder): 2 5/8 (6.7)

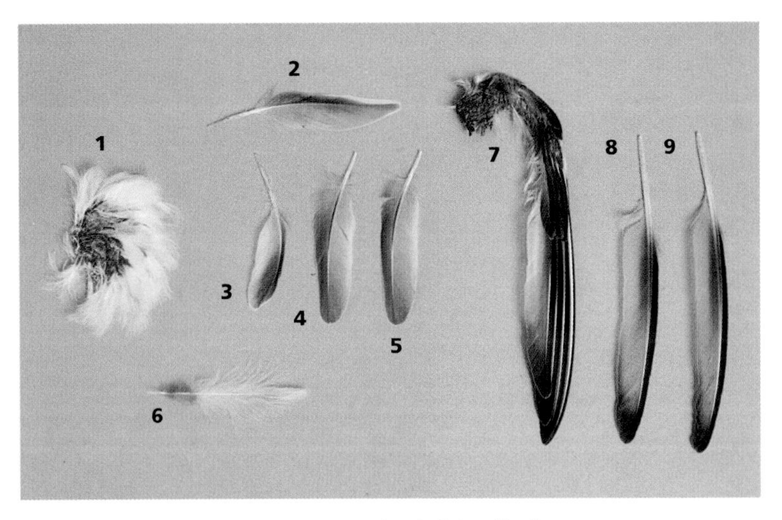

Sanderling—winter plumage *(Calidris alba)*

1 breast clump: n/a
2 flank contour: 1 1/4 (3.2)
3 S: 1 3/4 (4.5)
4, 5 S: 2 (5.1)
6 belly contour: 2 (5.1)
7 partial wing (Ps): 4 1/2 (11.4)
8 inner P: 3 1/2 (8.9)
9 outer P: 3 5/8 (9.2)

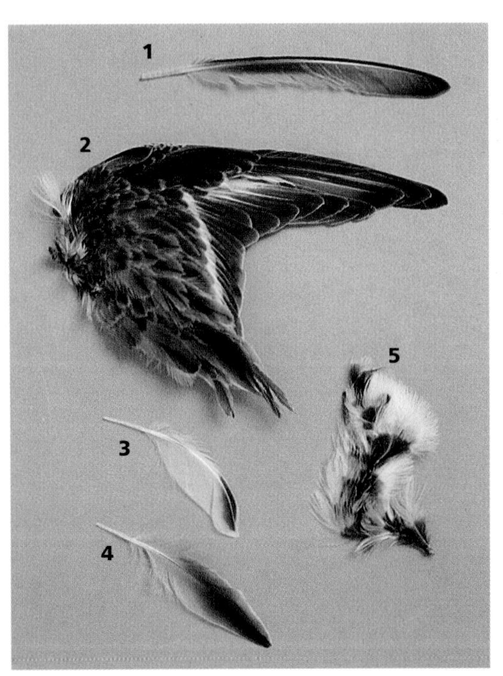

Dunlin *(Calidris alpina)*

1 P: 3 1/2 (8.9)
2 wing: 4 1/2 (11.4)
3 inner S: 2 (5.1)
4 Tert.: 2 1/2 (6.4)
5 belly clump: n/a

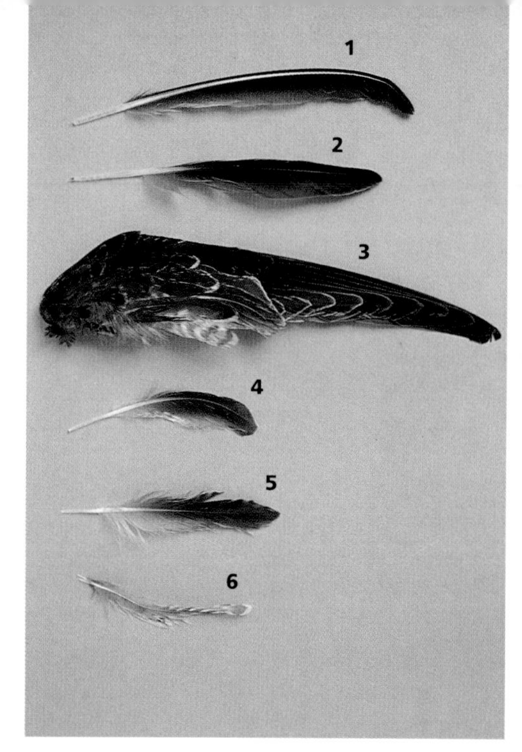

Upland Sandpiper
(Bartramia longicauda)

1 P: 4 1/2 (11.4)
2 mid P: 4 1/8 (10.5)
3 partial wing w/ coverts: 6 1/8 (15.5)
4 S: 2 1/2 (6.3)
5 Tert.: 2 7/8 (7.3)
6 underwing covert: 2 1/4 (5.7)

Short-billed Dowitcher
(Limnodromus griseus)

1 outer P: 4 3/8 (11.1)
2 wing: 5 1/4 (13.3)
3, 4 T covert: 2 3/8 (6.1)
5 S clump: up to 2 1/2 (6.4)
6 back contour: 3 3/4 (9.5)

Common Snipe *(Gallinago gallinago)*

1 outer P: $3\frac{1}{4}$ (8.3)
2 partial wing w/ coverts: $4\frac{1}{8}$ (10.5)
3 breast contour: $1\frac{3}{4}$ (4.5)
4 back contour: $1\frac{3}{4}$ (4.5)
5 wing covert: $2\frac{1}{4}$ (5.7)
6 wing covert clump: n/a
7 Tert.: $3\frac{1}{8}$ (7.9)
8 underwing covert: $2\frac{1}{4}$ (5.7)

American Woodcock *(Scolopax minor)*

1 back contour: $1\frac{7}{8}$ (4.8)
2 back contour: 2 (5.1)
3 head contour: 1 (2.5)
4, 5 belly down: $1\frac{1}{2}$ (3.8)
6 S covert: 3 (7.6)
7, 8 inner S: $3\frac{1}{4}$ (8.3)
9 outer P: $3\frac{1}{2}$ (8.9)
10 outermost P: $3\frac{1}{4}$ (8.3)

Skuas, Gulls, Terns, Skimmers *(family Laridae)*

Gulls

Franklin's Gull *(Larus pipixcan)*

1 partial wing (P to S): $9\frac{1}{2}$ (24.1)
2 P: $7\frac{3}{4}$ (19.7)
3 S: $4\frac{1}{8}$ (10.4)
4 underwing covert: $4\frac{3}{8}$ (11.1)

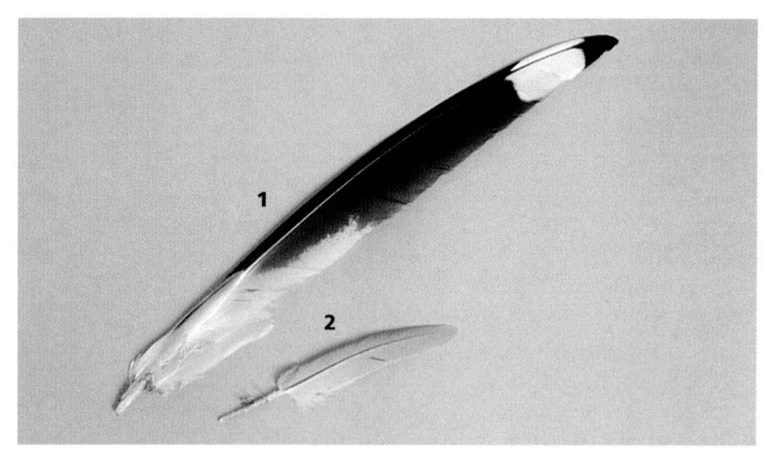

Ring-billed Gull (*Larus delawarensis*)

1 P: $11\frac{1}{8}$ (28.2)
2 P covert: $4\frac{1}{2}$ (11.4)

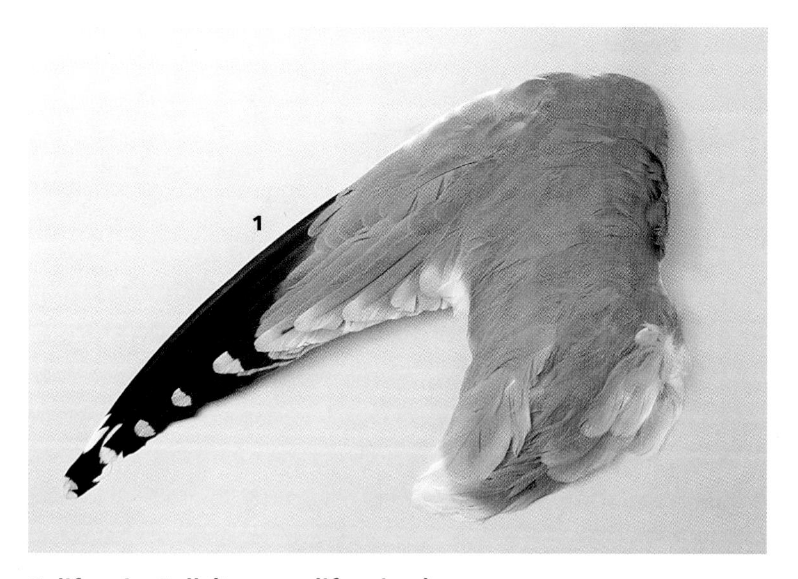

California Gull (*Larus californicus*)

1 outer P: 13¼ (33.7)

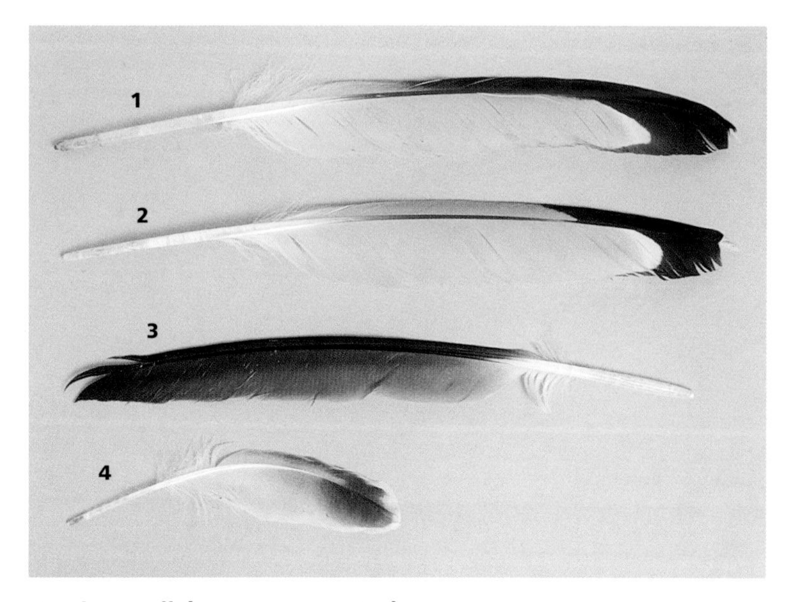

Herring Gull (*Larus argentatus*)

1 P: 11¼ (28.6)
2 P: 11⅛ (28.2)
3 P: 10¼ (26.0)
4 S: 5⅜ (13.7)

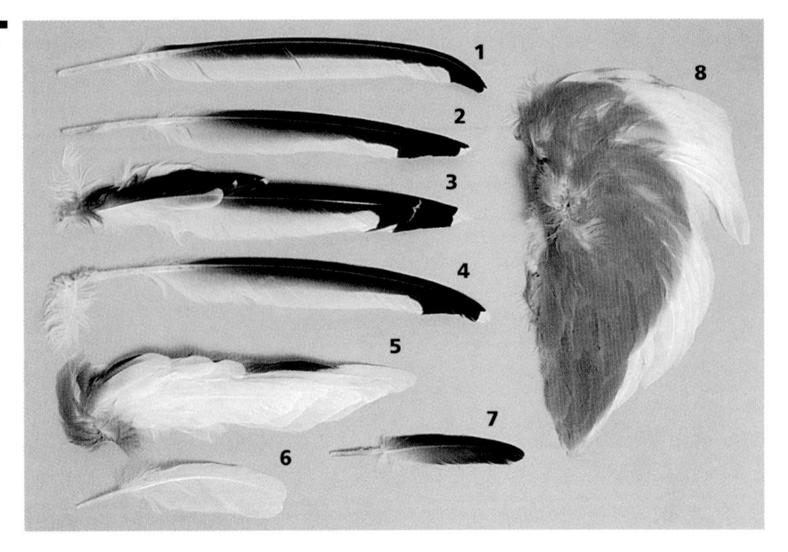

Sabine's Gull *(Xema sabini)*

1 outer P: $7\frac{3}{4}$ (19.7)
2 outer P: $7\frac{3}{8}$ (18.7)
3 partial wing (Ps w/ coverts): $7\frac{3}{8}$ (18.7)
4 outer P: $7\frac{3}{4}$ (19.7)
5 midwing (Ps): $6\frac{1}{8}$ (15.5)
6 S: $3\frac{3}{4}$ (9.5)
7 P covert: $3\frac{1}{2}$ (8.9)
8 shoulder clump: n/a

Terns

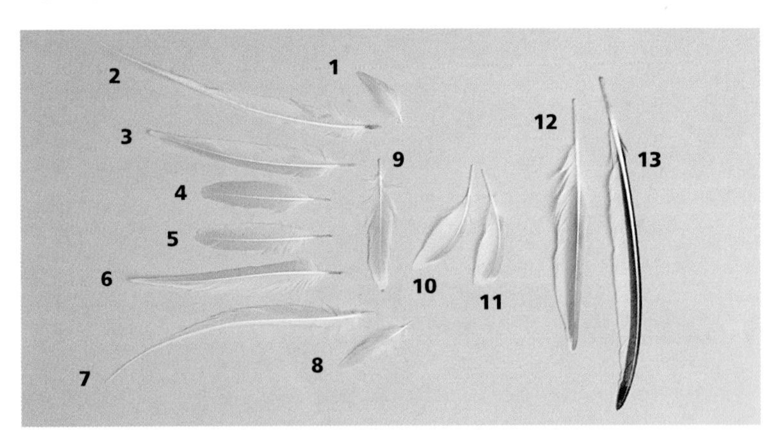

Roseate Tern *(Sterna dougallii)*

1 (crissum) under T: $1\frac{1}{2}$ (3.8)
2 outer T: $6\frac{3}{4}$ (17.1)
3 mid T: $4\frac{7}{8}$ (12.4)
4, 5 center T: 3 (7.6)
6 mid T: $4\frac{7}{8}$ (12.4)
7 outer T: $6\frac{3}{4}$ (17.1)
8 T covert: 2 (5.1)
9 Tert.: $2\frac{7}{8}$ (7.3)
10 S: $2\frac{5}{8}$ (6.6)
11 S: $2\frac{1}{2}$ (6.4)
12 mid P: $5\frac{1}{2}$ (14.0)
13 outer P: $7\frac{1}{4}$ (18.4)

Pigeons, Doves *(family Columbidae)*

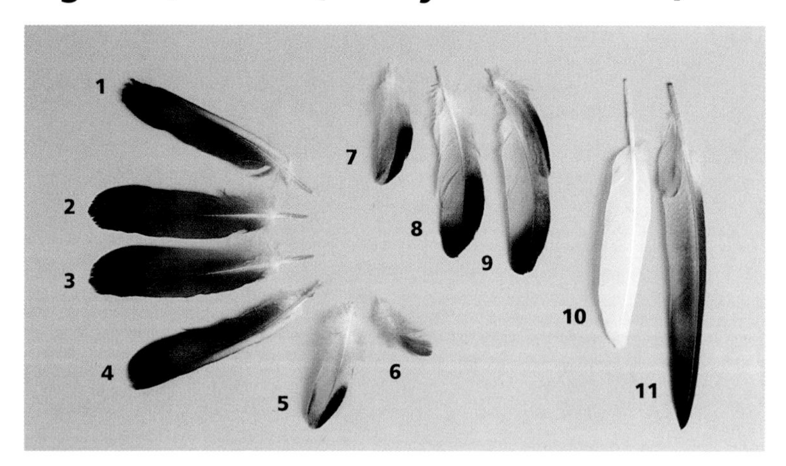

Rock Dove *(Columba livia)*

1 T: $5\frac{1}{8}$ (13.0)
2, 3 T: $5\frac{1}{4}$ (13.3)
4 T: $5\frac{1}{8}$ (13.0)
5 covert T: 3 (7.6)
6 back: n/a
7 Tert.: $2\frac{3}{4}$ (7.0)
8 S: $4\frac{1}{4}$ (10.8)
9 S w/ covert: $4\frac{5}{8}$ (11.7)
10 mid P: 6 (15.2)
11 outer P: $7\frac{3}{4}$ (19.7)

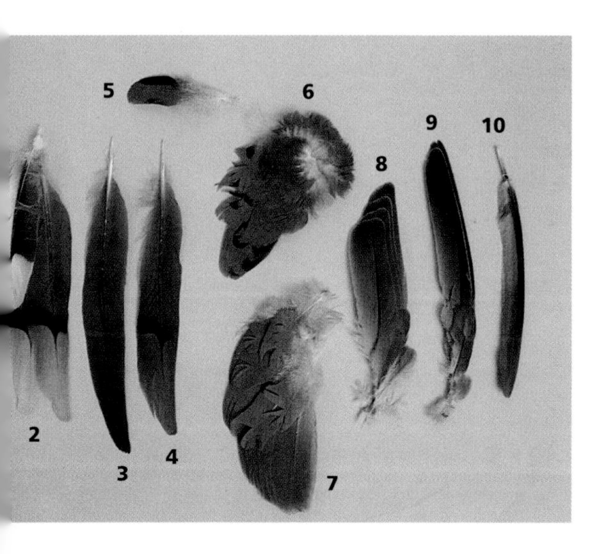

Mourning Dove *(Zenaida macroura)*

1 outer T: $3\frac{1}{4}$ (8.3)
2 mid T clump: n/a
3 center T: 6 (15.2)
4 center T: $5\frac{1}{2}$ (14.0)
5 shoulder contour: $2\frac{1}{4}$ (5.7)
6 shoulder clump: n/a
7 midwing (Ss w/coverts): $3\frac{1}{4}$ (8.3)
8 midwing (Ss): $4\frac{1}{4}$ (10.8)
9 P (2): 5 (12.7)
10 outer P: $4\frac{3}{4}$ (12.1)

Cuckoos, Roadrunners (family Cuculidae)

Yellow-billed Cuckoo (*Coccyzus erythropthalmus*)

1 partial wing (Terts., Ss): 3 1/2 (8.9)
2 breast contour clump: up to 2 3/8 (6.0)
3 back contour clump: 1 1/4 (3.2)
4 S: 4 1/8 (10.5)
5 S, P: up to 4 7/8 (12.4)
6 P: 4 7/8 (12.4)
7 T: 5 1/8 (13.0)
8 T: 5 3/4 (14.6)

Greater Roadrunner (*Geococcyx californianus*)

1 T: 8 (20.3)
2 T: 10 (25.4)
3 T: 11 (28.0)
4, 5 T: 11 3/8 (28.9)
6 T covert: 4 3/8 (11.1)

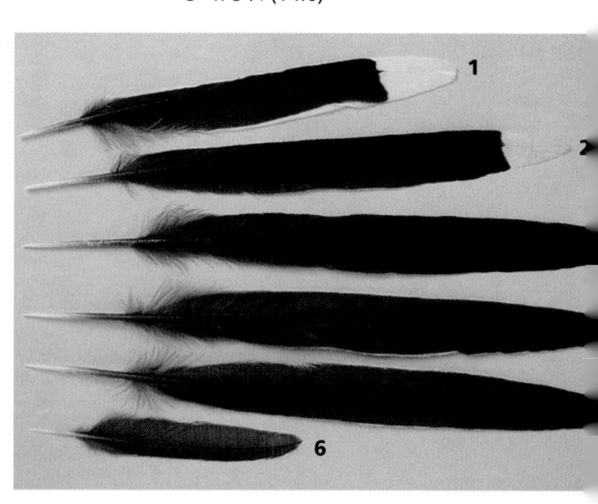

Greater Roadrunner

1 outer P: 4 1/2 (11.4)
2 P: 5 1/4 (13.3)
3 P: 5 5/8 (14.3)
4 inner P: 5 1/2 (14.0)
5, 6 S: 5 5/8 (14.3)
7 S: 5 1/2 (14.0)

Owls *(families Tytonidae and Strigidae)*

Barn Owl *(Tyto alba)*

1 partial wing (Ss w/ coverts): 7 (17.8)
2 P w/ covert: 9 3/8 (23.8)
3 inner P: 8 1/2 (21.6)
4 P: 10 5/8 (27.0)
5 wing covert: 2 1/4 (5.7)

Short-eared Owl (*Asio flammeus*)

1 outer P w/ covert: 7 7/8 (20.0)
2 partial wing (outer Ps): 7 3/8 (18.7)
3 back clump: n/a
4 S: 6 1/8 (15.5)
5 T (2): 6 (15.2)
6 outer T: 5 3/4 (14.6)

Great Horned Owl (*Bubo virginianus*)

1 outer T: 8 5/8 (21.9)
2 central T (2): 9 3/8 (23.8)
3 filoplume T: 7 5/8 (19.4)
4 belly contour: 5 1/4 (13.3)
5 neck clump: n/a
6 S: 7 1/4 (18.4)
7 S: 9 1/4 (23.5)
8 inner P: 10 3/8 (26.4)

Barred Owl (*Strix varia*)

1 T (2): 8 1/8 (20.6)
2 filoplume T: 6 3/8 (16.2)
3 breast clump: n/a
4 S: 8 1/2 (21.6)
5 partial wing (Ps): 11 (27.9)
6 outer P: 10 1/4 (26.0)
7 outer P: 7 1/8 (18.1)

Great Gray Owl (*Strix nebulosa*)

1 filoplume: 6 1/2 (16.5)
2 P: 14 1/4 (36.2)
3 belly filoplume clump: 7 1/2 (19.1)

Snowy Owl (*Nyctea scandiaca*)

1 tail section w/ undercovert: 10 1/2 (26.7)
2 T: 9 7/8 (25.1)
3 S: 8 3/4 (22.2)
4 partial wing w/ coverts: 10 (25.4)
5 outer P: 12 3/4 (32.4)

Long-eared Owl (*Asio otus*)

1 inner T: 6 3/8 (16.2)
2 outer T: 5 3/4 (14.6)
3 partial wing (Ss): 6 1/2 (16.5)
4 inner S: 6 (15.2)
5 partial wing (Ps): 8 1/2 (21.3)
6 P: 8 1/4 (21.0)
7 S covert: 2 (5.1)
8 breast down clump: n/a
9 T covert: 4 1/4 (10.8)
10 flank: 2 1/2 (6.4)

Northern Saw-whet Owl *(Aegolius acadicus)*

1 outer T: 3 1/8 (7.9)
2 inner T: 3 1/8 (7.9)
3–5 outer T: 3 1/8 (7.9)
6 belly clump: n/a
7 S covert: 2 (5.1)
8 inner S: 3 1/4 (8.3)
9 outer S: 3 3/4 (9.5)
10 mid P: 4 7/8 (12.4)
11 outer P: 3 1/2 (8.9)
12 alula: 1 7/8 (4.8)

Burrowing Owl *(Athene cunicularia)*

1 S: 9 1/8 (23.2)
2 S: 9 1/4 (23.5)
3 P: 8 5/8 (21.9)
4 P: 9 5/8 (24.5)
5 P: 8 1/2 (21.6)

Note “soundproofing” serrations on leading edges of primaries.

Nighthawks, Nightjars *(family Caprimulgidae)*

Common Nighthawk *(Chordeiles minor)*

1 outer T: 4 1/4 (10.8)
2 partial wing w/ coverts: 7 (17.8)
3 wing covert (2): 3 (7.6)
4 S: 2 7/8 (7.3)
5 partial wing: 4 5/8 (11.7)
6 Ps (2): 5 1/4 (14.6)
7 P: 6 3/8 (16.2)

Common Nighthawk

1 undertail covert: 1 5/8 (4.1)
2 undertail covert: 2 1/2 (6.4)
3, 4 T: 4 3/4 (12.0)
5 T: 4 1/2 (11.4)
6 T: 4 1/4 (10.8)
7 T: 4 (10.2)
8 undertail covert: 2 1/8 (5.4)
9 Tert.: 3 1/4 (8.3)
10 S: 2 7/8 (7.3)
11 S: 3 (7.6)
12 inner P: 4 1/8 (10.5)
13 mid P: 5 5/8 (14.3)
14 outer P: 6 3/8 (16.2)
15 S covert: 2 1/8 (5.4)
16 S covert: 2 1/4 (5.7)

Whip-poor-will *(Caprimulgus vociferus)*

1–4 T: 4 3/4 (12.0)
5 breast clump: n/a
6 wing (Ps): 5 1/2 (14.0)

Swifts *(family Apodidae)*

McCay

White-throated Swift *(Aeronautes saxatalis)*

1–3 wing covert: 1 3/4 (4.5)
4 T: 3 1/4 (8.2)
5 partial wing (Ps, Ss, Terts.): 6 1/4 (15.9)
6 P: 5 1/8 (13.0)

Hummingbirds *(family Trochilidae)*

Broad-tailed Hummingbird—female
(Selasphorus platycercus)

1 bird: 3½ (8.9)
2 T: 1⅛ (2.9)

Kingfishers *(family Alcedinidae)*

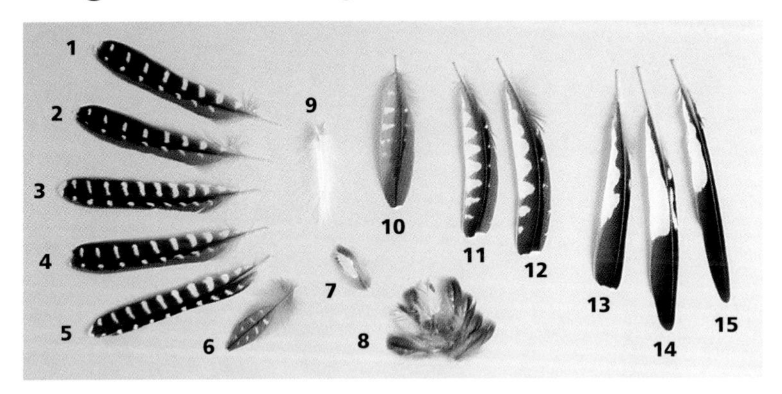

Belted Kingfisher *(Ceryle alcyon)*

1 outer T: $3\frac{3}{4}$ (9.5)
2 mid T: $3\frac{3}{4}$ (9.5)
3 center T: $3\frac{3}{4}$ (9.5)
4 mid T: $3\frac{3}{4}$ (9.5)
5 outer T: $3\frac{3}{4}$ (9.5)
6 rump: $1\frac{3}{4}$ (4.4)
7 breast: n/a
8 breast clump: n/a
9 (crissum) under T covert: $1\frac{3}{4}$ (4.4)
10 Tert.: $2\frac{3}{4}$ (7.0)
11 inner S: $3\frac{1}{8}$ (7.9)
12 outer S: $3\frac{5}{8}$ (9.2)
13 inner P: 4 (10.1)
14 outer P: $4\frac{3}{4}$ (12.0)
15 outer P: $4\frac{1}{2}$ (11.4)

Woodpeckers *(family Picidae)*

McCay

Lewis's Woodpecker *(Melanerpes lewis)*

1 breast down clump: n/a
2 breast contour: $1\frac{1}{4}$ (3.2)
3 inner T: $4\frac{3}{8}$ (11.2)
4 wing section (Ss): $4\frac{5}{8}$ (11.7)
5 inner P: $4\frac{3}{4}$ (12.0)
6 P: $5\frac{3}{4}$ (14.6)

McCay

Red-bellied Woodpecker *(Melanerpes carolinus)*

wing: 5 1/2 (14.0)

McCay

Red-bellied Woodpecker

tail: 3 1/8 (7.9)

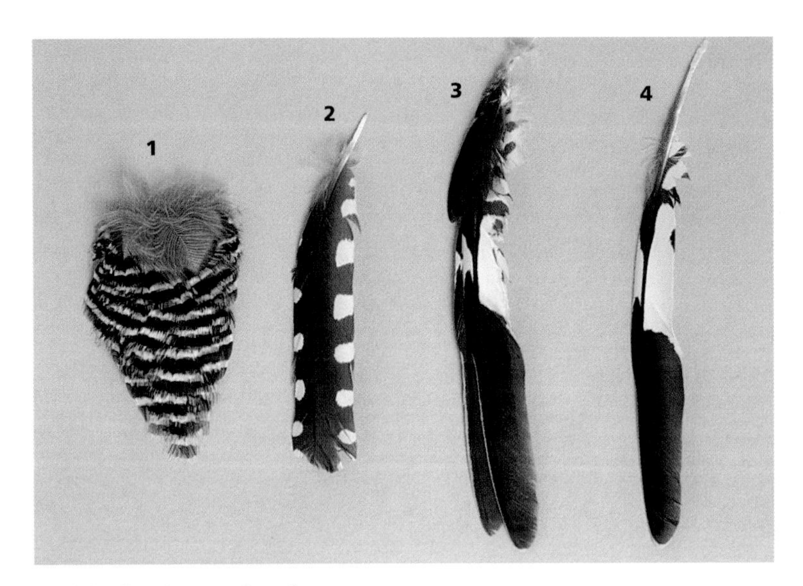

Red-bellied Woodpecker

1 upper-back section: n/a
2 S: 2 1/2 (6.4)
3 outer P (2): 4 3/4 (12.0)
4 outer P: 4 3/4 (12.0)

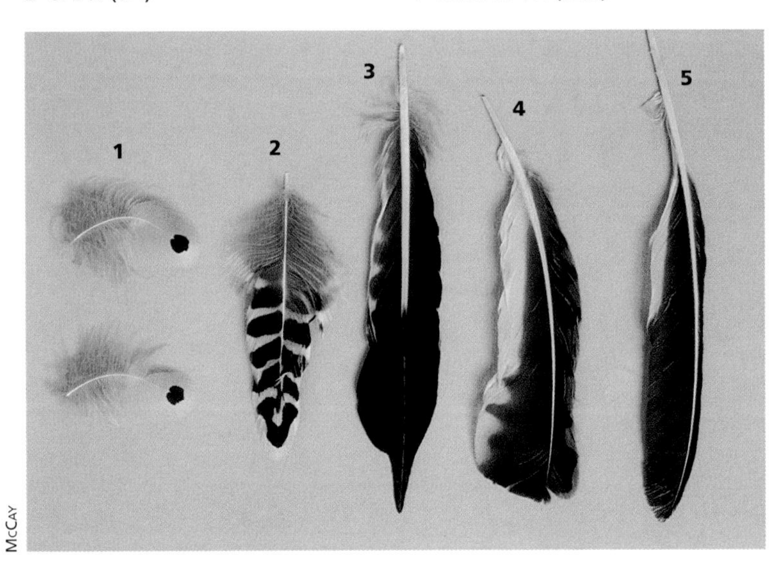

McCay

Northern Flicker—yellow-shafted *(Colaptes auratus)*

1 breast contours (2): 1 1/2 (3.8)
2 T covert: 2 3/4 (7.0)
3 T: 4 3/8 (11.1)
4 S: 3 3/4 (9.5)
5 P: 4 5/8 (11.7)

Northern Flicker—red-shafted

1 midwing (Ss w/ coverts): 4 1/4 (10.8)
2, 3 T: 4 5/8 (11.7)
4 S: 4 5/8 (11.7)
5 outer P: 5 1/2 (14.0)

Yellow-bellied Sapsucker *(Sphyrapicus varius)*

1 T covert: 1 1/2 (3.8)
2 center T: 3 1/8 (8.0)
3 outer T: 2 3/8 (6.0)
4 mid T: 2 7/8 (7.3)
5 center T: 3 1/8 (8.0)
6 outer T: 2 5/8 (6.6)
7 mid T: 2 7/8 (7.3)
8 breast clump: n/a
9 Tert.: 1 1/2 (3.8)
10 inner S: 2 1/2 (6.4)
11 outer S: 2 7/8 (7.3)
12 mid P: 3 1/4 (8.3)
13 outer P: 4 1/8 (10.5)
14 outer P: 3 1/2 (8.9)

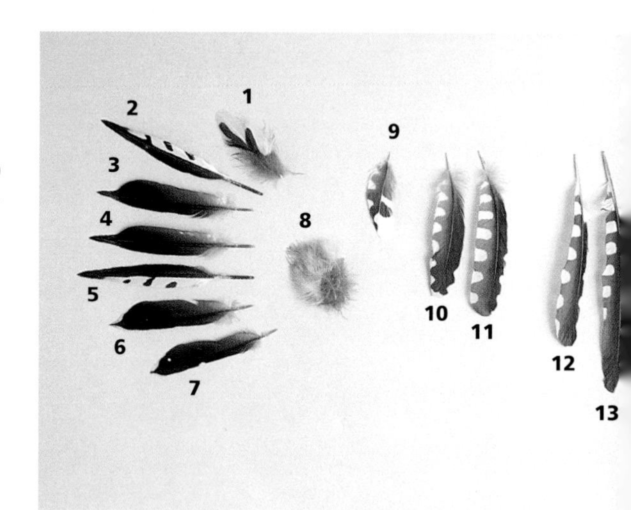

Yellow-bellied Sapsucker

1 breast contour: 1 1/4 (3.2)
2 midwing (Ss): 3 3/8 (8.5)
3 S: 3 1/8 (7.9)
4 P (2): 4 3/8 (11.1)
5 outer P: 4 1/4 (10.8)

Downy Woodpecker *(Picoides pubescens)*

1 partial wing (Ss): 2 1/2 (6.4)
2 partial wing (Ps): 3 7/16 (8.7)
3 outer T: 2 3/8 (6.0)
4 inner T: 2 7/8 (7.3)
5 S: 2 1/2 (6.3)
6 P: 3 1/8 (7.9)
7 nape contour (3): 3/8 (1.0)

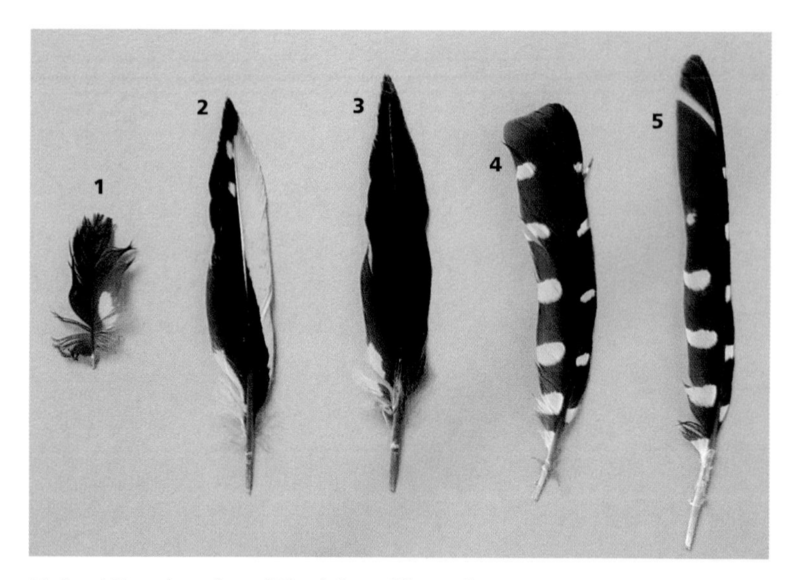

Hairy Woodpecker *(Picoides villosus)*

1 covert: 1 5/16 (3.3)
2 T: 3 3/8 (8.6)
3 T: 3 1/2 (8.7)
4 S: 3 5/16 (8.4)
5 outer P: 4 1/8 (10.5)

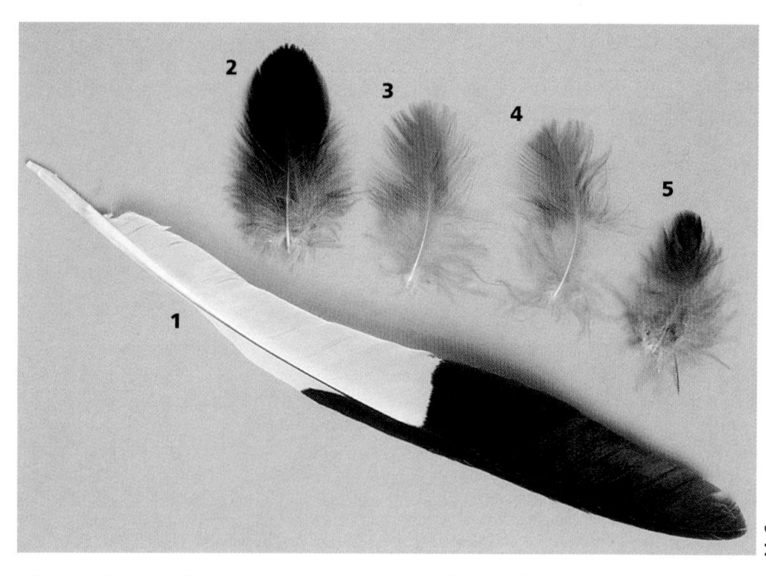

McCay

Pileated Woodpecker *(Dryocopus pileatus)*

1 outer P: 7 1/2 (19.1)
2–5 breast down: 1 5/8 (4.1) avg.

Tyrant Flycatchers *(family Tryannidae)*

Empidonax Flycatchers

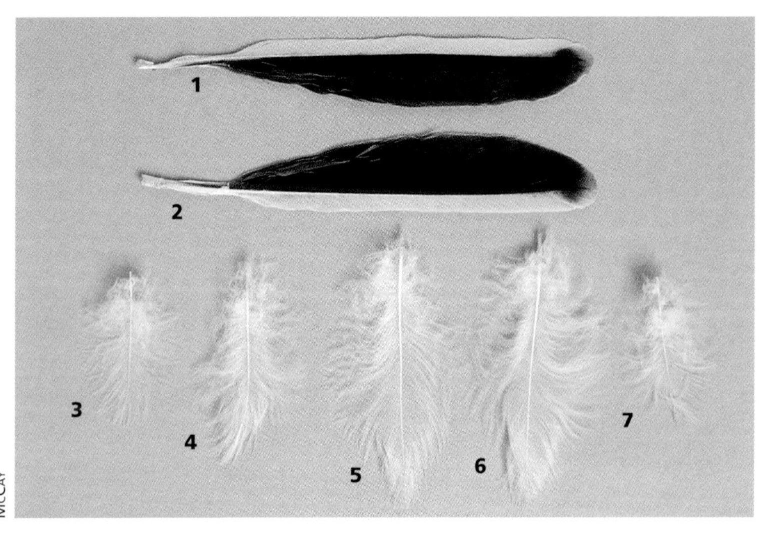

McCay

Western Kingbird *(Tyrannus verticalus)*

1, 2 T: $4\frac{1}{8}$ (10.5)

3–7 undertail coverts: range $2\frac{1}{8}$–$2\frac{3}{4}$ (5.4–7.0)

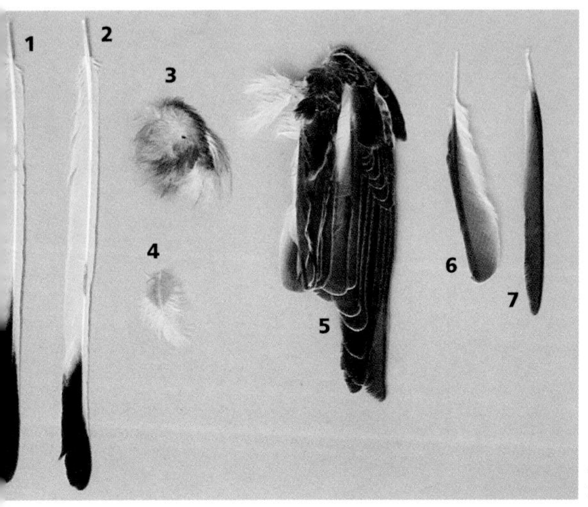

McCay

Scissor-tailed Flycatcher *(Tyrannus forficatus)*

1, 2 T: $5\frac{7}{8}$ (15.0)

3 breast contour: 1 (2.5)

4 underwing clump: n/a

5 wing: $4\frac{1}{2}$ (11.4)

6 S: 3 (7.6)

7 P: $3\frac{5}{8}$ (9.2)

Shrikes *(family Laniidae)*

Loggerhead Shrike *(Lanius ludovicianus)*

1, 2 P: 2 7/8 (7.3)
3 S: 2 3/8 (6.0)
4 partial wing (Ps, Ss, Terts.): 3 3/8 (8.6)

Vireos *(family Vireonidae)*

Blue-headed Vireo *(Vireo solitarius)*

1 P: 2 5/8 (6.7)
2 P: 2 1/2 (6.4)
3 S: 1 3/4 (4.4)
4 S covert: 3/4 (1.9)
5 partial wing (Ps, Ss, Terts.): 3 (7.6)

Crows, Jays (*family Corvidae*)

Blue Jay
(*Cyanocitta cristata*)

1 T: 4 7/8 (12.4)
2 T: 4 1/2 (11.4)
3 T: 5 (12.7)
4 P: 4 7/8 (12.4)
5 S: 4 3/4 (12.1)
6 S: 3 (7.6)
7, 8 Tert.: 1 3/8 (3.5)

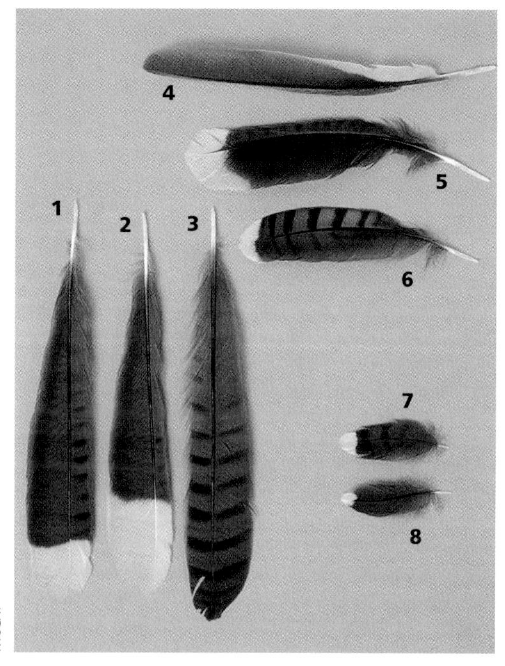

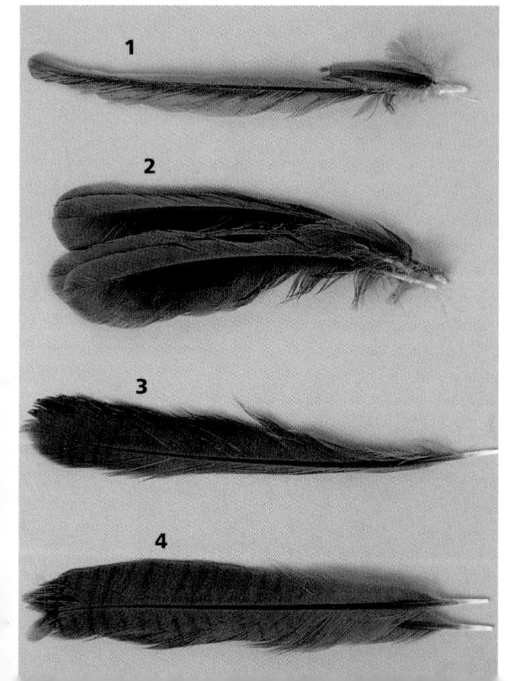

Steller's Jay
(*Cyanocitta stelleri*)

1 P: 5 1/4 (13.3)
2 wing section (Ss): 4 1/2 (11.4)
3 T: 5 3/4 (14.6)
4 T (2): 5 1/2 (14.0)

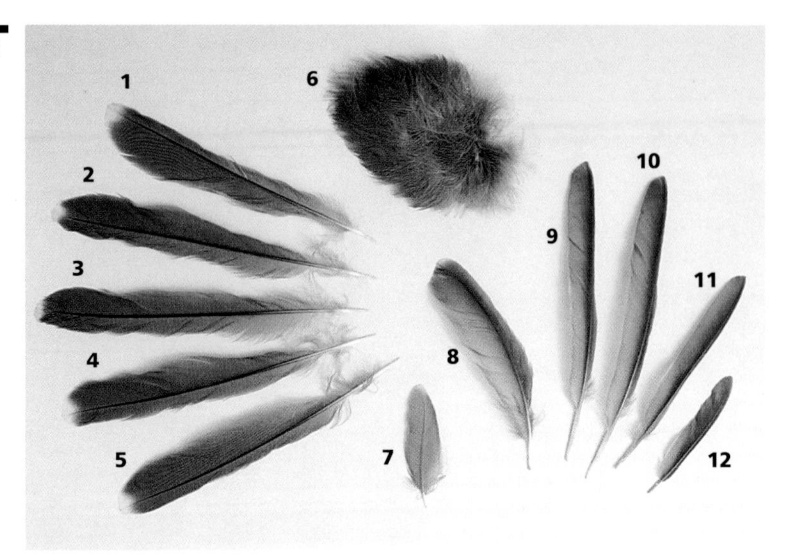

Gray Jay *(Perisoreus canadensis)*

1–5 T: $5\frac{5}{8}$ (14.3)
6 back clump: n/a
7 wing covert: 2 (5.1)
8 S: $3\frac{5}{8}$ (9.2)
9 P: $4\frac{3}{4}$ (12.1)
10 P: 5 (12.7)
11 P: $3\frac{3}{4}$ (9.5)
12 alula: $2\frac{1}{4}$ (5.7)

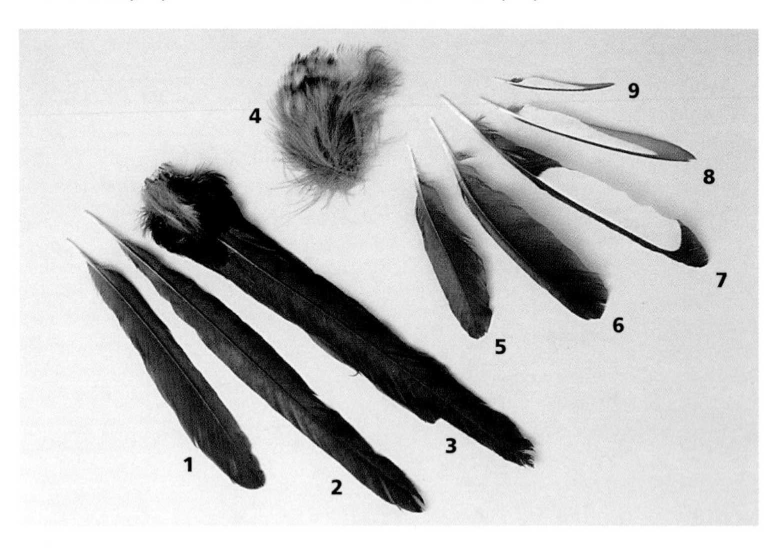

Black-billed Magpie *(Pica pica)*

1 outer T: $6\frac{5}{8}$ (16.8)
2 inner T: $9\frac{5}{8}$ (24.5)
3 partial tail: $10\frac{1}{8}$ (25.7)
4 lower back clump: n/a
5 S: $4\frac{1}{4}$ (10.8)
6 S: $5\frac{5}{8}$ (14.3)
7 inner P: $6\frac{7}{8}$ (17.5)
8 P: 5 (12.7)
9 outer P: $2\frac{3}{4}$ (7.0)

American Crow *(Corvus brachyrhynchos)*

1 P: $9\frac{3}{8}$ (23.8)
2 P: $8\frac{3}{4}$ (22.2)
3 S: $6\frac{1}{2}$ (16.5)

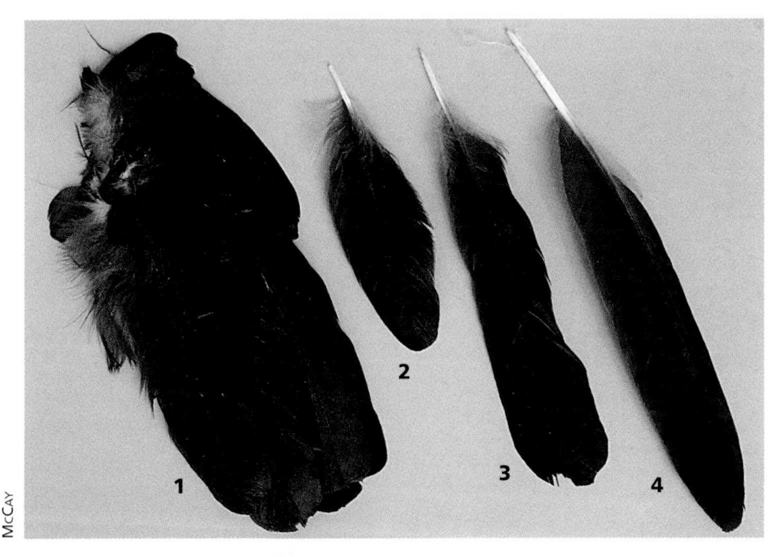

McCay

Northwestern Crow *(Corvus caurinus)*

1 partial wing (Ps w/ Ss): $7\frac{3}{4}$ (20.0)
2 Tert.: $4\frac{1}{4}$ (10.8)
3 S: $6\frac{1}{2}$ (16.5)
4 P: $7\frac{3}{4}$ (20.0)

Fish Crow
(Corvus ossifragus)

1 breast contours: 1 3/4 (4.4)
2 S: 7 (17.8)
3 P w/ covert: 9 (23.0)

Larks *(family Alaudidae)*

Horned Lark *(Eremophila alpestris)*

1 nail: 1/2 (1.3)
2 chin clump: n/a
3 breast clump: n/a
4 wing (Ps, Ss, Terts.): 3 7/8 (9.9)
5 P: 3 1/2 (8.9)

Note elongation of toe 1 due to long nail.

Swallows *(family Hirundinidae)*

Tree Swallow
(Tachycineta bicolor)

1 S: 2⅞ (7.3)
2 S: 3⅛ (7.9)
3 P: 3⅛ (7.9)
4 back contour (5): 1⅛ (2.8) avg.
5 back clump: n/a

Purple Martin
(Progne subis)

1 outer P: 4¾ (12.0)
2 P: 4⅜ (11.1)
3 Tert.: 2½ (6.4)
4 partial wing: 2½ (6.4)

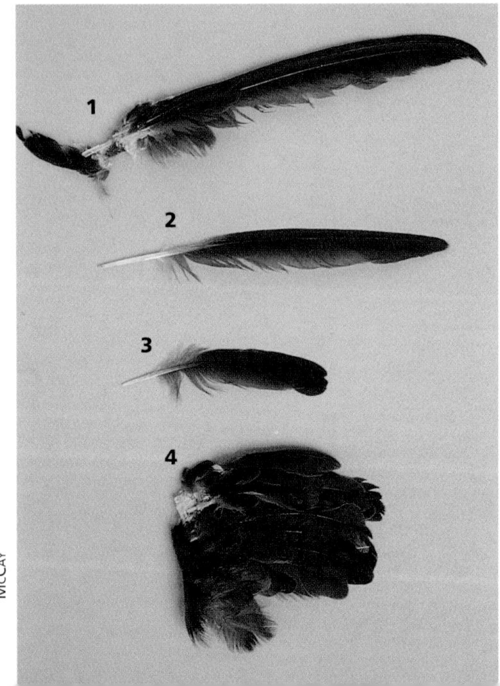

Barn Swallow *(Hirundo rustica)*

1 P: $3^{1}/_{2}$ (8.9)
2 Tert.: $1^{3}/_{4}$ (4.4)
3 T: $2^{7}/_{8}$ (7.3)
4 partial wing: $4^{1}/_{4}$ (10.8)
5 breast clump: n/a

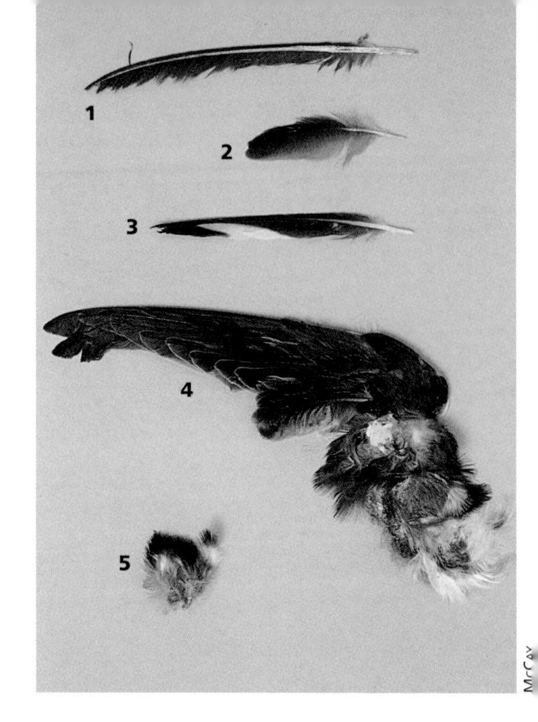

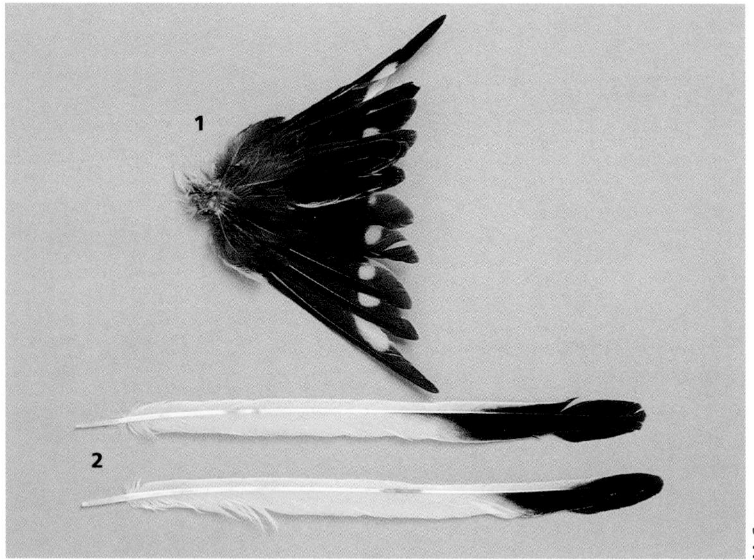

Barn Swallow *(Hirundo rustica)*

1 Tail: up to $5^{7}/_{8}$ (15.0)

Scissor-tailed Flycatcher (Tyrannus forficata)

2 T (2): $2^{7}/_{8}$ (7.3)

Chickadees, Titmice *(family Paridae)*

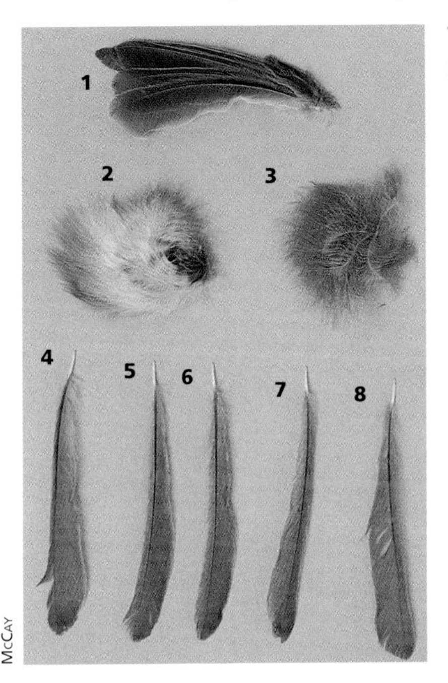

Tufted Titmouse *(Baeolophus bicolor)*

1 wing section (Ss): 2 1/2 (6.4)
2 flank contour clump: 1 1/2 (3.8)
3 nape contour clump: 1 1/8 (2.8)
4–8 T: 3 1/8 (7.9)

Nuthatches *(family Sittidae)*

White-breasted Nuthatch *(Sitta carolinensis)*

1 outer T: 2 1/8 (5.4)
2, 3 inner T: 2 1/8 (5.4)
4 outer T: 2 1/8 (5.4)
5 Tert.: 1 1/4 (3.2)
6 S: 2 3/8 (6.0)
7 P: 3 1/8 (7.9)
8 outer P: 2 3/4 (7.0)

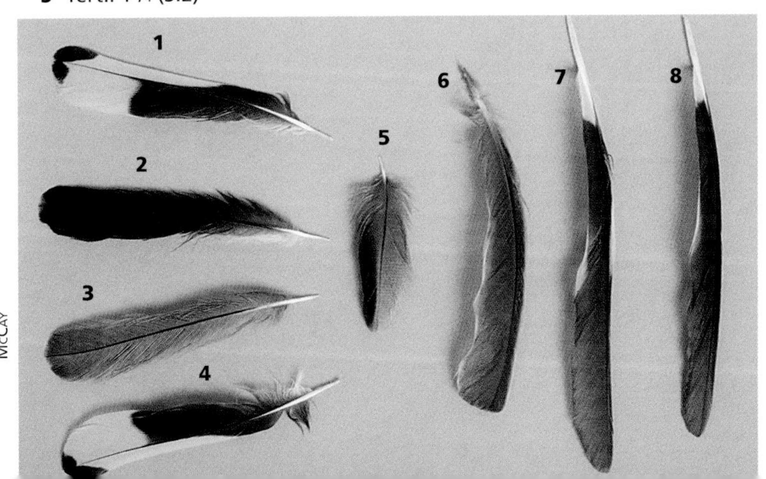

Wrens *(family Troglodytidae)*

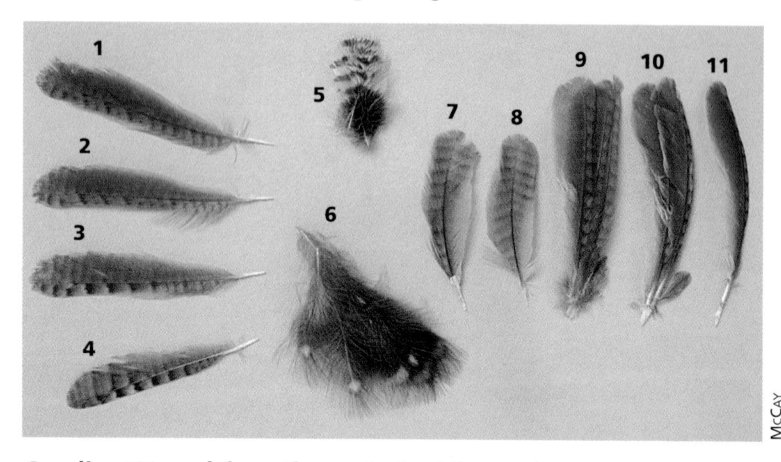

Carolina Wren *(Thryothorus ludoviciannus)*

1–4 T: 2¼ (5.7)
5 undertail covert: 1 (2.5)
6 lower back contours: 1¼ (3.2)
7, 8 Tert.: 1⅝ (4.1)
9 Ss: 2⅜ (6.0)
10 Ps: 2⅜ (6.0)
11 P: 2⅜ (6.0)

Kinglets *(family Regulidae)*

Golden-crowned Kinglet *(Regulus satrapa)*

1 P (2): 2⅛ (5.4)
2 P: 2 (5.1)
3 S: 1⅝ (4.1)
4 nape contour clump: n/a
5 partial wing: 2⅛ (5.4)
6–8 crown contour: ⅜ (1.0)

Thrushes *(family Turdidae)*

Eastern Bluebird *(Sialia sialis)*

1 partial wing (Terts., Ss): $2^1/_4$ (5.7)

2 S: $2^3/_8$ (6.0)

3 P: $3^1/_8$ (7.9)

4 breast contour clump: up to 1 (2.5)

Townsend's Solitaire *(Myadestes townsendi)*

1–6 T: $4^1/_4$ (10.8)

7 wing section (Terts., Ss): $2^5/_8$ (6.7)

8 S: $2^5/_8$ (6.7)

9, 10 belly contour: $1^1/_2$ (3.8)

11, 12 P: $3^7/_8$ (9.9)

13 wing section (outer Ps): $3^7/_8$ (9.9)

14 belly clump: n/a

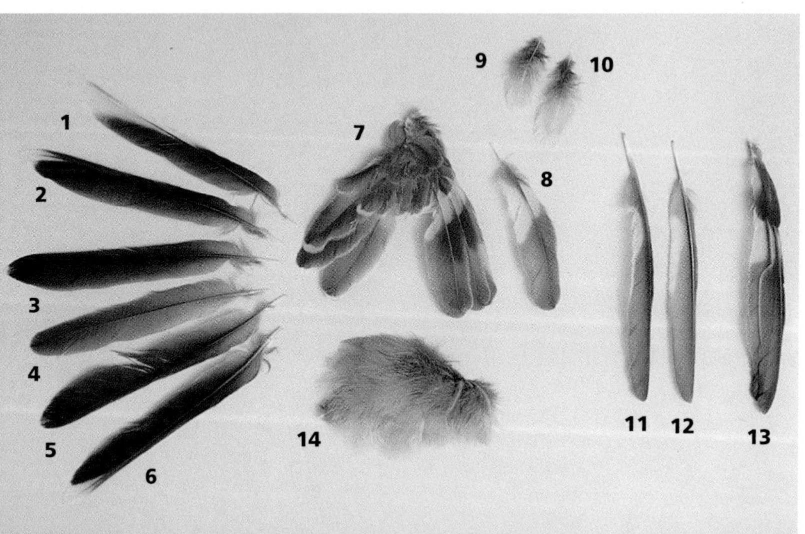

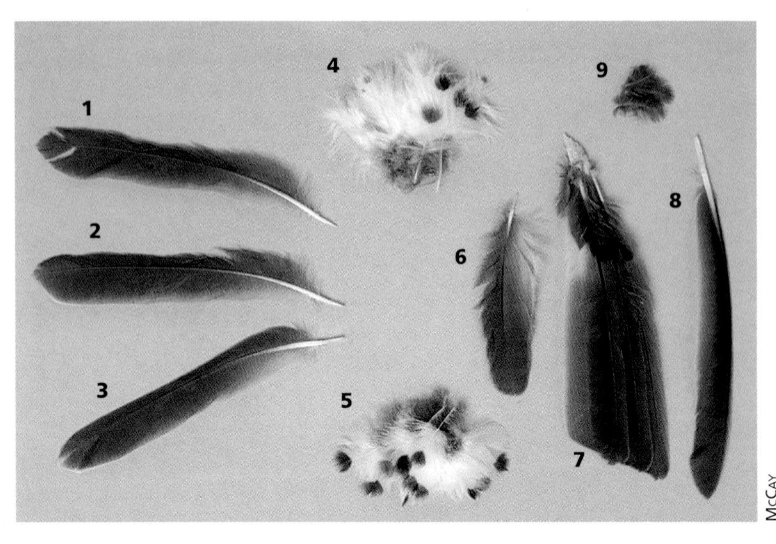

Wood Thrush *(Hylocichla mustelina)*

1–3 T: 3 (7.6)
4 breast contour clump: n/a
5 throat contour clump: n/a
6 Tert.: 1 7/8 (4.8)
7 partial wing (Ss): up to 3 1/8 (7.9)
8 outer P: 3 3/8 (8.9)
9 head contour clump: 5/8 (1.6)

Swainson's Thrush *(Catharus ustulatus)*

1 T: 3 (7.6)
2 Terts.: 2 1/4 (5.7)
3, 4 S: 2 7/8 (7.3)
5 P: 3 1/4 (8.3)
6 throat contours: 1/2 (1.3)

Hermit Thrush
(Catharus guttatus)

1 outer P: 2¾ (7.0)
2 S: 2¼ (5.7)
3 wing section: 3⅜ (8.6)
4 back contour clump: n/a
5 neck contours: ⅝ (1.6)
6 neck contour clump: n/a
7–9 T: 2¾ (7.0)

American Robin
(Turdus migratorius)

1 outer T: 4¼ (10.8)
2 inner T: 4⅜ (11.1)
3 breast contour clump: 1¼ (3.2)
4 Tert.: 2 (5.1)
5 S: 3½ (8.9)
6 P: 4½ (11.4)

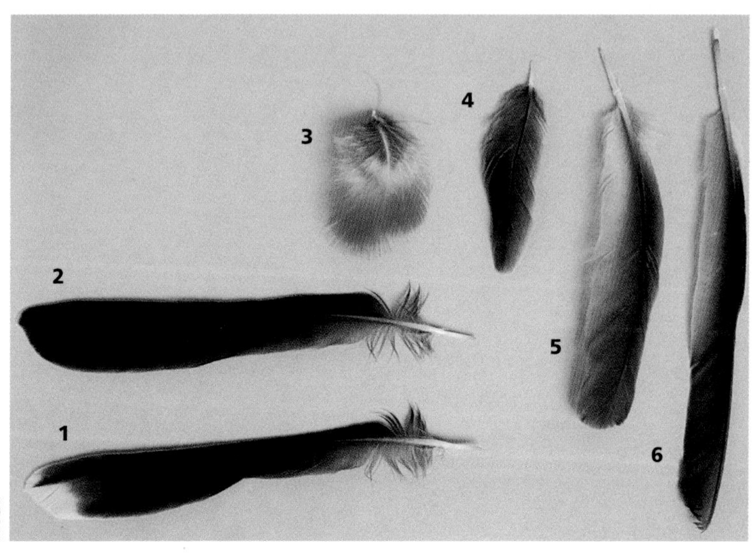

McCAY

Mockingbirds, Thrashers *(family Mimidae)*

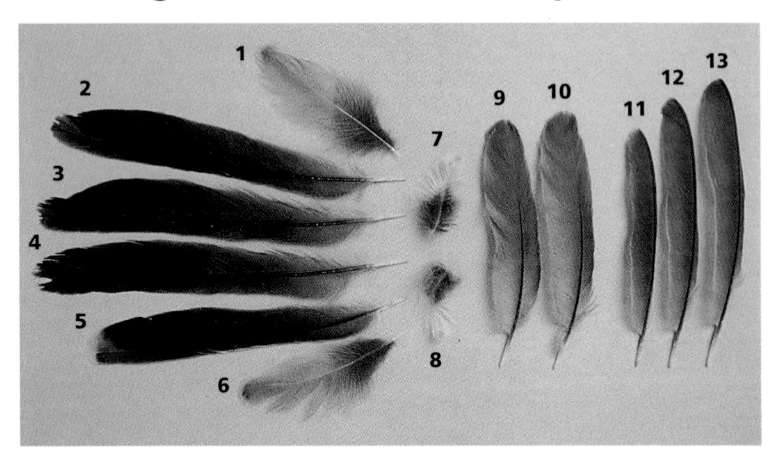

Curve-billed Thrasher *(Toxostoma curvirostre)*

1 undertail covert: $2\frac{1}{4}$ (5.7)
2 T: $4\frac{3}{16}$ (10.7)
3 T: $4\frac{7}{8}$ (12.4)
4 T: $4\frac{3}{4}$ (12.1)
5 T: $4\frac{5}{8}$ (11.7)
6 undertail covert: $2\frac{1}{4}$ (5.7)
7–8 breast contour: $1\frac{1}{8}$ (2.8)
9 S: 3 (7.6)
10 S: $3\frac{1}{8}$ (7.9)
11 P: $2\frac{7}{8}$ (7.3)
12 P: $3\frac{1}{4}$ (8.3)
13 P: $3\frac{1}{2}$ (8.9)

Starlings *(family Sturnidae)*

European Starling *(Sternus vulgaris)*

1 P: $4\frac{1}{2}$ (11.4)
2 S: $2\frac{5}{8}$ (6.7)
3 Terts.: $1\frac{5}{8}$ (4.1)
4 breast clump: $1\frac{5}{8}$ (4.1)
5 undertail covert (2): $2\frac{1}{4}$ (5.7)
6–8 T: $2\frac{7}{8}$ (7.3)

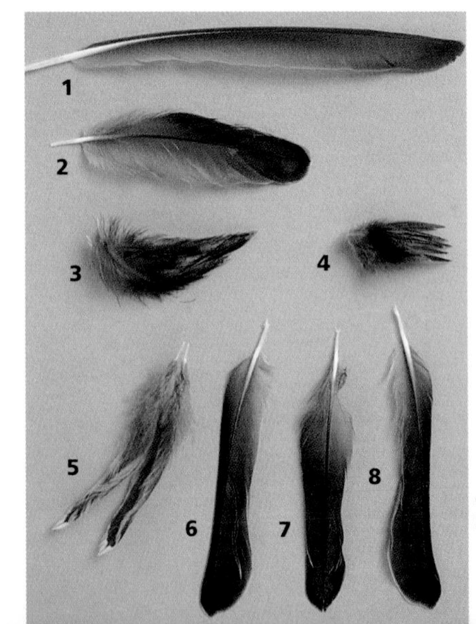

McCay

Waxwings *(family Bombycillidae)*

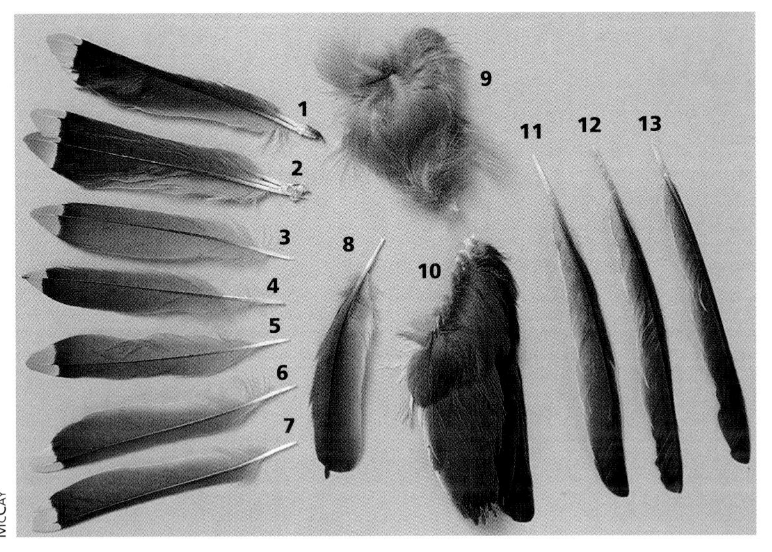

Cedar Waxwing *(Bombycilla cedrorum)*

1–7 T: $2\frac{3}{4}$ (7.0) avg.
8 Tert.: $2\frac{1}{4}$ (5.7)
9 nape contour clump: n/a
10 partial wing (Terts., Ss): $2\frac{3}{4}$ (7.0)
11, 12 P: $3\frac{1}{4}$ (8.3)
13 P: $3\frac{1}{8}$ (7.9)

Wood-Warblers *(family Parulidae)*

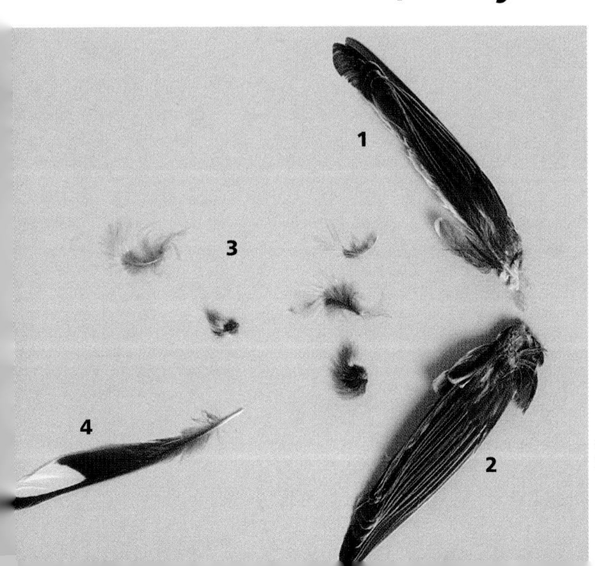

Yellow-rumped Warbler *(Dendroica coronata)*

1, 2 partial wing: up to $2\frac{1}{2}$ (6.4)
3 contours: $\frac{1}{2}$ (1.3) avg.
4 outer T: $2\frac{3}{8}$ (6.0)

Note: Paucity of feathers is representative of a typical kill site.

Yellow Warbler *(Dendroica petechia)*

1, 2 outer T: 1 7/8 (4.8)
3 partial tail: 1 7/8 (4.8)
4 throat contour clump: 1/2 (1.3)
5 S covert: 7/8 (2.2)
6 S: 1 1/2 (3.8)
7 P: 2 1/8 (5.4)

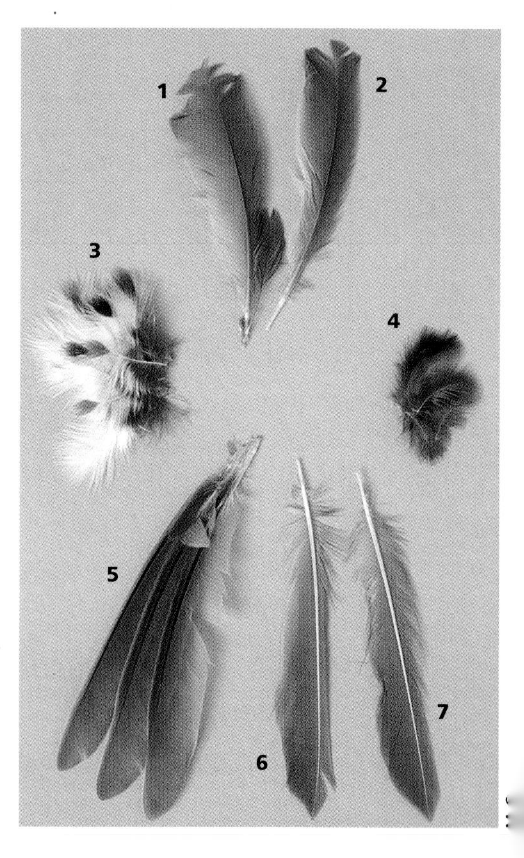

Ovenbird *(Seiurus aurocapillus)*

1 S: 1 13/16 (4.6)
2 S: 1 3/4 (4.4)
3 breast contour clump: 13/16 (2.1)
4 crown contour clump: 3/8 (1.0)
5 partial tail: 2 7/16 (6.2)
6, 7 T: 2 1/4 (5.7)

American Redstart—male *(Setophaga ruticilla)*

1–7 T: 2 3/8 (6.0) avg.
8 side contour clump: n/a
9 crown contour clump: n/a
10 S: 1 1/2 (3.8)
11 P: 2 (5.1)
12 P: 2 1/8 (5.4)

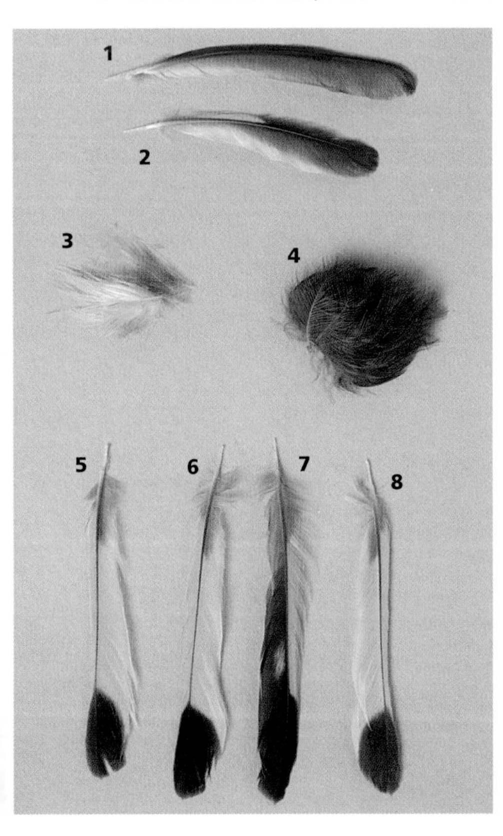

American Redstart—female

1 P: 2 1/8 (5.4)
2 S: 2 (5.1)
3 underwing covert: 3/4 (1.8)
4 back clump: up to 7/8 (2.2)
5–8 T: 2 1/4 (5.7) avg.

Emberizids *(family Emberizidae)*

Green-tailed Towhee *(Pipilo chlorurus)*

1, 2 wings (outer Ps): 5 1/8 (13.0)

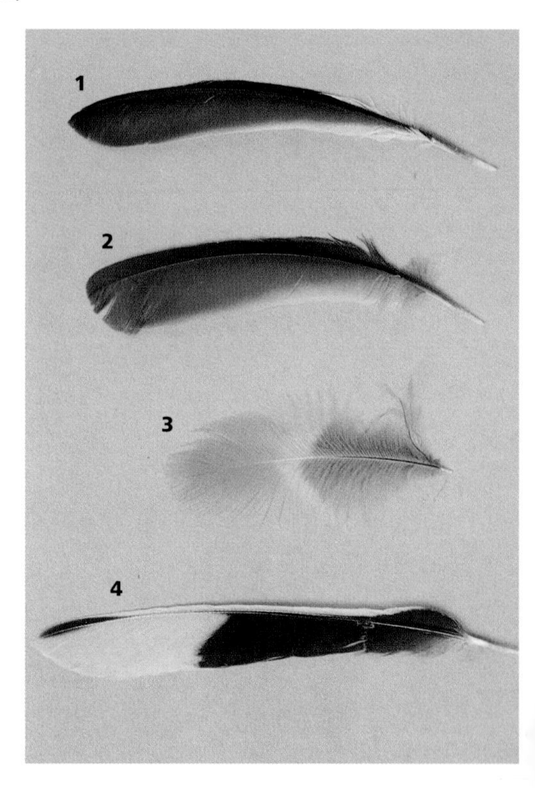

Eastern Towhee *(Pipilo erythrophthalmus)*

1 P: 2 7/8 (7.3)
2 S: 2 5/8 (6.7)
3 breast contour: 1 7/8 (4.8)
4 T: 3 1/4 (8.3)

Snow bunting—winter plumage *(Plectrophenax nivalis)*

1 P: $3\frac{1}{8}$ (7.9)
2 P: 3 (7.6)
3 back contour: 2 (5.1)
4 partial wing: 4 (10.2)

Cardinals *(family Cardinalidae)*

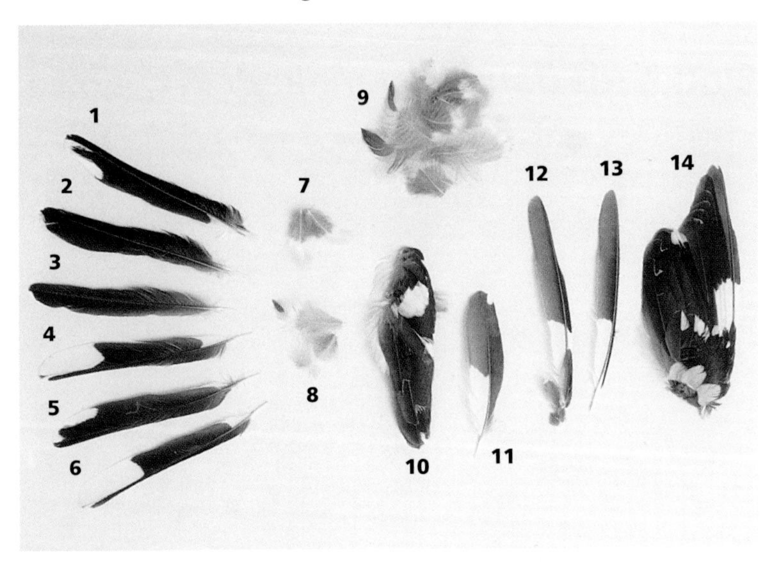

Black-headed Grosbeak *(Pheucticus melanocephalus)*

1–6 T: $3\frac{3}{8}$ (8.6)
7, 8 breast contours: 1 (2.5)
9 breast contour clump: n/a
10 wing section (Ss): $2\frac{3}{4}$ (7.0)
11 S: $2\frac{1}{2}$ (6.4)
12 P: $3\frac{3}{8}$ (8.6)
13 P: $3\frac{1}{4}$ (8.3)
14 partial wing (outer Ps): $3\frac{1}{2}$ (8.9)

Northern Cardinal *(Cardinalis cardinalis)*

1 female P: 3 1/4 (8.3)
2 female S w/ coverts: 2 5/8 (6.7)
3 female breast contour clump: n/a
4, 5 female T: 4 (10.2)
6 male crest: up to 1 1/8 (2.8)
7 male chin w/ breast contours: n/a

Note sex comparison.

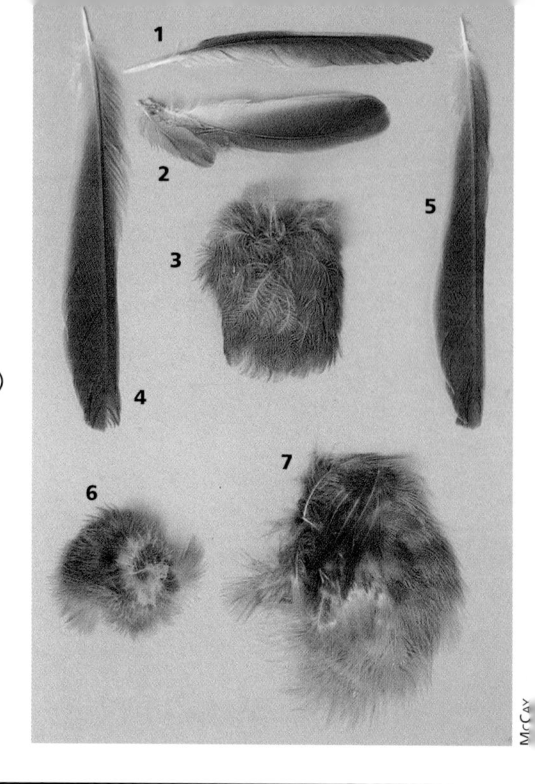

Blackbirds *(family Icteridae)*

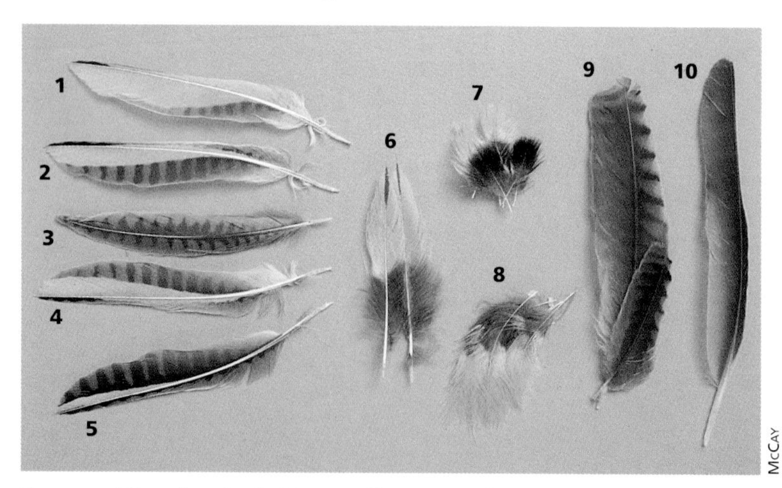

Eastern Meadowlark *(Sturnella magna)*

1–5 T: up to 3 (7.6)
6 undertail coverts: 2 1/4 (5.7)
7 throat contours: 7/8 (2.2)
8 breast contours: 1 3/4 (4.4)
9 S w/ covert: 3 3/8 (8.6)
10 P: 4 7/8 (12.4)

Baltimore Oriole *(Icterus galbula)*

1–4 T: 3 (7.6)
5 belly clump: n/a
6 belly contour: 5/8 (1.6)
7, 8 S: 2 3/8 (6.0)
9 P: 3 1/8 (7.9)

Finches *(family Fringillidae)*

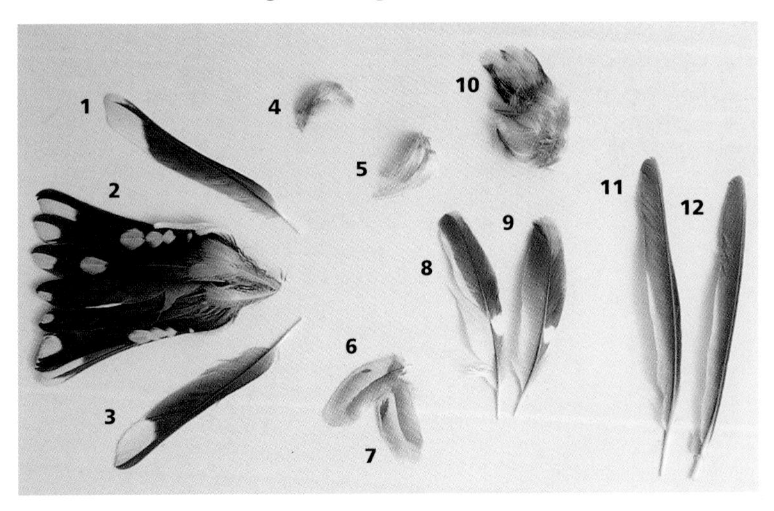

Evening Grosbeak *(Coccothraustes vespertinus)*

1 T: 2 3/4 (7.0)
2 partial tail: n/a
3 T: 2 3/4 (7.0)
4, 5 wing covert: 1 1/4 (3.2)
6 underwing clump: n/a
7 underwing covert: 3/4 (1.9)
8, 9 S: 2 1/4 (5.7)
10 back down clump: n/a
11 P: 3 5/8 (9.2)
12 P: 3 1/2 (8.9)

Old World Sparrows *(family Passeridae)*

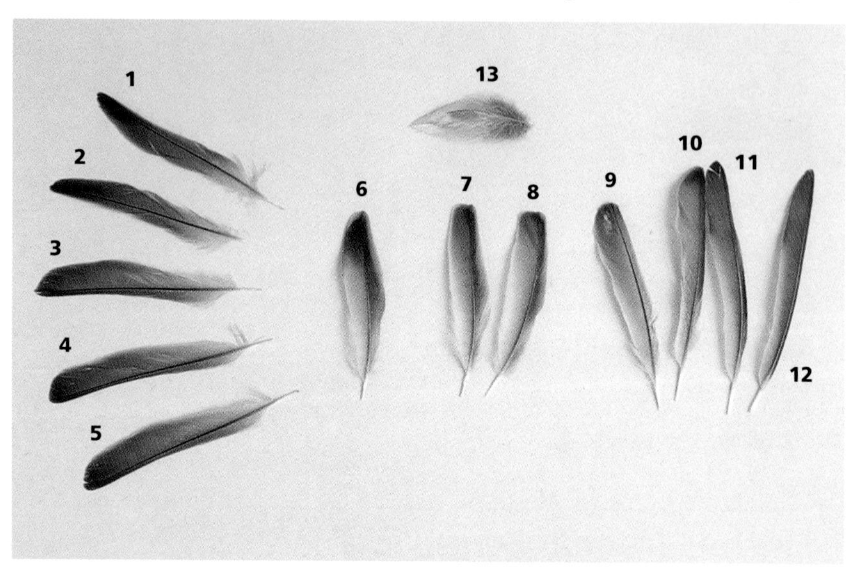

House Sparrow—male *(Passer domesticus)*

1–5 T: 2 1/2 (6.4)
6 Tert.: 2 (5.1)
7, 8 S: 2 (5.1)
9 P: 2 1/8 (5.4)
10 P: 2 3/8 (6.0)
11, 12 P: 2 5/8 (6.7)
13 T covert: 1 1/2 (3.8)

Skulls

7

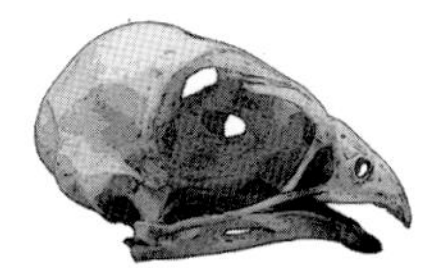

M.E./E.M./C.D.B.

Because the anatomical forms of skulls are so diverse, they provide yet another means of identifying bird species. Skulls and other bones are very useful signs for the tracker and can be just as helpful as pellets or dust baths. And they provide much information when taken in an ecological context. Have you discovered bones or a skull in association with a kill site, a roost, or a nest? How many bones are present, or does only the skull remain? Is the skull still completely intact, or has the brain cavity been broken by small mammals that discovered the carcass or the predator that killed it? Skulls always have a story to tell.

When you come upon a bird skull, first take the time to study the surrounding environment and all the information it has to offer. Then approach the skull, focusing on its particular site. Next, study the bill, which differs among bird species, each of which specializes and thrives in its own feeding niche. Taxonomists organize and separate birds into groups by bill shape and structure. Birding field guides are a great asset in the study of bills, as the head and bill are typically illustrated in superb detail because both are crucial to bird identification in the field.

We have separated skulls into four categories based on bill characteristics, not taxonomical order. These categories will help you begin to narrow down the possibilities: Is it a typical bird bill, an obvious duck bill, a hooked bill, or an unusual bill? Once you have made this decision, refer to the appropriate section to check measurements. Not all species are represented, but the range of illustrations will help you narrow down the possibilities considerably.

The particular skulls illustrated here were chosen for various reasons: They represent a bird family especially well; they are important to researchers in the field; they tend to be more common than others; or they are exceptionally beautiful. If you'd like to get deeper into the study of bird skulls, the bibliography lists sources that are more comprehensive overviews of bird skulls for North America. Add your observations of the habitat and the immediate site where the skull was found and you will certainly be able to narrow down the possibilities, often to the likely bird family or even to the species.

Measurements are provided for bill length, cranium length, and the combination of the two. All measurements are made in a linear fashion; do not measure along the curve of the bill, as this will skew the measurements drastically and your numbers will no longer be a useful comparison to what we have provided. Brown et al. (1987) believe that the ratio of bill length to overall length is diagnostic for a species (that is, bill length ÷ overall length = a percentage). Therefore, we have also provided this data, as it may be useful to the tracker, field worker, and birder. Skulls are arranged within each category from shortest bill length to longest.

We have also provided some technical terms for bones of the skull, which are useful to know as you get started. But since this

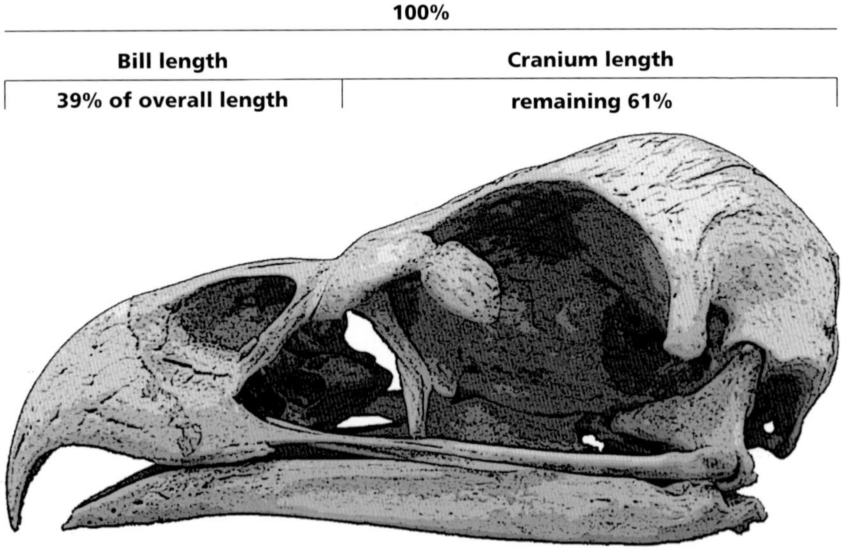

Bald eagle skull—diagram of measurements.

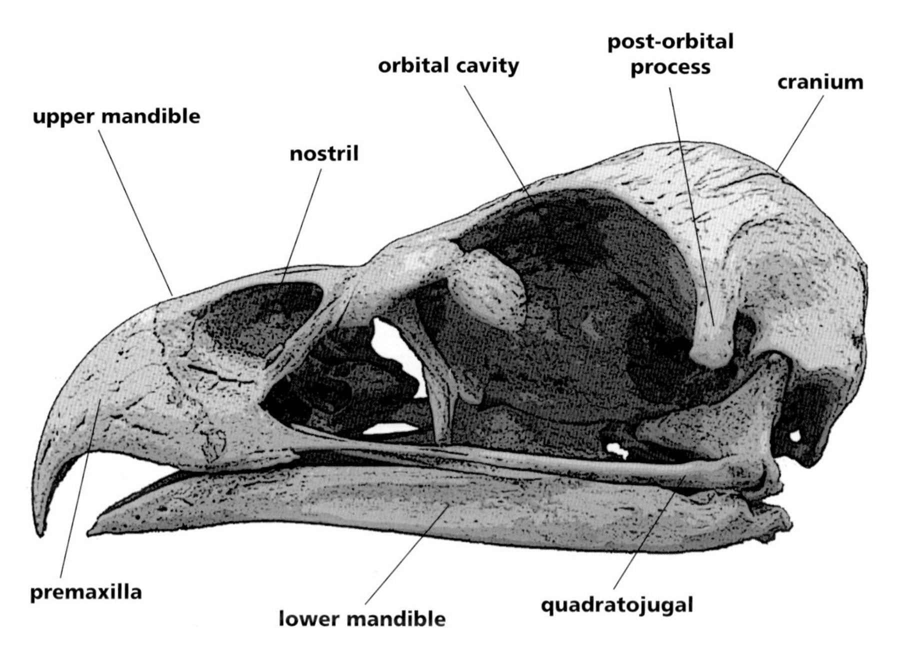

The parts of a bald eagle skull.

chapter serves only as an introduction, we will not be delving comprehensively into skull osteology, the study of bones. For the practical goals of field identification, we are using a pragmatic approach: Certain technical terms and bones are worth knowing, such as the *sclerotic ring* found in intact owl skulls, as they help in species identification.

Bird skulls are made up of over thirty bones, most of which are fused together so well that on casual inspection, the skull appears seamless and created from only several bones. The bones of all species except loons are filled with air cavities, which provide a light but strong structure for flight. And bills are covered by a sheath, called a *ramphothea,* which may be entirely or partially present when you find a skull. This sheath may alter the shape of the bill drastically; if you were to find a crossbill skull without the bill sheaths present, you might be surprised to see that the bones of the bill are as straight as those of other finches. Study the diagrams provided to get a clear visual understanding of measurements and an introduction to technical terms.

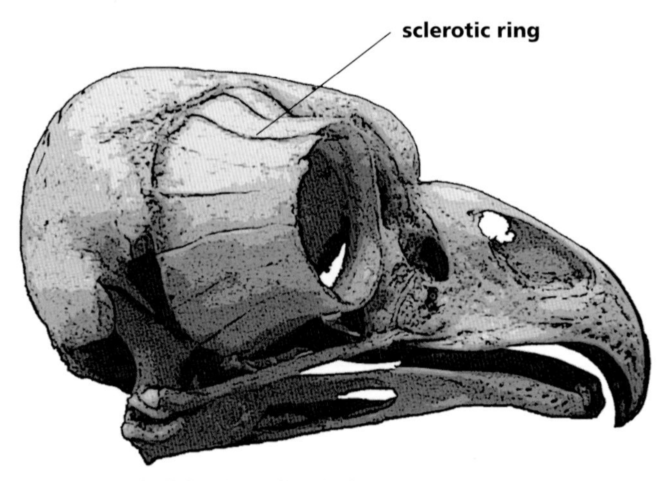

Great horned owl skull showing sclerotic ring.

Typical Bills

Typical bills include a wide variety of bird bills, regardless of length. Stout, thin, long, and short bills all fall within this category.

Black-capped Chickadee *(Poecile atricapillus)*

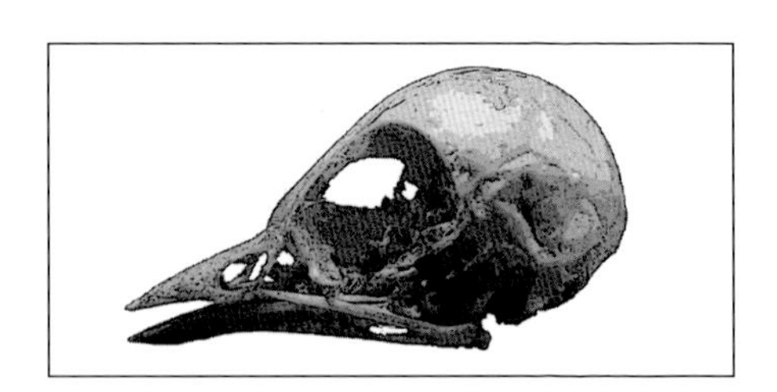

Total: 7/8–1 in., 2.2–2.6 cm
Bill: 5/16–3/8 in., 0.7–1.0 cm
(32–38% of skull length)
Cranium: 5/8–11/16 in., 1.5–1.7 cm

Yellow Warbler *(Dendroica petechia)*

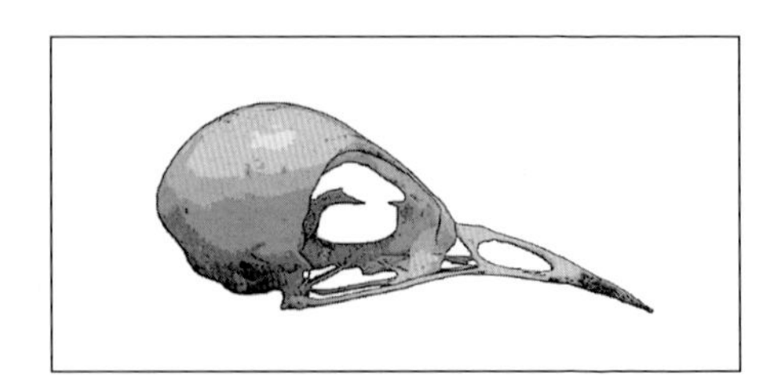

Total: 1–1 1/16 in., 2.6–2.7 cm
Bill: 7/16–1/2 in., 1.2–1.3 cm
(46–48% of skull length)
Cranium: 9/16–5/8 in., 1. 4–1.6 cm

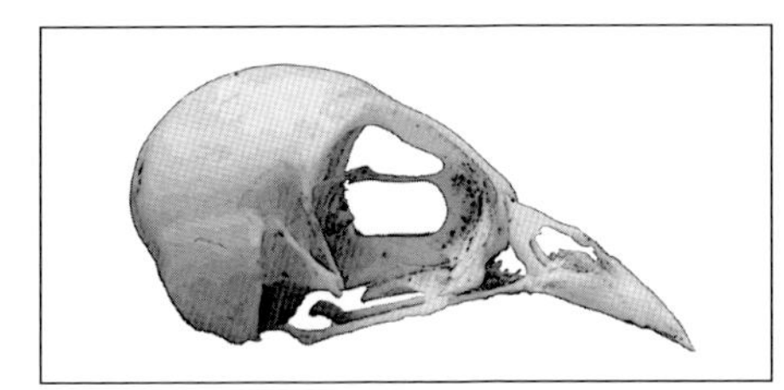

Dark-eyed Junco
(Junco hyemalis)

Total: 1 1/8 in., 2.8 cm
Bill: 3/8–9/16 in., 1.0–1.4 cm
(36–50% of skull length)
Cranium: 11/16–3/4 in., 1.8–1.9 cm

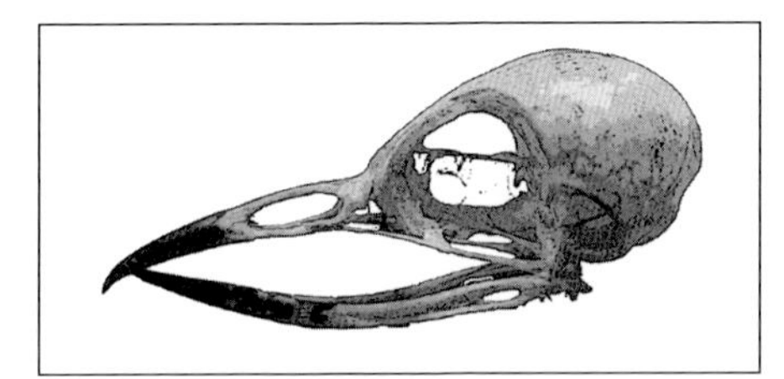

House Wren
(Troglodytes aedon)

Total: 1–1 1/4 in., 2.5–3.1 cm
Bill: 3/8–9/16 in., 1.1–1.4 cm
(39–45% of skull length)
Cranium: 9/16–5/8 in., 1.4–1.6 cm

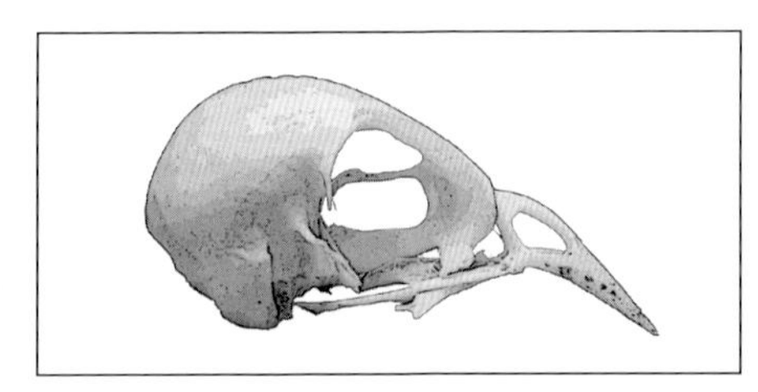

Song Sparrow
(Melospiza melodia)

Total: 1 1/8–1 3/16 in., 2.8–3.0 cm
Bill: 7/16–1/2 in., 1.1–1.3 cm
(37–43% of skull length)
Cranium: 11/16–13/16 in., 1.8–2.0 cm

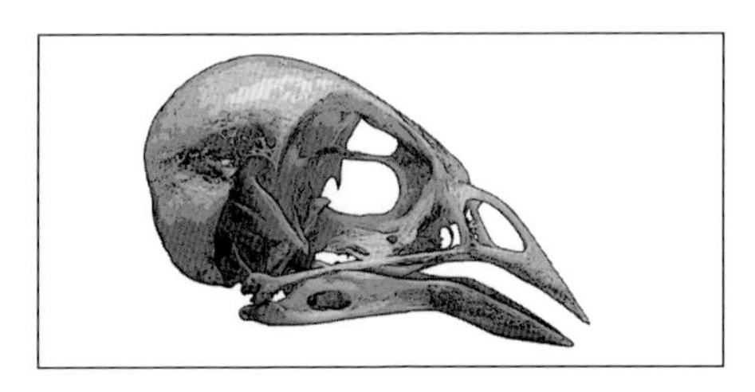

Green-tailed Towhee
(Pipilo chlorurus)

Total: 1 3/16–1 1/4 in., 3.1–3.2 cm
Bill: 9/16 in., 1.4 cm
(44–45% of skull length)
Cranium: 3/4–13/16 in., 1.9–2.0 cm

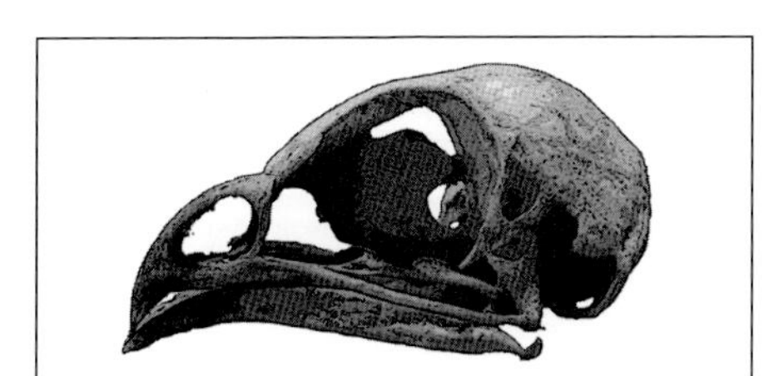

Northern Bobwhite
(Colinus virginianus)

Total: 1 7/16 in., 3.7 cm
Bill: 7/16 in., 1.1 cm
(30% of skull length)
Cranium: 1 in., 2.5 cm

Piping Plover
(Charadrius melodus)

Total: $1\frac{7}{16}$–$1\frac{1}{2}$ in., 3.7–3.8 cm
Bill: $\frac{5}{8}$ in., 1.6 cm
(36–42% of skull length)
Cranium: $\frac{13}{16}$–$\frac{7}{8}$ in., 2.1–2.2 cm

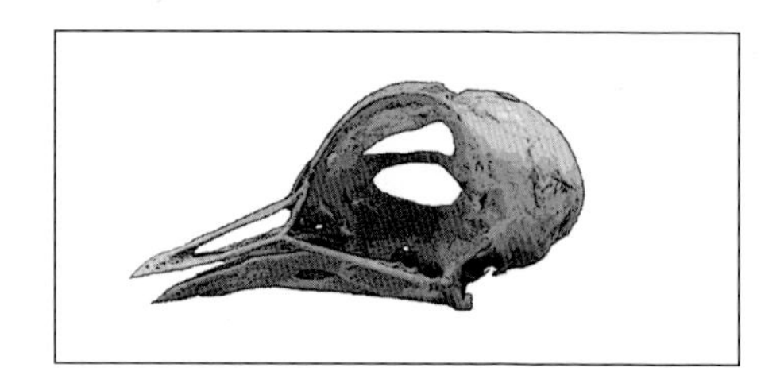

Baltimore Oriole
(Icterus galbula)

Total: $1\frac{3}{8}$–$1\frac{7}{16}$ in., 3.5–3.7 cm
Bill: $\frac{7}{16}$–$\frac{3}{4}$ in., 1.2–1.9 cm
(34–51% of skull length)
Cranium: $\frac{13}{16}$–$\frac{7}{8}$ in., 2.1–2.2 cm

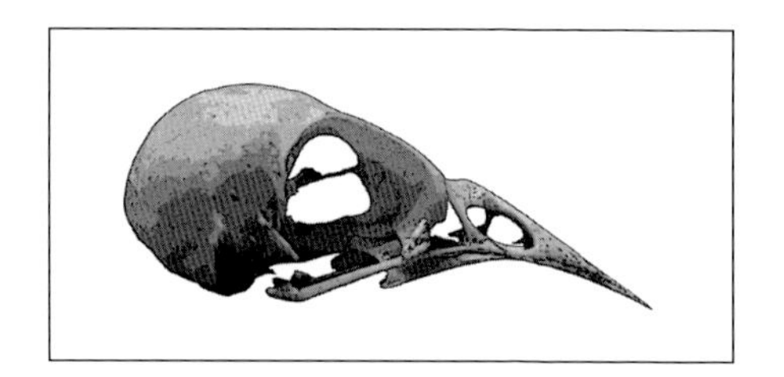

Northern Cardinal
(Cardinalis cardinalis)

Total: $1\frac{5}{16}$–$1\frac{7}{16}$ in., 3.4–3.6 cm
Bill: $\frac{3}{4}$ in., 1.9 cm
(53–56% of skull length)
Cranium: $\frac{13}{16}$–$\frac{7}{8}$ in., 2.1–2.3 cm

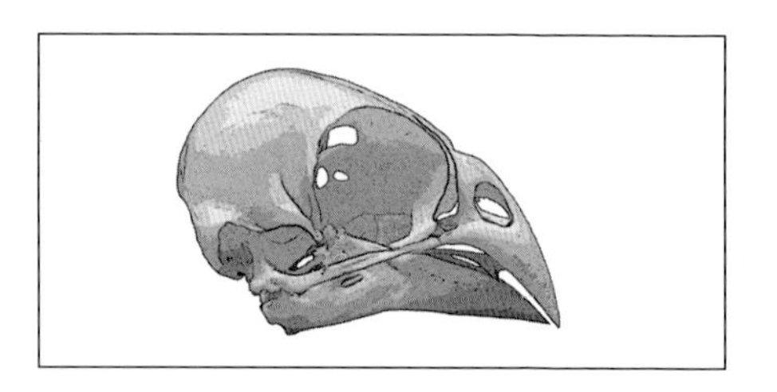

Evening Grosbeak
(Coccothraustes vespertinus)

Total: $1\frac{7}{16}$–$1\frac{1}{2}$ in., 3.6–3.8 cm
Bill: $\frac{5}{16}$–$\frac{3}{8}$ in., 0.8–1.0 cm
(21–26% of skull length)
Cranium: $1\frac{1}{8}$ in., 2.9 cm

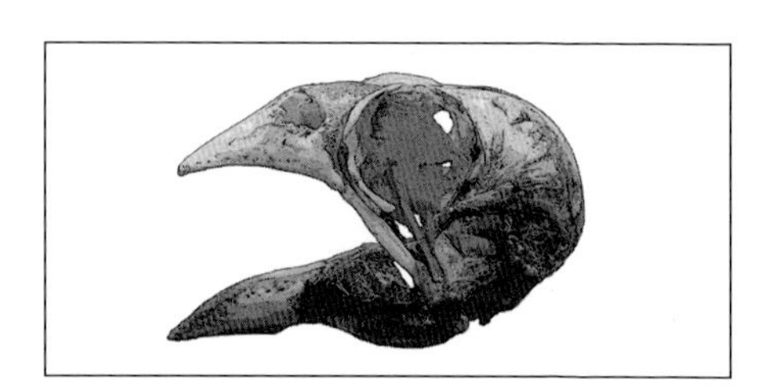

Red-winged Blackbird
(Agelaius phoeniceus)

Total: $1\frac{3}{8}$–$1\frac{9}{16}$ in., 3.5–4.0 cm
Bill: $\frac{5}{8}$–$\frac{3}{4}$ in., 1.6–1.9 cm
(46–48% of skull length)
Cranium: $\frac{13}{16}$–$\frac{15}{16}$ in., 2.1–2.3 cm

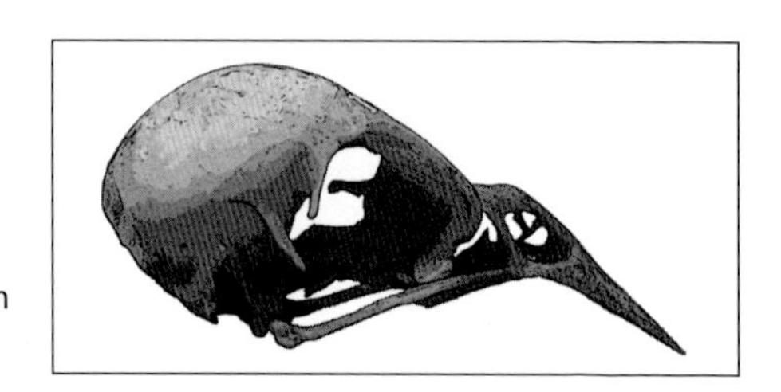

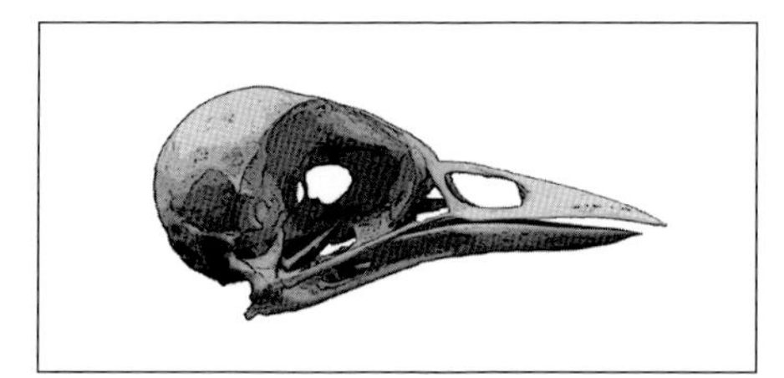

Eastern Kingbird
(Tyrannus tyrannus)

Total: 1 9/16–1 5/8 in., 4.0–4.1 cm

Bill: 11/16 in., 1.7 cm
(41–43% of skull length)

Cranium: 7/8–15/16 in., 2.2–2.4 cm

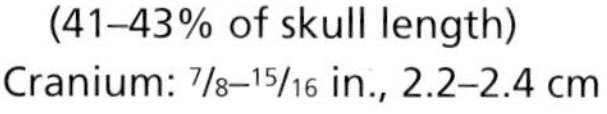

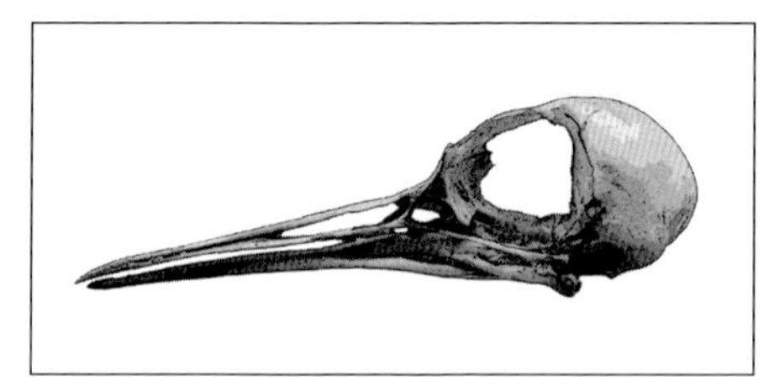

Spotted Sandpiper
(Actitis macularia)

Total: 1 1/2–1 13/16 in., 3.8–4.6 cm

Bill: 13/16–1 1/16 in., 2–2.7 cm
(52–59% of skull length)

Cranium: 11/16–3/4 in., 1.7–1.9 cm

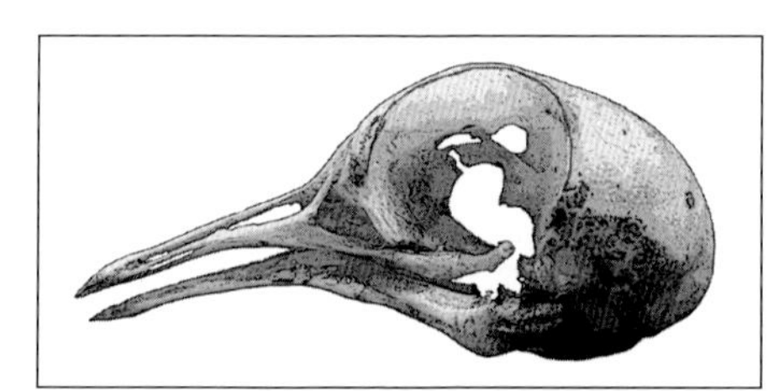

Mourning Dove
(Zenaida macroura)

Total: 1 9/16–1 3/4 in., 4–4.4 cm

Bill: 5/8–3/4 in., 1.6–1.9 cm
(40–43% of skull length)

Cranium: 15/16–1 in., 2.4–2.5 cm

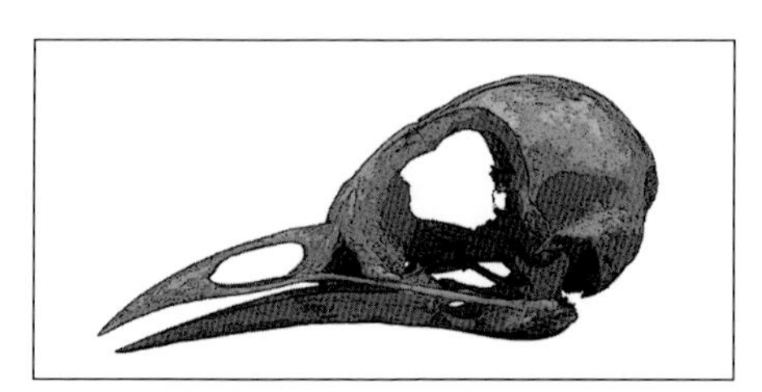

American Robin
(Turdus migratorius)

Total: 1 11/16–1 7/8 in., 4.3–4.7 cm

Bill: 13/16 in., 2.0 cm
(43–47% of skull length)

Cranium: 1–1 1/8 in., 2.7–2.9 cm

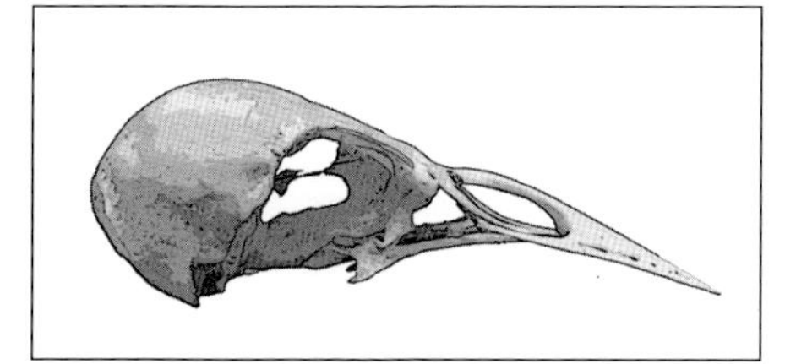

European Starling
(Sturnus vulgaris)

Total: 1 7/8–1 15/16 in., 4.8–4.9 cm

Bill: 15/16 in., 2.4 cm
(49–50% of skull length)

Cranium: 1 in., 2.5 cm

Brown Thrasher
(Toxostoma rufum)

Total: $1\frac{15}{16}$–$2\frac{1}{16}$ in., 4.9–5.2 cm
Bill: $\frac{7}{8}$–$\frac{15}{16}$ in., 2.2–2.4 cm
(44–46% of skull length)
Cranium: $1\frac{1}{16}$–$1\frac{1}{8}$ in., 2.7–2.8 cm

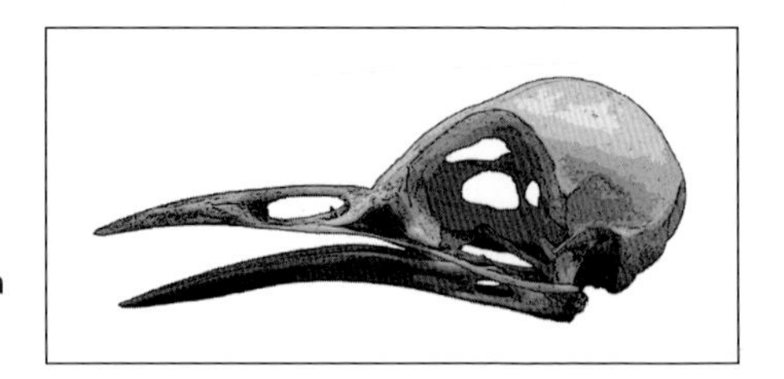

Least Tern
(Sterna antillarum)

Total: 2–$2\frac{1}{16}$ in., 5.1–5.2 cm
Bill: $\frac{15}{16}$–1 in., 2.3–2.6 cm
(46–50% of skull length)
Cranium: $1\frac{1}{8}$ in., 2.8 cm

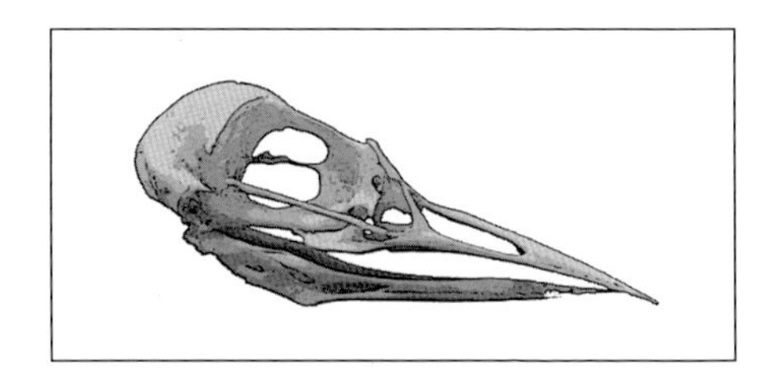

Ruffed Grouse
(Bonasa umbellus)

Total: 2–$2\frac{3}{16}$ in., 5.1–5.5 cm
Bill: $\frac{11}{16}$–1 in., 1.7–2.5 cm
(33–45% of skull length)
Cranium: $1\frac{5}{16}$–$1\frac{7}{16}$ in., 3.3–3.6 cm

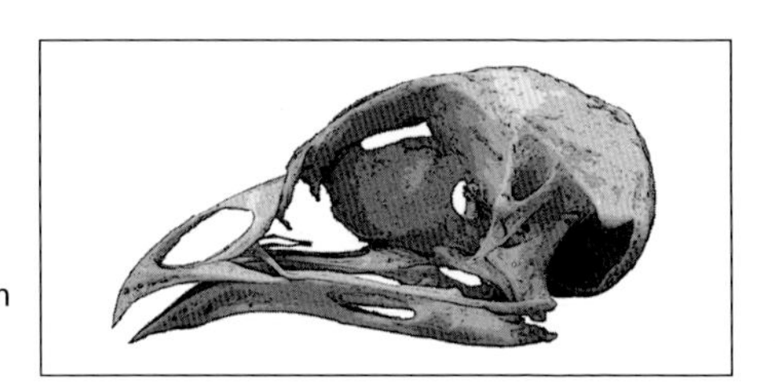

Blue Jay
(Cyanocitta cristata)

Total: $2\frac{1}{8}$ in., 5.4 cm
Bill: $\frac{7}{8}$–$\frac{15}{16}$ in., 2.2–2.4 cm
(41–44% of skull length)
Cranium: $1\frac{3}{16}$–$1\frac{1}{4}$ in., 3.0–3.2 cm

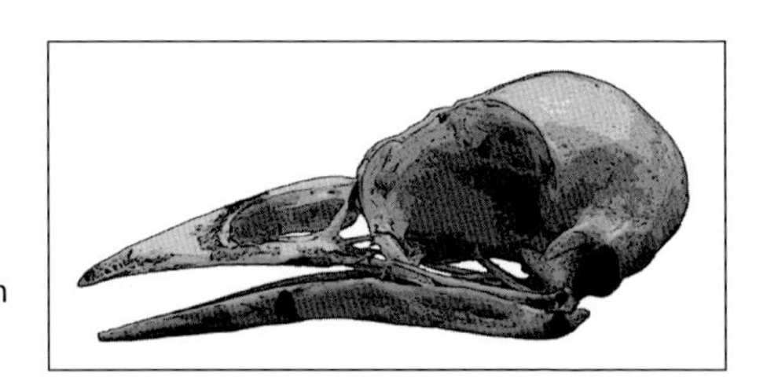

Common Grackle
(Quiscalus quiscula)

Total: $1\frac{3}{4}$–$2\frac{1}{4}$ in., 4.4–5.7 cm
Bill: $\frac{3}{4}$–$1\frac{1}{16}$ in., 1.9–2.7 cm
(43–47% of skull length)
Cranium: 1–$1\frac{3}{16}$ in., 2.5–3.0 cm

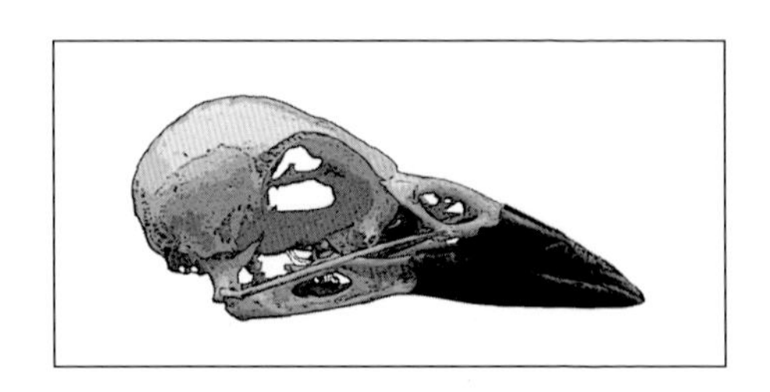

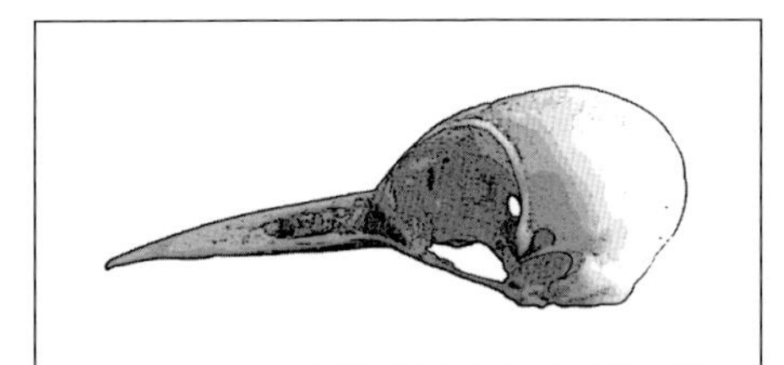

Northern Flicker
(Colaptes auratus)

Total: $2\frac{3}{8}$– $2\frac{7}{16}$ in., 6.0–6.1 cm

Bill: $1\frac{3}{16}$–$1\frac{5}{16}$ in., 3.0–3.4 cm (50–55% of skull length)

Cranium: $1\frac{1}{8}$–$1\frac{3}{16}$ in., 2.9–3.0 cm

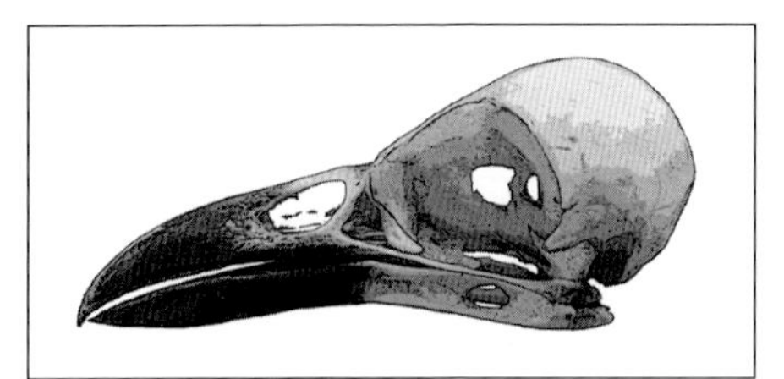

Black-billed Magpie
(Pica pica)

Total: $2\frac{3}{4}$–$2\frac{15}{16}$ in., 7.0–7.5 cm

Bill: $1\frac{3}{8}$–$1\frac{5}{8}$ in., 3.5–4.1 cm (50–54% of skull length)

Cranium: $1\frac{5}{16}$–$1\frac{3}{8}$ in., 3.3–3.5 cm

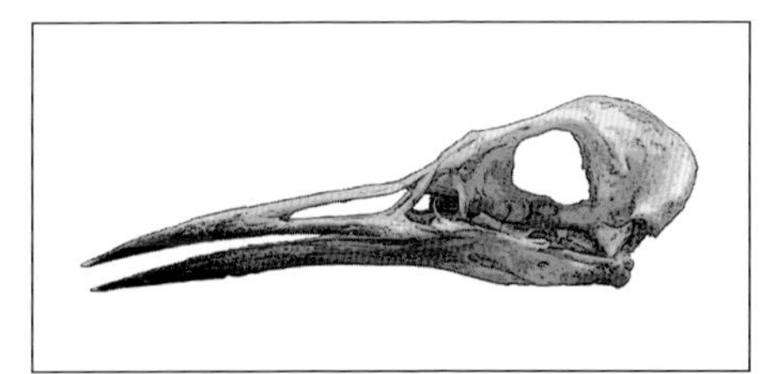

Common Tern
(Sterna hirundo)

Total : $2\frac{7}{8}$–$3\frac{1}{8}$ in., 7.2–7.8 cm

Bill: $1\frac{1}{2}$–$1\frac{11}{16}$ in., 3.8–4.3 cm (53–56% of skull length)

Cranium: $1\frac{3}{8}$–$1\frac{7}{16}$ in., 3.5–3.6 cm

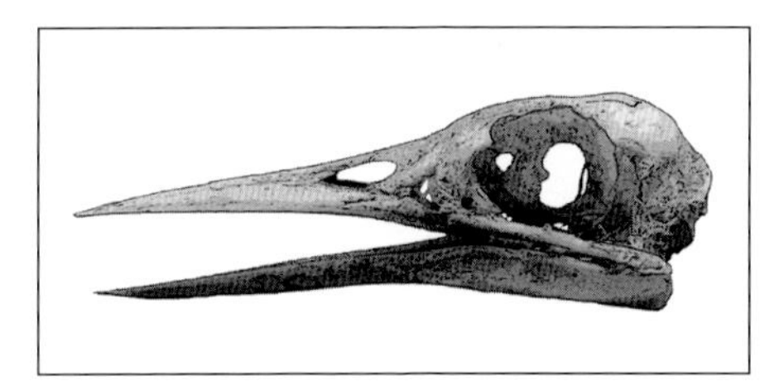

Belted Kingfisher
(Ceryle alcyon)

Total: $3\frac{3}{16}$–$3\frac{3}{8}$ in., 8.1–8.5 cm

Bill: $1\frac{13}{16}$–$1\frac{7}{8}$ in., 4.6–4.7 cm (55–57% of skull length)

Cranium: $1\frac{3}{8}$–$1\frac{1}{2}$ in., 3.5–3.8 cm

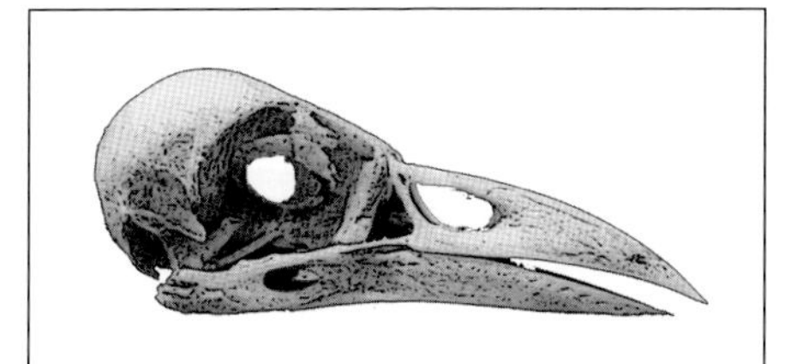

American Crow
(Corvus brachyrhynchos)

Total: $3\frac{3}{16}$–$3\frac{9}{16}$ in., 8.1–9.0 cm

Bill: $1\frac{3}{4}$–$1\frac{15}{16}$ in., 4.4–5.0 cm (54–56% of skull length)

Cranium: $1\frac{9}{16}$–$1\frac{11}{16}$ in., 4.0–4.1 cm

Pileated Woodpecker
(Dryocopus pileatus)

Total: $3\frac{3}{16}$–4 in., 8.1–10.2 cm
Bill: $1\frac{11}{16}$–$2\frac{1}{2}$ in., 4.7–6.3 cm
(58–62% of skull length)
Cranium: $1\frac{1}{2}$–$1\frac{5}{8}$ in., 3.7–4.1 cm

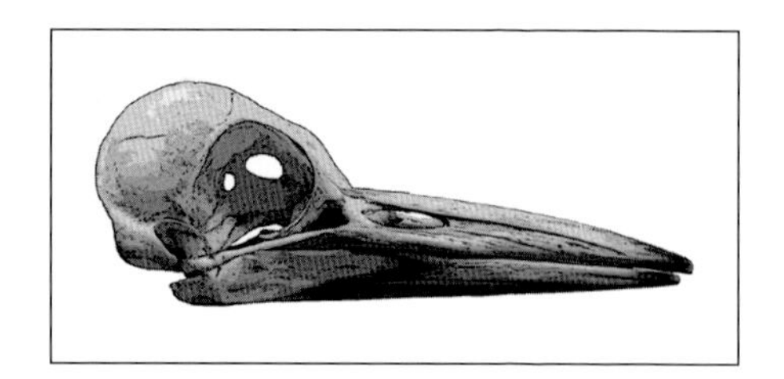

Wild Turkey
(Meleagris gallopavo)

Total: $3\frac{7}{8}$–$4\frac{1}{8}$ in., 10.0–10.5 cm
Bill: $1\frac{11}{16}$–$1\frac{7}{8}$ in., 4.3–4.6 cm
(43–44% of skull length)
Cranium: $2\frac{1}{8}$–$2\frac{7}{16}$ in., 5.3–6.2 cm

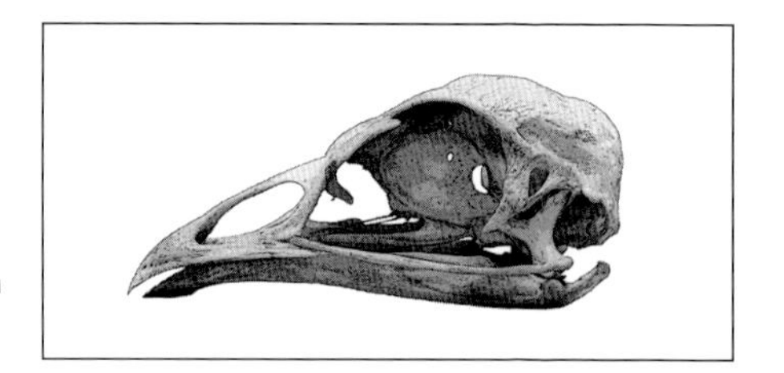

American Woodcock
(Philohela minor)

Total: $3\frac{13}{16}$–$4\frac{5}{16}$ in., 9.7–11.0 cm
Bill: $2\frac{3}{4}$–$3\frac{1}{4}$ in., 7.0–8.2 cm
(72–75% of skull length)
Cranium: 1–$1\frac{1}{16}$ in., 2.5–2.7 cm

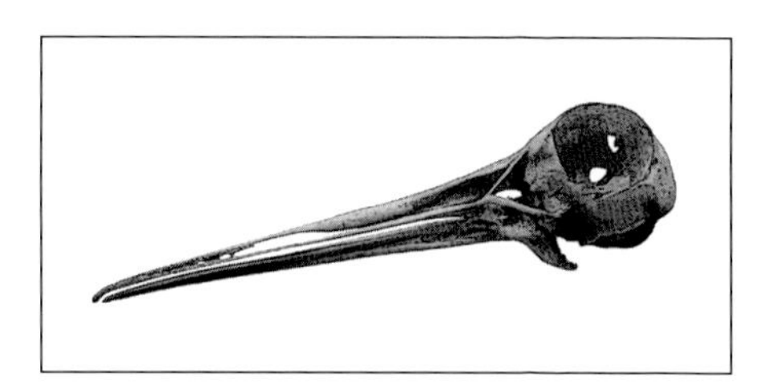

Common Raven
(Corvus corax)

Total: $4\frac{7}{16}$–$4\frac{9}{16}$ in., 11.3–11.7 cm
Bill: $2\frac{11}{16}$–3 in., 6.8–7.7cm
(60–66% of skull length)
Cranium: $1\frac{5}{8}$–$1\frac{13}{16}$ in., 3.0–4.2 cm

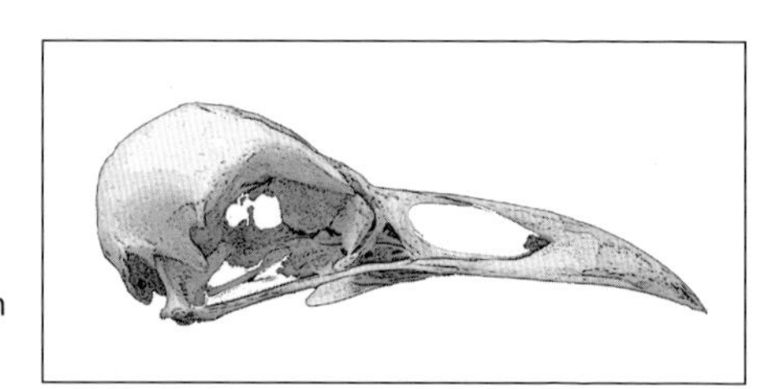

Herring Gull
(Larus argentatus)

Total: $4\frac{3}{4}$–$4\frac{15}{16}$ in., 12.0–12.3 cm
Bill: $2\frac{3}{4}$–$2\frac{7}{8}$ in., 7.0–7.3 cm
(58–59% of skull length)
Cranium: 2–$2\frac{1}{8}$ in., 5.0–5.4 cm

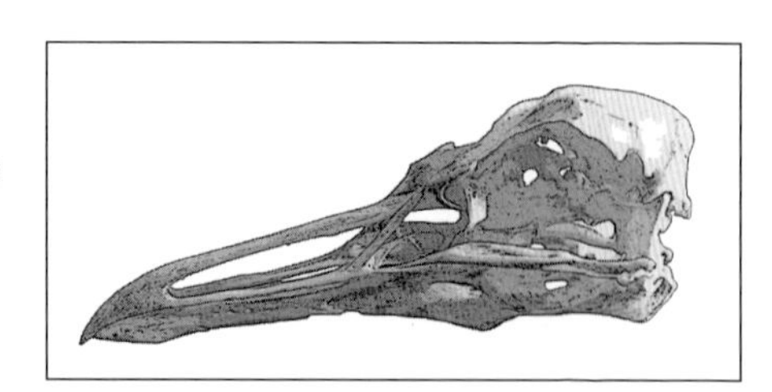

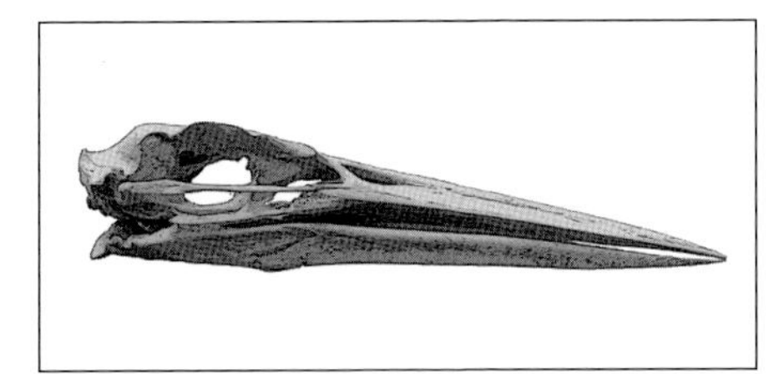

Great Blue Heron
(Ardea herodias)

Total: $7\frac{3}{8}$–8 in., 19.0–20.3 cm
Bill: $4\frac{3}{4}$–$5\frac{7}{16}$ in., 12.1–13.9 cm (64–68% of skull length)
Cranium: $2\frac{11}{16}$–$3\frac{3}{8}$ in., 6.8–7.9 cm

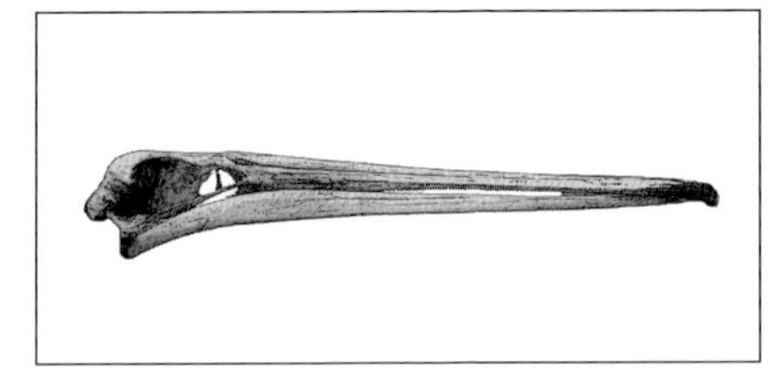

Brown Pelican
(Pelecanus erythrorhynchos)

Total: $15\frac{1}{8}$–16 in., 38.4–40.0 cm
Bill: 12–$12\frac{1}{2}$ in., 30.6–31.8 cm (79–80% of skull length)
Cranium: 3–$3\frac{1}{4}$ in., 7.8–8.2 cm

Duck Bills

Duck bills are wider than they are high and have a hard tip at the center of the upper mandible, referred to as the nail, which aids in grazing on tough sedges and grasses. This nail is a permanent fixture, rather than temporary, as is the egg tooth.

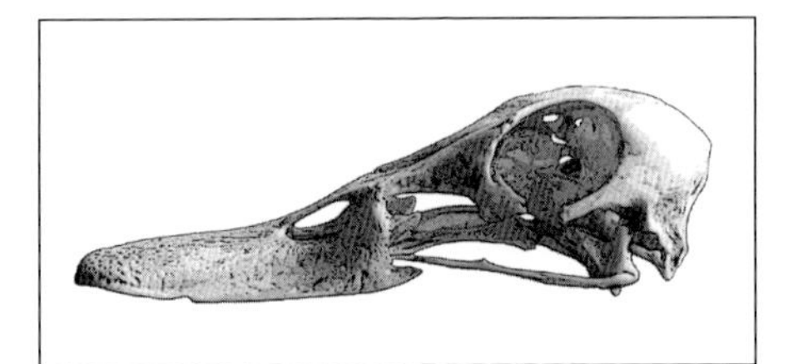

Mallard
(Anas platyrhynchos)

Total: $4\frac{1}{4}$–$4\frac{7}{16}$ in., 10.8–11.3 cm
Bill: $1\frac{15}{16}$–$2\frac{1}{8}$ in., 4.9–5.4 cm (45–49% of skull length)
Cranium: $2\frac{5}{16}$ in., 5.8 cm

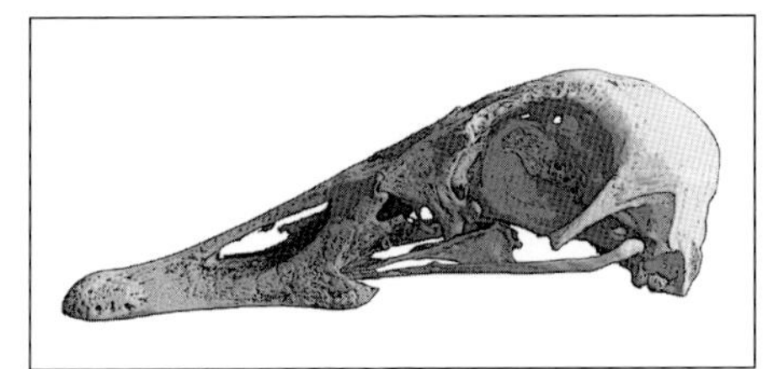

Canada Goose
(Branta canadensis)

Total: $4\frac{9}{16}$ in., 11.6 cm
Bill: $2\frac{1}{16}$ in., 5.2 cm (45% of skull length)
Cranium: $2\frac{1}{2}$ in., 6.3 cm

Hooked Bills

Hooked bills are distinguished by their decurved upper mandibles, as in hawks.

American Kestrel
(Falco sparverius)

Total: $1\frac{5}{8}$ in., 4.1 cm
Bill: $\frac{7}{16}$–$\frac{9}{16}$ in., 1.2–1.4 cm
(29–34% of skull length)
Cranium: $1\frac{1}{8}$–$1\frac{3}{16}$ in., 2.8–3.0 cm

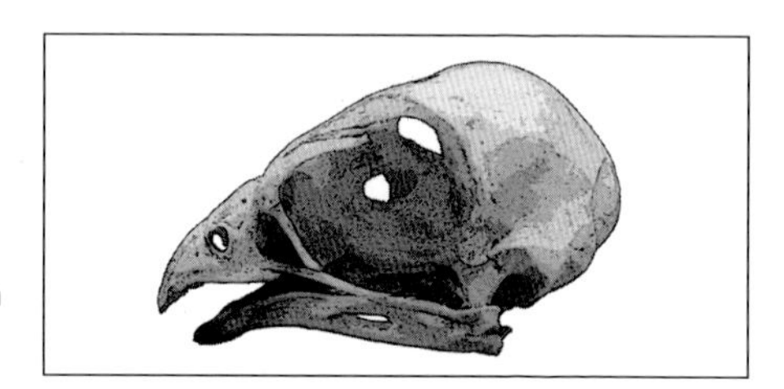

Eastern Screech-owl
(Otus asio)

Total: 2–$2\frac{1}{8}$ in., 5.1–5.4 cm
Bill: $\frac{5}{8}$–$\frac{11}{16}$ in., 1.6–1.8 cm
(31–33% of skull length)
Cranium: $1\frac{3}{8}$ in., 3.5 cm

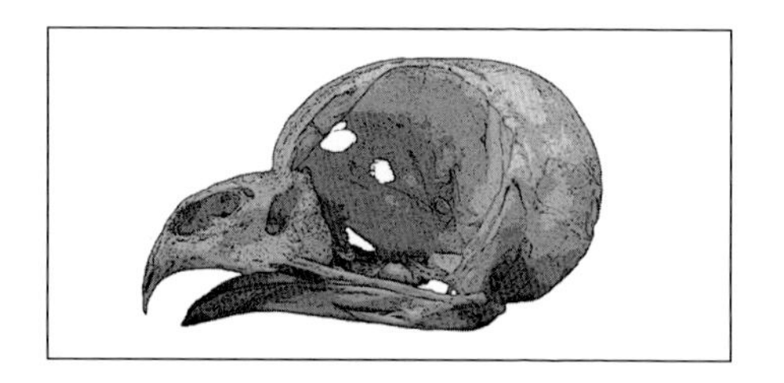

Cooper's Hawk
(Accipiter cooperii)

Total: $2\frac{7}{16}$–$2\frac{1}{8}$ in., 6.2–6.3 cm
Bill: $1\frac{1}{8}$–$1\frac{3}{8}$ in., 2.7–3.4 cm
(44–54% of skull length)
Cranium: $1\frac{9}{16}$–$1\frac{5}{8}$ in., 4.0–4.2 cm

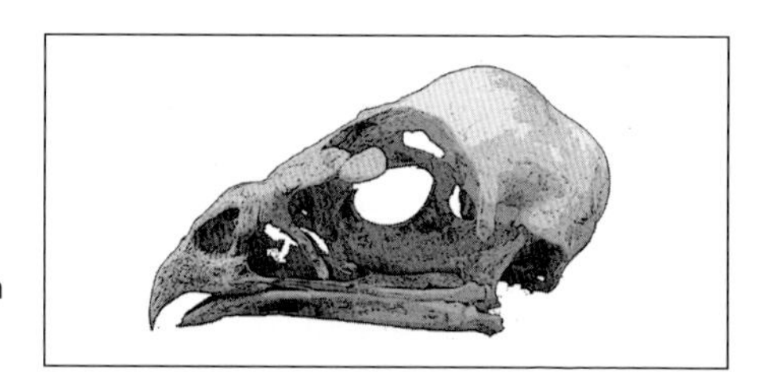

Osprey
(Pandion haliaetus)

Total: $2\frac{7}{8}$–$3\frac{1}{4}$ in., 7.3–8.3 cm
Bill: $1\frac{1}{4}$–$1\frac{5}{8}$ in., 3.2–4.2 cm
(44–51% of skull length)
Cranium: $1\frac{5}{8}$–$1\frac{7}{8}$ in., 4.2–4.7 cm

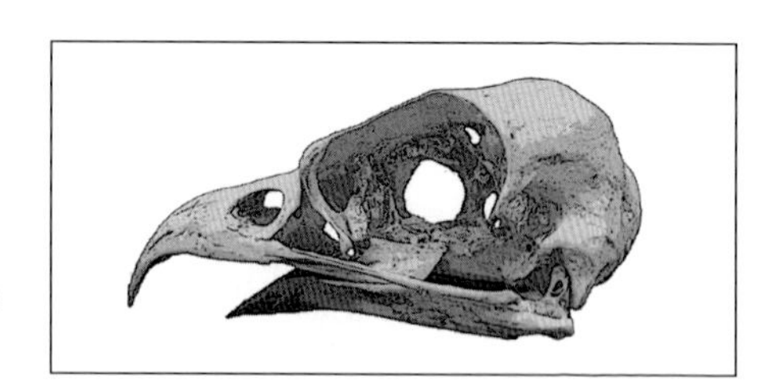

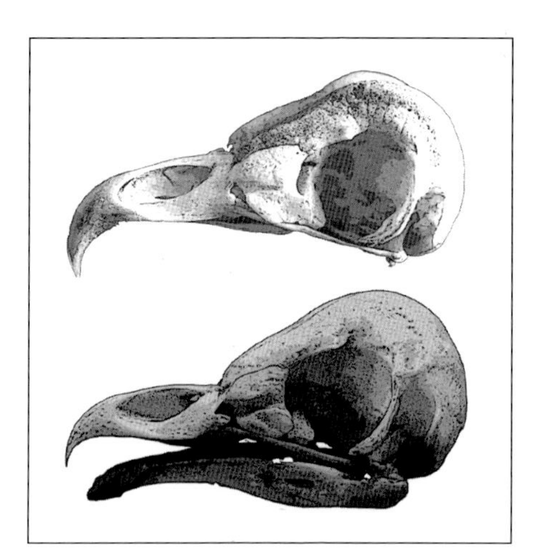

Barn Owl
(Tyto alba)

Total: $2\frac{5}{8}$–$3\frac{3}{16}$ in., 6.6–8.1 cm

Bill: $\frac{15}{16}$–$1\frac{1}{4}$ in., 2.4–3.2 cm
(32–39% of skull length)

Cranium: $1\frac{5}{8}$–$1\frac{7}{8}$ in., 4.2–4.7 cm

Bottom illustration shows lower mandible.

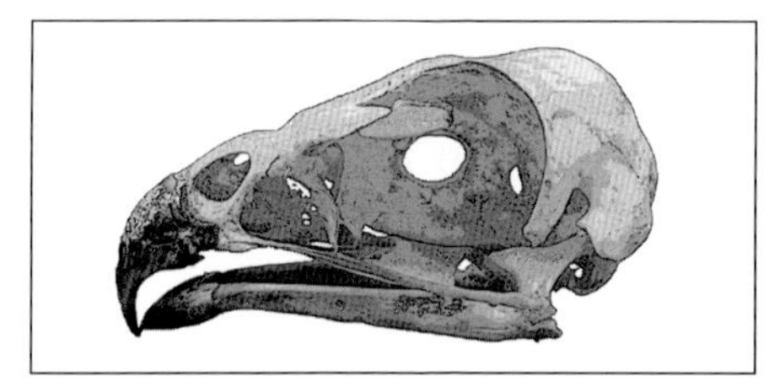

Red-tailed Hawk
(Buteo jamaicensis)

Total: $3\frac{1}{4}$–$3\frac{3}{8}$ in., 8.2–8.5 cm

Bill: $1\frac{1}{8}$–$1\frac{1}{2}$ in., 2.8–3.7 cm
(43–44% of skull length)

Cranium: $2\frac{1}{8}$–$2\frac{1}{4}$ in., 5.4–5.7 cm

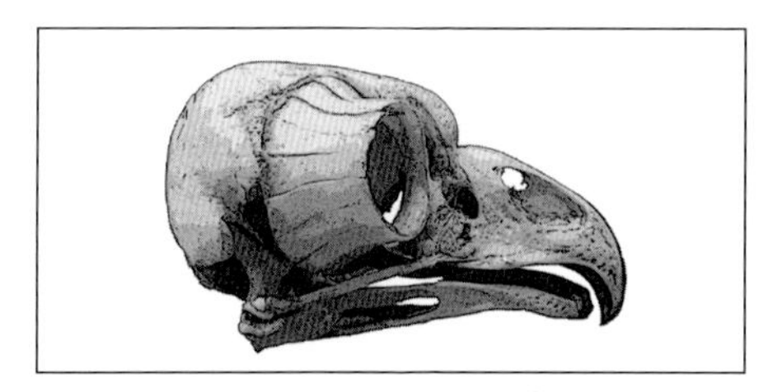

Great Horned Owl
(Bubo virginianus)

Total: $3\frac{5}{16}$–$3\frac{5}{8}$ in., 8.4–9.2 cm

Bill: $1\frac{5}{8}$–$1\frac{3}{4}$ in., 4.1–4.4 cm
(47–48% of skull length)

Cranium: $2\frac{1}{16}$–$2\frac{1}{8}$ in., 5.2–5.4 cm

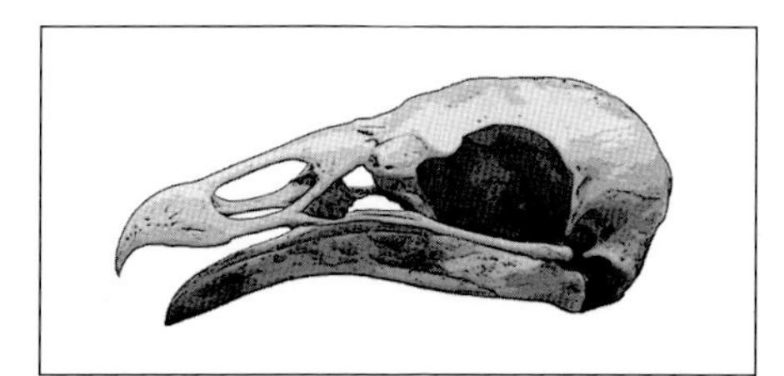

Turkey Vulture
(Cathartes aura)

Total: $3\frac{3}{4}$–$3\frac{7}{8}$ in., 9.5–9.8 cm

Bill: $1\frac{5}{8}$–$1\frac{11}{16}$ in., 4.1–4.3 cm
(43–44% of skull length)

Cranium: $2\frac{1}{8}$–$2\frac{3}{16}$ in., 5.4–5.5 cm

Red-breasted Merganser *(Mergus serrator)*

Total: 4 3/8–4 7/16 in., 11.1–11.3 cm

Bill: 2 9/16 in., 6.5 cm (57–59% of skull length)

Cranium: 1 13/16–1 7/8 in., 4.6–4.7 cm

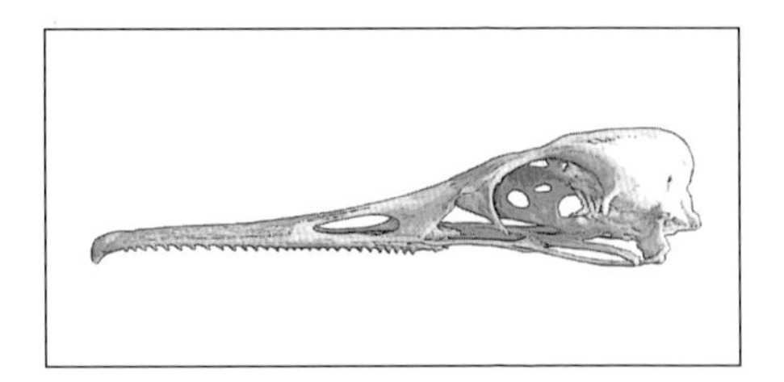

Bald Eagle *(Haliaeetus leucocephalus)*

Total: 4 1/2 in., 11.5 cm

Bill: 2 1/16–2 5/16 in., 5.1–6.0 cm (44–52% of skull length)

Cranium: 2 3/8–2 1/2 in., 6.0–6.3 cm

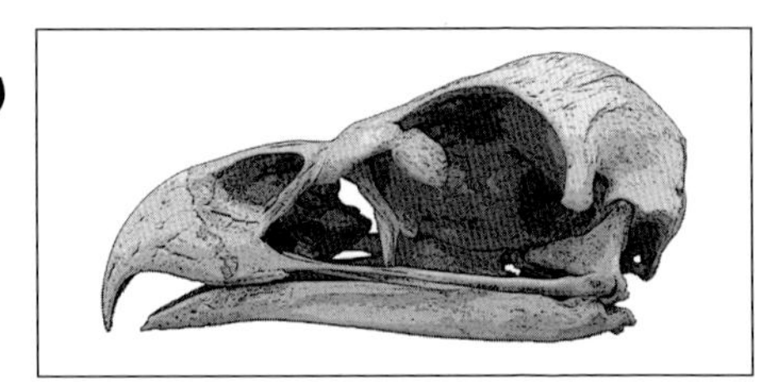

Double-crested Cormorant *(Phalacrocorax auritus)*

Total: 4 7/8–5 1/4 in., 12.4–13.3 cm

Bill: 2 5/16–2 9/16 in., 5.8–6.5 cm (47–49% of skull length)

Cranium: 2 9/16–2 11/16 in., 6.5–6.8 cm

Bottom illustration shows lower mandible.

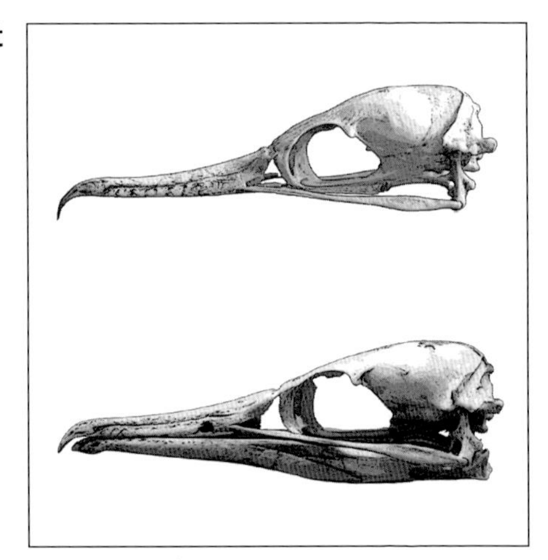

Unusual Bills

Unusual bills are those that are unique for some reason: They may be recurved, decurved, or spoon-billed. These are often some of the more beautiful skulls, ones that seem more like sculptures rather than mere signs.

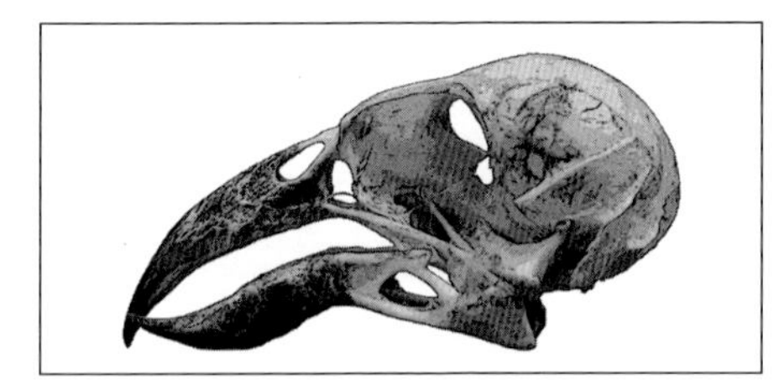

Red Crossbill
(Loxia curvirostra)

Total: $1\frac{5}{16}$–$1\frac{3}{8}$ in., 3.3–3.5 cm

Bill: $\frac{7}{16}$–$\frac{1}{2}$ in., 1.1–1.3 cm
(33–37% of skull length)

Cranium: $\frac{7}{8}$ in., 2.2 cm

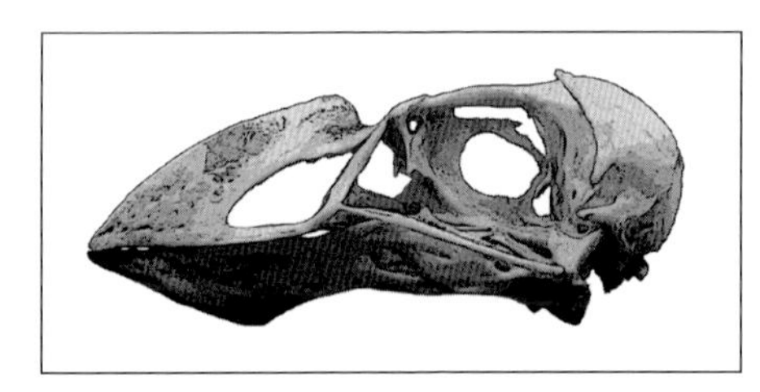

Atlantic Puffin
(Fratercula arctica)

Total: $2\frac{7}{8}$ in., 7.3 cm

Bill: $1\frac{3}{16}$–$1\frac{1}{4}$ in., 3.0–3.2 cm
(41–44% of skull length)

Cranium: $1\frac{5}{8}$–$1\frac{11}{16}$ in., 4.1–4.3 cm

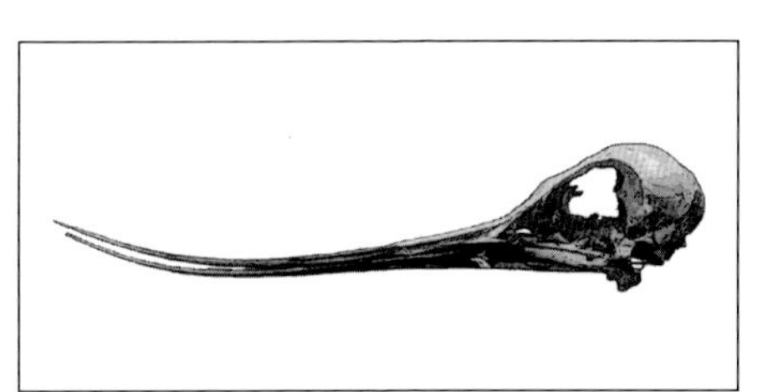

American Avocet
(Recurvirostra americana)

Total: $4\frac{3}{8}$–$4\frac{9}{16}$ in., 11.1–11.6 cm

Bill: $3\frac{1}{4}$–$3\frac{7}{16}$ in., 8.2–8.7 cm
(75–78% of skull length)

Cranium: $1\frac{1}{8}$ in., 2.8–2.9 cm

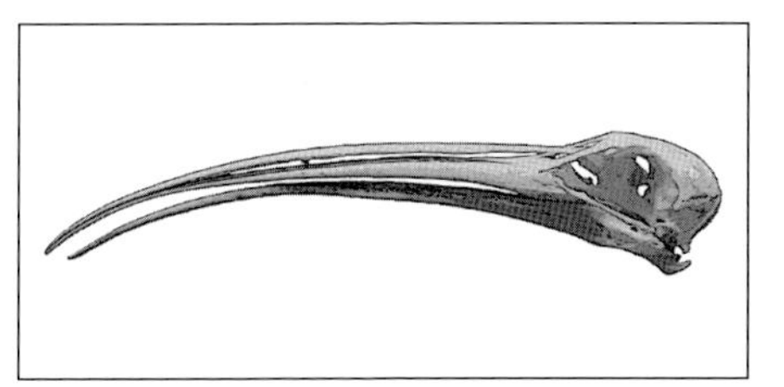

White Ibis
(Eudocimus albus)

Total: $6\frac{5}{8}$–$7\frac{7}{8}$ in., 16.8–20.0 cm

Bill: $4\frac{7}{8}$–$6\frac{1}{8}$ in., 12.4–15.5 cm
(74–78% of skull length)

Cranium: $1\frac{3}{4}$ in., 4.5 cm

Black Skimmer
(Rynchops niger)

Total: $4\frac{1}{16}$–$5\frac{7}{16}$ in., 10.3–13.8 cm

Bill: $2\frac{1}{4}$–$3\frac{7}{16}$ in., 5.7–8.7 cm (55–63% of skull length)

Cranium: $1\frac{13}{16}$–2 in., 4.6–5.1 cm

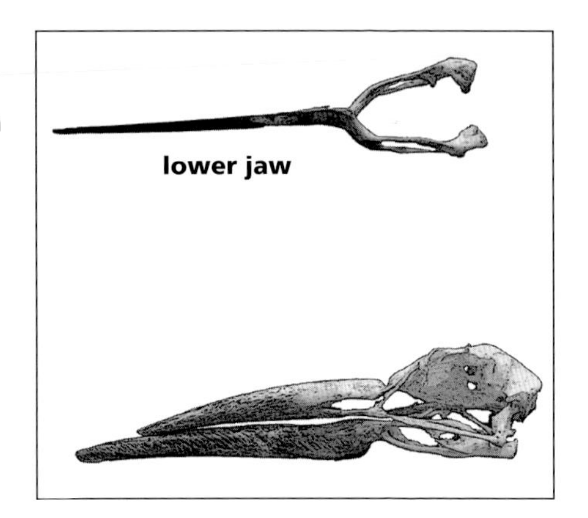

Long-billed Curlew
(Numenius americanus)

Total: $7\frac{13}{16}$–$9\frac{3}{16}$ in., 19.9–23.4 cm

Bill: $6\frac{7}{16}$–$7\frac{11}{16}$ in., 16.4–19.5 cm (82–84% of skull length)

Cranium: $1\frac{3}{8}$–$1\frac{1}{2}$ in., 3.5–3.8 cm

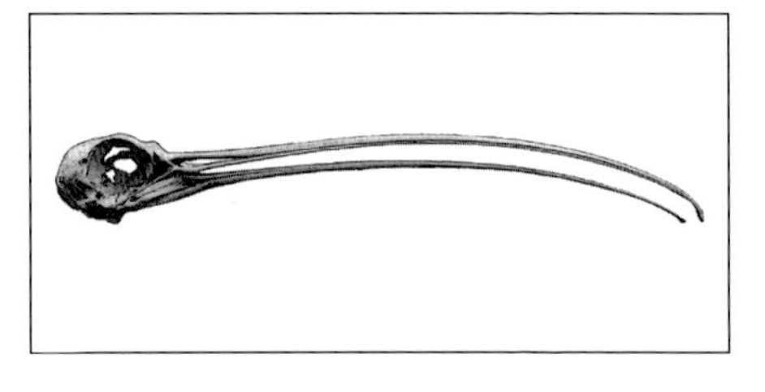

APPENDIX: RESOURCES

American Bird Conservancy
1250 24th St., NW, Suite 220
Washington, DC 20037
(888) BIRD-MAG

American Birding Association
P.O. Box 6599
Colorado Springs, CO 80934
(800) 835-2473

Avian Science and
Conservation Centre
Macdonald Campus
of McGill University
21, 111 Lakeshore Rd.
Ste. Anne de Bellevue,
PQ H9X 3V9
Canada
(514) 398-7760

Birds of Vermont Museum
900 Sherman Hollow Rd.
Huntington, VT 05462
(802) 434-2167

Diane Boretos
Call of the Wild
Environmental Services
P.O. Box 572
West Falmouth, MA 02574

Boulder Outdoor Survival School
Josh Bernstein, David Wescott
P.O. Box 1590
Boulder, CO 80306
(800) 335-7404
(303) 444-9779

International Crane Foundation
E-11376 Shady Lane Rd.
Baraboo, WI 53913

Keeping Track
Sue Morse
P.O. Box 848
Richmond, VT 05477
(802) 434-7000
fax: (802) 434-5383

Laboratory of Ornithology
Cornell University
159 Sapsucker Woods Rd.
Ithaca, NY 14850
(607) 254-2473

Clare Walker Leslie
(naturalist, artist, illustrator, and author, offering classes in field sketching, painting, and nature journaling)
76 Garfield St.
Cambridge, MA 02138
(617) 547-9128

National Audubon Society
(central bureau)
700 Broadway
New York, NY 10003
(212) 979-3000
www.audubon.org

The National Wildlife Federation
8925 Leesburg Pike
Vienna, VA 22184
(703) 790-4100
www.nwf.org

The Natural Resources Defense Council
40 West 20th St.
New York, NY 10011
(212) 727-2700

A Naturalist's World
(includes tracking classes)
Jim Halfpenny
P.O. Box 989
Gardiner, MT 59030
(406) 848-9458
www.tracknature.com

Nature and Vision Tracking School
(includes tracking classes)
Charles Worsham
760 Thomas Rd.
Madison Heights, VA 24572
(804) 846-1987

Ndakinna Wilderness Project
(includes tracking classes)
Jim Bruchac
P.O. Box 308
23 Middle Grove Rd.
Greenfield Center, NY 12833
(518) 583-9980
www.ndakinna.com

Paul Rezendes Programs in Nature (includes tracking classes)
Paul Rezendes, Paulette Roy, Mark Elbroch, John McCarter, Alcott Smith
3933 Bearsden Rd.
Royalston, MA 01368-9400
(978) 249-8810
www.paulrezendes.com

Alcott Smith, D.V.M.
(tracking, forest, and nature studies program)
P.O. Box 897
Hanover, NH 03755
(603) 448-6352

John Stokes
The Tracking Project
P.O. Box 266
Corrales, NM 87048

The Teton Science School
P.O. Box 68
1 Ditch Creek Rd.
Kelly, WY 83011
(307) 733-4765
www.tetonscience.org

The Tracker School
Tom Brown, Jr.
P.O. Box 173
Asbury, NJ 08802-0173
(908) 479-4681

WildEarth: The Journal of Wildlands Recovery and Protection
P.O. Box 455
Richmond, VT 05477
(802) 434-4077

Wilderness Awareness School (includes tracking classes)
Jon Young
P.O. Box 5000, #5-137
Duvall, WA 98019
(425) 788-1301
www.NatureOutlet.com

The Wildlands Project
1955 W. Grant Rd., Suite 145
Tucson, AZ 85745-1147
(520) 884-0875
www.twp.org

The Yellowstone to Yukon Conservation Initiative
710 9th St., Studio B
Canmore, AB T1W 2V7
Canada
y2y@banff.net

BIBLIOGRAPHY

Abbey, Edward. *Desert Solitaire: A Season in the Wilderness.* New York: Simon and Schuster, 1968.

Allaby, Michael. *A Dictionary of Zoology.* 2nd ed. New York: Oxford University Press, 1999.

Askins, Robert A. *Restoring North America's Birds: Lessons from Landscape Ecology.* New Haven, CT: Yale University Press, 2000.

Attenborough, D. *The Life of Birds.* Princeton, NJ: Princeton University Press, 1998.

Atwater, Sally, and Judith Schnell, eds. *The Wildlife Series: Ruffed Grouse.* Harrisburg, PA: Stackpole Books, 1989.

Baicich, Paul J., and Colin J. O. Harrison. *A Guide to the Nests, Eggs, and Nestlings of North American Birds.* 2nd ed. Boston: Academic Press, 1997.

Bang, Preben, and Preben Dahlstrom. *Collins Guide to Animal Tracks and Signs: A Guide to the Tracking of All British and European Mammals and Birds.* English translation. St. James Place, London: Collins Sons and Co., Ltd., 1974.

Barnard, Ellsworth. *In a Wild Place: A Natural History of High Ledges.* Lincoln, MA: Massachusetts Audubon Society, 1998.

Bent, Arthur. *Life Histories of North American Woodpeckers.* New York: Dover, 1964.

Benyus, Janine M. *The Field Guide to Wildlife Habitats of the Eastern United States.* New York: Fireside, 1989.

Bird, David. *The Bird Almanac.* Buffalo: Firefly Books, 1999.

Bouchner, Miroslav. *Animal Tracks.* English edition. Prague: Aventinum Publishing House, 1998.

Brown, R., J. Ferguson, M. Lawrence, and D. Lees. *Tracks and Signs of the Birds of Britain and Europe: An Identification Guide*. Kent: Christopher Helm, 1987.

Brown, Tom. *Tom Brown's Field Guide to Nature Observation and Tracking*. New York: Berkeley Publishing, 1983.

Butler, Robert W. *The Great Blue Heron*. Vancouver, BC: University of British Columbia Press, 1999.

Caduto, Michale J., and Joseph Bruchac. *Keepers of the Animals: Native American Stories and Wildlife Activities for Children*. Golden, CO: Fulcrum Publishing, 1991.

Cajete, Gregory. *Look to the Mountain: An Ecology of Indigenous Education*. Skyland, NC: Kivaki Press, 1994.

Carson, Rachel. *Silent Spring*. Boston: Houghton Mifflin Company, 1962.

Chadwick, Douglas H., and Joel Sartore. *The Company We Keep: America's Endangered Species*. Washington, DC: National Geographic Society, 1996.

Cruickshank, Allen, and Helen Cruickshank. *1001 Questions Answered about Birds*. New York: Grosset and Dunlap, 1958.

Dickson, James. *The Wild Turkey: Biology and Management*. Harrisburg, PA: Stackpole Books, 1992.

Donahue, Brian. *Reclaiming the Commons: Community Farms and Forests in a New England Town*. New Haven, CT: Yale University Press, 1999.

Drengson, Alan, and Duncan Taylor, eds. *Ecoforestry: The Art and Science of Sustainable Forest Use*. Gabriola Island, BC: New Society Press, 1997.

Dunn, Jon, and Kimball Garrett. *A Field Guide to Warblers of North America*. Boston: Houghton Mifflin Company, 1997.

Dunne, Pete. *The Feather Quest: A North American Birder's Year*. Boston: Houghton Mifflin Company, 1992.

Dunning, Joan. *Secrets of the Nest: The Family Life of North American Birds*. Boston: Houghton Mifflin Company, 1994.

Eastman, John. *Birds of Forest, Yard, and Thicket*. Mechanicsburg, PA: Stackpole Books, 1997.

———. *Birds of Lake, Pond, and Marsh*. Mechanicsburg, PA: Stackpole Books, 1999.

———. *Birds of Field and Shore*. Mechanicsburg, PA: Stackpole Books, 2000.

Ehrlich, P., and G. Daily. "Sapsuckers at Work." *Whole Earth Magazine* (Summer 1998): 24–26.

Ehrlich, Paul, David Dobkin, and D. Wheye. *The Birder's Handbook: A Field Guide to the Natural History of North American Birds.* New York: Simon and Schuster, 1988.

Field Guide to the Birds of North America. 2nd ed. Washington, DC: National Geographic Society, 1987.

Forrest, Louise. *A Field Guide to Tracking Animals in Snow.* Mechanicsburg, PA: Stackpole Books, 1988.

Franck, Frederick. *The Zen of Seeing.* New York: Vintage Books, 1973.

Goodnow, David. *How Birds Fly.* Columbia, MD: Periwinkle Books, 1992.

Grambo, Rebecca L. *Eagles.* Stillwater, MN: Voyageur Press, 1999.

Griffin, Donald R. *Animal Minds.* Chicago: University of Chicago Press, 1992.

Grooms, Steve. *The Cry of the Sandhill Crane.* Minnetonka, MN: Northword Press, 1991.

Halfpenny, James. *Scats and Tracks of the Rocky Mountains.* Helena, MT: Falcon Publishing, 1998.

Hancock, James. *Herons and Egrets of the World: A Photographic Journey.* San Diego: Academic Press, 1999.

Hanzak, J. *The Pictorial Encyclopedia of Birds.* New York: Crown Publications, 1967.

Harrison, Hal H. *A Field Guide to the Birds' Nests: United States East of the Mississippi River.* Boston: Houghton Mifflin Company, 1975.

Harrison, Kit, and George Harrison. *The Birds of Winter.* New York: Random House, 1990.

Hawke, Paul. *The Ecology of Commerce: A Declaration of Sustainability.* New York: HarperCollins, 1993.

Hay, John. *The Bird of Light.* New York: W. W. Norton and Company, 1991.

Heinrich, Bernd. *The Mind of the Raven.* New York: HarperCollins Publishers, 1999.

———. *Ravens in Winter.* New York: Vintage Press, 1989.

———. *The Trees in My Forest.* New York: Vintage Press, 1997.

Heintzelman, Donald S. *The Illustrated Bird Watcher's Dictionary.* Tulsa, OK: Winchester Press, 1980.

Hiller, Ilo. *Introducing Birds to Young Naturalists.* College Station, TX: Texas A & M University Press, 1989.

Hinchman, Hannah. *A Trail through Leaves: The Journal as a Path to Place.* New York: W. W. Norton and Company, 1997.

Jaeger, Ellsworth. *Tracks and Trailcraft.* New York: Macmillan Co., 1948.

Jorgensen, Neil. *A Sierra Club Naturalist's Guide: Southern New England.* San Francisco: Sierra Club Books, 1978.

Kaufman, Kenn. *Advanced Birding.* Boston: Houghton Mifflin, 1990.

———. *Focus Guide to Birds of North America.* Boston: Houghton Mifflin, 2000.

Kellert, Stephen R., and Edward O. Wilson, eds. *The Biophilia Hypothesis.* Washington, DC: Island Press, 1993.

Kerlinger, Paul. *How Birds Migrate.* Mechanicsburg, PA: Stackpole Books, 1995.

Kilham, Lawrence. *On Watching Birds.* College Station, TX: Texas A & M University Press, 1997.

Koch, Maryjo. *The Nest: An Artist's Sketchbook.* New York: Stewart, Tabori & Chang, 1999.

Kochan, Jack B. *Bills and Mouths.* Mechanicsburg, PA: Stackpole Books, 1996.

———. *Feet and Legs.* Mechanicsburg, PA: Stackpole Books, 1996.

———. *Heads and Eyes.* Mechanicsburg, PA: Stackpole Books, 1996.

———. *Wings and Tails.* Mechanicsburg, PA: Stackpole Books, 1996.

Kricher, John. *A Neotropical Companion.* Princeton, NJ: Princeton University Press, 1997.

Kricher, John, and Gordon Morrison. *Peterson Field Guides: California and Pacific Northwest Forests.* New York: Houghton Mifflin Co., 1993.

———. *Peterson Field Guides: Ecology of Eastern Forests.* New York: Houghton Mifflin Co., 1988.

———. *Peterson Field Guides: Rocky Mountain and Southwest Forests.* New York: Houghton Mifflin Co., 1993.

Laughlin, Sarah B., and Diane M. Pence. *A Guide to Bird Education Resources: Migratory Birds of the Americas: An Annotated Bibliography.* Washington, DC: National Fish and Wildlife Foundation, 1997.

Lawrence, Gale. *A Field Guide to the Familiar: Learning to Observe the Natural World.* Hanover, NH: University of New England Press, 1998.

Leahy, Christopher. *The Birdwatcher's Companion.* New York: Gramercy Books, 1982.

Lefranc, N. *Shrikes: A Guide to the World.* New Haven, CT: Yale University Press, 1997.

Leopold, Aldo. *A Sand County Almanac.* London: Oxford University Press, 1968.

Lerner, Steve. *Eco-pioneers: Practical Visionaries Solving Today's Environmental Problems.* Cambridge, MA: MIT Press, 1998.

Leslie, Clare Walker. *The Art of Field Sketching.* Dubuque, IA: Kindall/Hunt Publishing Company, 1984.

———. *Nature All Year Long.* New York: Greenwillow Books, 1991.

Leslie, Clare Walker, and Charles E. Roth. *Nature Journaling.* Pownal, VT: Storey Books, 1998.

Levin, Ted. *Backtracking: The Way of a Naturalist.* Chelsea, VT: Chelsea Green Publishing Company, 1987.

Liebenberg, Louis. *The Art of Tracking: The Origin of Science.* Cape Town, South Africa: David Philip, 1990.

Luoma, Jon R. *The Hidden Forest: The Biography of an Ecosystem.* New York: Henry Holt and Company, 1999.

MacMahon, James A. *Deserts.* New York: Alfred A. Knopf, 1985. (One of seven excellent field guides, each focused on a major ecosystem in the United States.)

Marchand, J. *Life in the Cloud: An Introduction to Winter Ecology.* Hanover, NH: University Press of New England, 1987.

Martin, Alexander C., Herbert S. Zim, and Arnold L. Nelson. *American Wildlife and Plants: A Guide to Wildlife Food Habits.* New York: Dover Publications, 1951.

Marzluff, John M., and Russel Balda. *The Pinyon Jay: Behavioral Ecology of a Colonial and Cooperative Corvid.* London: T. & A. D. Poyser, 1992.

Maslow, Jonathan Evan. *The Owl Papers.* New York: Vintage Press, 1983.

Matthews, Downs. *Skimmers.* New York: Simon and Schuster Books for Young Readers, 1990.

Matthiessen, Peter. *Wildlife in America.* New York: Penguin Press, 1987.

———. *The Windbirds: Shorebirds of North America.* Shelburne, VT: Chapters Publishing Ltd., 1994.

Meadows, Donella H., Dennis L. Meadows, and Jorgen Randers. *Beyond the Limits.* Post Mills, VT: Chelsea Green Publishing Company, 1992.

Merlin, Pinau. *A Field Guide to Desert Holes.* Tucson: Arizona Desert Museum Press, 1999.

Meyer, Christine, and Faith Moosang, eds. *Living with the Land: Communities Restoring the Earth.* Philadelphia: New Society Publishers, 1992.

Middleton, A. *Wild Bird Guides: American Goldfinch.* Mechanicsburg, PA: Stackpole Books, 1998.

Miller, Brian. "Using Focal Species in the Design of Nature Reserve Networks." *Wild Earth* 8 (Winter 1998–99).

Mitchel, John Hanson. *A Field Guide to Your Own Back Yard.* Woodstock, VT: W. W. Norton and Company, 1999.

Mosby, Henry, and Charles O. Handley. *The Wild Turkey in Virginia: Its Status, Life History and Management.* Virginia Division of Game, 1943.

Murie, O. *A Field Guide to Animal Tracks.* New York: Houghton Mifflin Co., 1954.

Nabhan, Gary Paul. *Cultures of Habitat: On Nature, Culture, and Story.* Washington, DC: Counterpoint, 1997.

Nelson, Richard. *The Island Within.* New York: Vintage Press, 1989.

Patent, Dorothy Hinshaw. *The Whooping Crane: A Comeback Story.* New York: Clarion Books, 1988.

Peck, Sheila. *Planning for Biodiversity: Issues and Examples.* Washington, DC: Island Press, 1998.

Plant, Christopher, and Judith Plant. *Turtle Talk: Voices for a Sustainable Future.* Philadelphia: New Society Publishers, 1990.

Proctor, N., and P. Lynch. *Manual of Ornithology: Avian Structure and Function.* New Haven, CT: Yale University Press, 1993.

Pyle, Peter, Steve N. G. Howell, Robert P. Yunick, and David F. DeSante. *Identification Guide to North American Passerines.* Bolinas, CA: Slate Creek Press, 1987.

Raynes, Bert. *Birds of Grand Teton National Park and the Surrounding Area.* Grand Teton Natural History Association, 1984.

Rezendes, Paul. *Tracking and the Art of Seeing: How to Read Animal Tracks and Sign.* 2nd ed. New York: HarperCollins Publishers, 1999.

Rezendes, Paul, and Paulette Roy. *Wetlands: The Web of Life.* San Francisco: Sierra Club, 1996.

Sanders, Jack. *Internet Guide to Birds and Birding.* Camden, ME: Ragged Mountain Press, 2000.

Savage, Candace. *Bird Brains: The Intelligence of Crows, Ravens, Magpies, and Jays.* San Francisco: Sierra Club, 1995.

Shalaway, Scott. *Building a Backyard Bird Habitat.* Mechanicsburg, PA: Stackpole Books, 2000.

Sheldon, I., T. Hartson, and M. Elbroch. *Animal Tracks of New England.* Renton, WA: Lone Pine Publishing, 2000.

Sibley, David. *NAS Sibley Guide to Birds.* New York: Alfred A. Knopf, 2000.

Smith, Richard. *Animal Tracks and Signs of North America.* Harrisburg, PA: Stackpole Books, 1982.

Snyder, Gary. *The Practice of the Wild.* New York: Farrar, Straus and Giroux, 1990.

Soule, Michael E., and John Terborgh, eds. *Continental Conservation: Scientific Foundations of Regional Reserve Networks.* Washington, DC: Island Press, 1999.

Stein, Bruce A., Lynn S. Kutner, and Jonathan S. Adams. *Precious Heritage: The Status of Biodiversity in the United States.* New York: Oxford University Press, 2000.

Stokes, John, and Kanawahienton (David Benedict), Rokwaho (Dan Thompson), and Tekaronianekon (Jake Swamp). *Thanksgiving Address.* Six Nations Indian Museum and the Tracking Project, Corrales, NM: 1993.

Sutton, Patricia, and Clay Sutton. *How to Spot Hawks and Eagles.* New York: Houghton Mifflin, 1996.

———. *How to Spot an Owl.* Shelburne,VT: Chapters Publishing Ltd., 1994.

Teale, Edwin Way. *A Naturalist Buys an Old Farm.* Storrs, CT: Bibliopola Press, 1998.

Voitkevich, A. A. *The Feathers and Plumage of Birds.* London: Sidgwick and Jackson, 1966.

Wackernagel, Mathis, and William Rees. *Our Ecological Footprint: Reducing Human Impact on the Earth.* Gabriola Island, BC: New Society Publishers, 1996.

Waldbauer, Gilbert. *The Birder's Bug Book.* Cambridge, MA: Harvard University Press, 1998.

Wessels, Tom. *Reading the Forested Landscape: A Natural History of New England.* Woodstock, VT: Countryman Press, 1997.

Wheeler, Brian K., and William S. Clark. *A Photographic Guide to North American Raptors.* San Diego: Academic Press, 1999.

Wilcove, David S. *The Condor's Shadow: The Loss and Recovery of Wildlife in America.* New York: Anchor Books, 1999.

Wilson, Don E., F. Russell Cole, James D. Nichols, Rasanayagam Rudran, and Mercedes S. Foster, eds. *Measuring and Monitoring Biodiversity: Standard Methods for Mammals.* Washington, DC: Smithsonian Institution, 1996.

Wilson, Edward O. *The Diversity of Life.* New York: W. W. Norton and Company, 1992.

ACKNOWLEDGMENTS

First and foremost, we must thank the birds, which shared their stories with us so that we might share them with others. I have loved birds since I was very small; they have brought joy to my life for as long as I can remember. And I thank the bears and everything wild for nourishing me throughout my life, and certainly through this project.

I'd also like to recognize and celebrate all the people who share tracking with others across the country. Your work is so important! A special recognition for Tom Brown, Jr., Jon Young, and Paul Rezendes for sharing their insights into tracking with me. Tom also role modeled vision, passion, and focus, Jon helped me become a better teacher and demonstrated the "peace-maker principles," and Paul taught me photography, gave me greater confidence in what I knew, and allowed me to teach and grow in his school. They are good people, indeed. Thanks also to Roy Brown, Michael Lawrence, John Fergusen, and David Lees for their fine published works on birds and tracking in the United Kingdom—they taught me a great deal. And to any other person who has written anything on tracking, I greatly appreciate your efforts and experience, and that you were kind enough to document them for the benefit of others.

I have tracked with countless people over the years, and I thank all of them for sharing their time, energy, and passion for living. Each of them taught me much about tracking, awareness, and purpose. Among them were Kayla Sanford, Mike Pewtherer, Jonathan Talbott, Frank Grindrod, Keith Badger, John McCarter, Ricardo Sierra, Walker Korby, Eleanor Marks, Nancy Birtwell, Fred Vanderbeck, and Diane Boretos. And I thank my students, who have all taught me so much.

A number of people aided me in the research for this book. Greg Levandoski, an excellent birder, shared freely his expertise, enthusiasm, photography, and apartment in Big Bend National Park. Sue Thorpe of the New Hampshire Science Center graciously aided me on several occasions in gaining further data on raptor tracks and pileated woodpeckers. Her love for her work and wildlife, her generosity, and her good nature are contagious. Diane Boretos shared her piping plover research, measured many skulls, collected robin scat in the field, and has tirelessly worked for decades to protect wild creatures and natural places. Mike Cox of the Vermont Institute of Natural Science provided raptor pellets for comparison with those we were finding in the field, and Joe Delveccio and Julia Lankton helped collect them. Bill Evans of Cornell University, who works continually to educate others about the terrible effects of communication towers on birds, lent us slides. The Teton Science School happily opened the Murie Collection and allowed us to touch and handle casts made by Olaus Murie and to photograph pellets in the collection. Maureen McConnell of the Boston Museum of Science provided literature for the introduction; Maureen continually works to promote awareness of tracking within the scientific community. And thanks to Keith Badger, who is not only a tracker, but also a phenomenal educator, for his continuous enthusiasm and support of the project from start to finish.

This project would not have happened without the full support of my family. My grandfather introduced me to nature and birds many years ago among the hedgerows and fields of Suffolk, England. Since then, every member of my family has not only suffered with but also supported and encouraged my obsessive interests in wildlife, birding, and tracking. My parents aided in every way possible and provided so much for this project—the pens for the illustrations, the use of their scanner and art programs, and a space above their garage to set up a desk. My grandmother cheered me on, and my great-uncle, Robert Cross, a wonderful writer and editor, reviewed the contract and generously shared his experiences. Thanks to them all; it is a gift to be a part of such a wonderful family.

I also owe Eleanor Marks tremendous thanks. Somehow, Eleanor managed to juggle her dissertation, personal commitments, and this book with incredible energy, vision, and heart. She spent weeks among museum collections, months connecting with people, countless hours photographing, and endless days writing and editing to make this book a better, stronger resource. And Eleanor has truly celebrated all the people who have contributed to this work, even those who have passed on and those who have yet to appear in this world. May we all be so generous.

And none of this would have happened without the faith, vision, and support of Mark Allison of Stackpole Books. Thank you very much for this opportunity.

MARK ELBROCH

I want to begin by thanking Aldo Leopold for changing both my professional life and my avocation; he influenced me to go beyond being just a typical political scientist to becoming an ecologist studying international environmental problems and the use of negotiations to solve them. He also inspired me to learn tracking. The miracle here is that he accomplished these transformations in my life despite having died decades ago.

I also would like to thank my dear friend Dr. Emine Kiray for believing in this project and generously contributing ideas, inspiration, and editorial feedback, for her sheer delight in the scope and vision of this book, and for her open-hearted giving of the resources necessary to fund a major part of the fieldwork required for a work of this dimension. I say "sağ ol" ("thank you" in Turkish).

Profound thanks to photographer Sally McCay of the University of Vermont for designing and lab-testing our indoor lighting, conducting the first feather shoot at the Smithsonian, and giving support and professional problem-solving throughout the year; and most of all, for having a great sense of humor when recalcitrant feathers decided to go their own way and bolt for freedom after painstaking arrangement in the (apparently) unbearable confines of the frame.

I thank my dear friend Dr. Ellin Reisner for introducing me to Ron Naveen, noted author, photographer, and researcher of Antarctica, penguins, and boobies and founder of the conservation organization Oceanities, who networked with us in regard to conducting museum research.

Thanks to Bob Spear, founder and director of the Birds of Vermont Museum, for helping us connect with naturalist researchers in the field, from Bernd Heinrich to folks at the Vermont Institute of Science.

Deep, heart-felt thanks to Dr. Carla Dove of the Smithsonian's National Museum of Natural History, Division of Birds, for her spirited support, frank enthusiasm, fielding of obscure technical questions, introducing us to other ornithologists (in particular, the eminent Roxie Laybourne), her careful reading of the feather chapter herein, and her encouragement to build cooperation among ornithologists, museum folks, field researchers, and trackers to promote conservation. Carla connected us with other Smithsonian ornithologists, who were also wonderfully supportive to our work, including Dr. Gary Graves, Dr. Helen James, Dr. Storrs Olson, Dr. Claudia Angle, Jim Dean, and, especially, Marcy Heacker-Skeans.

Thanks are also due to Terry Goodhue and John Drury, two gifted field researchers, for allowing me to stay at their offshore research base. I was so moved and impressed by the back-breaking commitment of these birder-ecologists (the true experts of their Gulf of Maine ecosystem) who ingeniously juggle all kinds of multiple work schedules in order to make themselves available for weeks at a time to conduct the painstaking observational and quantitative research that is absolutely essential to long-term university-based research projects.

I also want to thank Alison Pirie, Assistant to the Curator, Department of Ornithology at the Museum of Comparative Zoology at Harvard University, for her wonderful laugh, her warm welcome, for answering technical questions, and guiding us through the unique MCZ avian skull classification system. She has encyclopedic knowledge of the MCZ collections, tends them with great care, expertise, and love, and is unabashedly passionate and knowledgeable about birds. It was a joy to work with her. We also thank Dr. Kevin J. McGowan, Curator of Birds and Mammals at Cornell University Museum of Vertebrates, for allowing us generous access to their avian skull collection.

I thank the author, painter, and naturalist Clare Walker Leslie for her enthusiastic support of this book, and especially for her absolute confidence that a work bringing together trackers, birders, ornithologists, and naturalists would make an important and unique contribution not only to the conduction of field studies, but also to inspiring birders, trackers, professionals, and laypeople alike to think as ecologists and work cooperatively on research. Also, I thank Julie Trachtenberg D'Amours of the Teton Science School in Jackson Hole, Wyoming, for sharing the Olaus, Adolph, and Mardie Murie Collection with us; and Dr. Mark Pokras, Dawn Kelly, Dr. Flo Tseng, Rose Miconi, Dr. Inga Sidor, Robin Shearer, and Dr. Carolyn Corsiglia for generously sharing the resources of the Tufts University Veterinary School of Medicine's New England Wildlife Clinic, so that I could complete our photo documentation of feathers. It was a sheer delight to be in their impassioned and eloquent presence. These folks' technical expertise, their devotion to wildlife, their research on environmental risks to birds (especially Dr. Pokras' work on the toxicity of lead sinkers to loons), and their incredibly hard work and relentless commitment to wildlife medicine, research, and outreach to educate the public were truly inspiring.

I also want to thank Dawn Kelly for introducing me to Tom French of the Massachusetts Division of Fisheries and Wildlife Federal Endangered Species and Natural Heritage Program, who shared his impressive feather collection gathered from over a decade of research on peregrine predation; Brad G. Blodget, Massachusetts State Ornithologist with the Division of Fisheries and Wildlife, for providing information and answering technical questions; and the delightful Dr. Russell P. Balda of Northern Arizona University–Flagstaff for his warm support and for sharing his original slides, his expert book, and references covering over forty years of research on corvids. His commitment to education and to understanding the habitat needs of pinyon jays and other corvids is wonderful to experience.

I would also like to thank tracker, naturalist, and educator Bob Metcalfe for sharing the wisdom of his expertise in the field and for introducing me to Massachusetts wildlife rehabilitators, especially teachers David Taylor and Liz Glass, and their enthusiastic students Scott Harmer and Rebecca Conley.

Of course I wish to wholeheartedly thank the indefatigable Amber Santangelo, director of "Wild Again" and past president of the Wildlife Rehabilitators Association of Massachusetts. I honor the HUGE commitment she has made out of her profound caring to rehabilitate injured wildlife. She epitomizes what I have experienced over and over again when working with wildlife rehabbers: devotion, "tough love," and the ability to work and carry on despite exhaustion and the seemingly endless assault of automobiles and power lines on the physical integrity of birds and other wild animals.

Many, many thanks to all the devoted folks who taught me tracking and helped to launch me as a naturalist: Paul Rezendes, Paulette Roy, and John McCarter, the very first to teach mammal tracking with great dedication to thousands of people throughout New England; Jeff Kunz, Lydia Rogers, Lyn Heubner, and Bob Metcalfe—all members of Paul's Apprenticeship Program, and the first ones to go "puzzling things out" in the forest with me; Paul Wanta and Heather

Lenz, devoted trackers, naturalists, and inspired teachers committed to working with Native youth and women (and who introduced me to John Stokes); Bob Leverett, who has done ground-breaking work on discovering, documenting, and defending tracts of old-growth forest throughout the eastern United States, and is one of the most heart-felt and articulate teachers I have ever met; Susan Morse, who through her visionary program "Keeping Track," teaches tracking to local conservation commissions throughout the United States; Dr. Alcott Smith, a forest "deep ecologist" who teaches tracking and nature studies with passionate and eloquent insight—he was the first person to take me seriously as a tracker and taught me to be a forest naturalist by "reading the landscape"; Jim Bruchak and his open-hearted family of accomplished teachers, writers, and artists, who shared their ancestral land and knowledge of Native American and Abnaki skills, science, and philosophy in regard to tracking and ecology; the deeply inspiring and generous teachers at the Boulder Outdoor Survival School in Boulder, Utah, especially David Holladay, Allyson Vance, and Andie Pitcher, true geniuses of the desert; and professional wetland scientist Diane Boretos, for introducing me to conservation biology, to the multinational Wildlands Project, and for her sheer delight in fieldwork, and decades of inspired conservation work.

A special thanks to my co-learners in our tracking mentoring group, Kayla Sanford, Nancy Birtwell, Diane Boretos, Frank Grindrod, and Fred Vanderbeck, a group initiated and taught by Mark, in which I received my most intensive and transformative learning experiences while tracking with others. And to Mark Elbroch himself, a heartfelt thank you for his generous mentoring, his devotion, passion, and brilliance, for his "24-7" original, audacious, and unrelenting field research, and for the way he inspires others to fall in love with a place by tracking, birding, naturalizing, and simply poking about (as his friend Mike calls it) but with profound attention and awareness, for hours and hours, in the deep trance of the tracker.

Finally, I thank Chris Counihan for his extraordinary day-in and day-out support of this project. He believed in it from its inception and did everything imaginable to help, from financial and logistical contributions to the far more important necessities of endlessly sharing emotional support, patience, a gentle sense of humor, wonderful food from his garden, and unshakable faith in this book.

ELEANOR MARKS

INDEX

ABOUT THE AUTHORS

Mark Elbroch has tracked throughout the United States, Canada, and the United Kingdom and has been involved in numerous tracking projects: surveying Glacier National Park's bear sign, detecting marten and fisher in Washington State, gathering track parameters of southern bobcats and gray foxes, and studying the distinction between northern and southern flying squirrel trail parameters. He conducts wildlife inventories for land owners and managers using tracks and sign and currently teaches tracking courses throughout the New England area; he is a staff instructor for Paul Rezendes Programs. He coauthored *Animal Tracks of New England* and has written or contributed to many other articles on tracks and sign. Mark Elbroch can be reached with questions or comments at markelbroch@yahoo.com.

Eleanor Marks, Ph.D., is a teacher, writer, naturalist, and environmental advocate. Her work has focused on the environment for twenty years using a variety of media, from poetry to teaching to cowriting and being senior producer for the television documentary "PCB's: Pollution, Clean-up and Beyond." She took her first tracking class with Paul Rezendes in 1987 and joined his apprenticeship program in 1995. She recently finished a book on global environmental organizing by women in the Southern Hemisphere for ecologically sustainable economic development at the village level; locally, she focuses her research and teaching on tracking and habitat and ecosystems dynamics within her home region of New England.

Diane Boretos has worked as a wetlands biologist for seventeen years in the field of wetland protection, both for the Massachusetts Department of Environmental Protection and local conservation commissions. Currently, Diane is a senior scientist with the environmental consulting firm "Call of the Wild," where she conducts botanical and wildlife inventories as well as creates land-management plans. Diane teaches wetlands ecology, coastal ecology, and tracking classes for the New England Wildflower Society and the Nature Conservancy.